W9-CRB-392

BIOORGANIC CHEMISTRY

Volume I   Enzyme Action

CONTRIBUTORS

MICHAEL J. ARSLANIAN

D. S. AULD

MYRON L. BENDER

JOHN H. CHALMERS, JR.

EUGENE E. DEKKER

N. C. DENO

THOMAS H. FIFE

RONALD A. GALAWAY

MORTON J. GIBIAN

ELIZABETH J. JEDZINIAK

V. C. JOSHI

E. T. KAISER

MAKOTO KOMIYAMA

TOYOKI KUNITAKE

ROBERT M. McGRATH

G. M. MAGGIORA

LAUREN A. MESSER

Y. NAKAGAWA

MARION H. O'LEARY

ALFONSO L. POGOLOTTI, JR.

DANIEL V. SANTI

R. L. SCHOWEN

THOMAS A. SPENCER

JAMES K. STOOPS

EDWARD S. TOMEZSKO

SALIH J. WAKIL

JOHN WESTLEY

BIOORGANIC CHEMISTRY

Edited by

E. E. van Tamelen

Department of Chemistry
Stanford University
Stanford, California

Volume I

ENZYME ACTION

A treatise to supplement Bioorganic Chemistry:
An International Journal

Edited by
E. E. van Tamelen

ACADEMIC PRESS   New York   San Francisco   London   1977

A Subsidiary of Harcourt Brace Jovanovich, Publishers

CHEMISTRY

6159-796X

ACADEMIC PRESS, INC.
111 Fifth Avenue, New York, New York 10003

United Kingdom Edition published by
ACADEMIC PRESS, INC. (LONDON) LTD.
24/28 Oval Road, London NW1

Library of Congress Cataloging in Publication Data

Main entry under title:

Bioorganic chemistry.

    Includes bibliographies and index.
    CONTENTS: v. 1.   Enzyme action.
    1.   Biological chemistry.   2.   Chemistry,
Organic.   I.   Van Tamelen, Eugene E., Date
QP514.2.B58   574.1'92   76-45994
ISBN 0–12–714301–7 (v. 1)

PRINTED IN THE UNITED STATES OF AMERICA

Contents

Chapter 12   The Catalytic Mechanism of Thymidylate Synthetase

ALFONSO L. POGOLOTTI, JR., AND DANIEL V.
SANTI

Chapter 13   Studies of Amine Catalysis via Iminium Ion Formation

THOMAS A. SPENCER

Chapter 14   Fatty Acid Synthetase Complexes

JAMES K. STOOPS, MICHAEL J. ARSLANIAN, JOHN
H. CHALMERS, JR., V. C. JOSHI, AND SALIH J.
WAKIL

Chapter 15   Sulfane-Transfer Catalysis by Enzymes

JOHN WESTLEY

List of Contributors

Numbers in parentheses indicate the pages on which the authors' contributions begin.

MICHAEL J. ARSLANIAN (339), Marrs McLean Department of Biochemistry, Baylor College of Medicine, Houston, Texas

D. S. AULD (1), Biophysics Research Laboratory, Department of Biological Chemistry, Harvard Medical School, and the Division of Medical Biology, Peter Bent Brigham Hospital, Boston, Massachusetts

MYRON L. BENDER (19), Northwestern University, Evanston, Illinois

JOHN H. CHALMERS, JR. (339), Marrs McLean Department of Biochemistry, Baylor College of Medicine, Houston, Texas

EUGENE E. DEKKER (59), Department of Biological Chemistry, The University of Michigan, Ann Arbor, Michigan

N. C. DENO (79), Department of Chemistry, The Pennsylvania State University, University Park, Pennsylvania

THOMAS H. FIFE (93), Department of Biochemistry, University of Southern California, Los Angeles, California

RONALD A. GALAWAY (117), Department of Chemistry, University of California, Riverside, California

MORTON J. GIBIAN (117), Department of Chemistry, University of California, Riverside, California

ELIZABETH J. JEDZINIAK (79), Department of Chemistry, The Pennsylvania State University, University Park, Pennsylvania

V. C. JOSHI (339), Marrs McLean Department of Biochemistry, Baylor College of Medicine, Houston, Texas

E. T. KAISER (137), Departments of Chemistry and Biochemistry, University of Chicago, Chicago, Illinois

MAKOTO KOMIYAMA (19), Northwestern University, Evanston, Illinois

TOYOKI KUNITAKE (153), Department of Organic Synthesis, Faculty of Engineering, Kyushu University, Fukuoka, Japan

ROBERT M. MCGRATH (231), Sorghum Beer Unit, National Food Research Institute, Council for Scientific and Industrial Research, Pretoria, Republic of South Africa

G. M. MAGGIORA (173), Department of Biochemistry, University of Kansas, Lawrence, Kansas

LAUREN A. MESSER (79), Department of Chemistry, The Pennsylvania State University, University Park, Pennsylvania

Y. NAKAGAWA (137), Departments of Chemistry and Biochemistry, University of Chicago, Chicago, Illinois

MARION H. O'LEARY (259), Department of Chemistry, University of Wisconsin, Madison, Wisconsin

ALFONSO L. POGOLOTTI, JR. (277), Department of Pharmaceutical Chemistry, University of California, San Francisco, California

DANIEL V. SANTI (277), Department of Biochemistry and Biophysics and Department of Pharmaceutical Chemistry, University of California, San Francisco, California

R. L. SCHOWEN (173), Department of Chemistry, University of Kansas, Lawrence, Kansas

THOMAS A. SPENCER (313), Department of Chemistry, Dartmouth College, Hanover, New Hampshire

JAMES K. STOOPS (339), Marrs McLean Department of Biochemistry, Baylor College of Medicine, Houston, Texas

EDWARD S. TOMEZSKO (79), Department of Chemistry, The Pennsylvania State University, University Park, Pennsylvania

SALIH J. WAKIL (339), Marrs McLean Department of Biochemistry, Baylor College of Medicine, Houston, Texas

JOHN WESTLEY (371), Department of Biochemistry, University of Chicago, Chicago, Illinois

Foreword

What is bioorganic chemistry? It is the field of research in which organic chemists interested in natural product chemistry interact with biochemistry. For many decades the natural product chemist has been concerned with the way in which Nature makes organic molecules. In the absence of any information other than that provided by structure, conclusions had necessarily to be derived from structural analysis. Broad groups of natural products could be recognized, such as alkaloids, isoprenoids, and polyketides (acetogenins), which clearly had elements of structure indicating a common biosynthetic origin. Indeed, for alkaloids and terpenoids, structural work was greatly helped by such biogenetic hypothesis. Similarly, after A. J. Birch had made an extensive analysis of polyketides, the repeating structural element postulated also helped in the determination of structure.

The alternative, and complement, to the above analysis is to consider the chemical mechanisms whereby the units of structure are assembled into the final natural product. For example, alkaloid structure can often be analyzed in terms of anion–carbonium ion combination. Also, the later stages of biosynthesis of many alkaloids can be analyzed by the concept of phenolate radical coupling. In polyisoprenoids the critical mechanism for carbon–carbon bond formation is the carbonium ion–olefin interaction to give a carbon–carbon bond and regenerate a further carbonium ion.

The analysis of natural product structures in terms of either structural units or mechanisms of bond formation has been subjected to rigorous tests since radioactively labeled compounds became generally available. It is gratifying that, on the whole, the theories developed from structural and mechanistic analysis have been fully confirmed by *in vivo* experiments.

Organic chemists have always been fascinated by the possibility of imitating in the laboratory, but without the use of enzymes, the precise steps of a biosynthetic pathway. Such work may be called biogenetic-type, or biomimetic, synthesis. This type of synthesis is a proper activity for the bioorganic chemist and undoubtedly deserves much attention. Nearly all such efforts are, however, much less successful than Nature's synthetic activities using enzymes. It is well appreciated that Nature has solved the outstanding problem of synthetic chemistry, viz., how to obtain 100% yield and complete stereospecificity in a chemical synthesis. It, therefore, remains a major task for bioorganic chemists to understand the mechanism of enzyme action and the precise reason why an enzyme is so efficient. We are still far from the day when we can construct an organic molecule which will be as efficient a catalyst as an enzyme but which will not be based on the conventional polypeptide chain.

Much of contemporary bioorganic chemistry is presented in these volumes. It will be seen that much progress has been made, especially in the last two decades, but that there are still many fundamental problems left of great intellectual challenge and practical importance.

The world community of natural product chemists and biochemists will be grateful to the editor and to all the authors for the effort that they have expended to make this work an outstanding success.

DEREK BARTON
Chemistry Department
Imperial College of Science and Technology
London, England

Preface

Although natural scientists have always been concerned with the development and behavior of living systems, only in the twentieth century have investigators been in a position to study on a molecular level the intimate behavior of organic entities in biological environments. By mid-century, the form and function of various natural products were being defined, and complex biosynthetic reactions were even being simulated in the nonenzymatic laboratory. As the cinematographic focus on biomolecules sharpened, one heard increasingly the adjective *bioorganic* applied to the interdisciplinary area into which such activity falls.

In 1971, publication of a new journal, *Bioorganic Chemistry*, was begun. As a follow-up, what could be more timely and useful than a well-planned, multivolume collection of bioorganic review articles, solicited from carefully chosen professionals, surveying the entire field from all possible vantage points? This four-volume work contains a collection, but it did not originate in this manner.

As the journal *Bioorganic Chemistry* developed, the number and quality of regular, original research articles were maintained at an acceptable level. However, comprehensive review articles appeared only sporadically, despite their intrinsic value at a time when general interest in bioorganic chemistry was burgeoning. In order to enhance this function of the journal, as well as to mark the fifth anniversary of its birth, we originally planned to publish in 1976 a special issue comprised entirely of reviews by active practitioners. After contact with a handful of stalwart bioorganic chemists, about two hundred written invitations for reviews were mailed during late 1975 to appropriate, diverse scientists throughout

the world. The response was overwhelming! More than seventy preliminary acceptances were received within a few months, and it soon became
evident that the volume could not be handled adequately through publication by journal means. After consultation with representatives of Academic
Press, we agreed to publish the manuscripts in book form.

Although the stringency of journal deadlines disappeared, the weightier
matter of editorial treatment had to be reconsidered. Should contributions
be published in the same, piecemeal, random fashion as received? Such
practice would be acceptable for journal dissemination, but for book
purposes, broader, more orderly, and inclusive treatment might be desirable and also expected. Partly because of editorial indolence, but mostly
because of a predilection for maintaining the candor and spontaneity
which might be lost with increased editorial control, we decided not to
attempt coverage of all identifiable areas of bioorganic chemistry, not to
seek out preferentially the recognized leaders in particular areas, and
even not to utilize outside referees. Consequently, we present reviews
composed by scientists who were not coerced or pressured, but who
wrote freely on subjects they wanted to write about and treated them as
they wanted to, at the cost perhaps of a certain amount of objectivity and
restraint as well as proper coverage of some important bioorganic areas.

We turn now to the results of this publication project. Because of the
inevitable attrition for the usual reasons, fewer than the promised number
of reviews materialized: fifty-seven manuscripts were received in good
time and accepted by this office. Eight countries are represented by the
entire collection, which emanates almost entirely from academia, as
would be expected. A great variety of topics congregated—greater than
we had foreseen. Inclusion of all papers in one volume was impractical,
and thus the problem arose of logically dividing the heterogeneous material into several unified subsections, each suitable for one volume, a
problem compounded by the fact that an occasional author elected to
treat, in one manuscript, several unconnected topics happening to fall in
his purview. Therefore, perfect classification without discarding or dis-
secting bodies of material as received was simply not possible.

After some reflection and a few misconceptions, we evolved a plan for
division into four more or less scientifically integral sections; these, happily, also constitute approximately equal volumes of written material, an
aspect of some importance to the publisher. The enzyme–substrate interaction was expected to be a well-represented subject, and, in fact, too
many manuscripts on this subject for one proportionally sized volume
were received. Although the separation of enzyme action and substrate
behavior is contrived and not basically justifiable, it turned out that, for
the most part, a group of authors heavily emphasized the former, while
another concentrated on the latter. Accordingly, Volume I was entitled

"Enzyme Action," and Volume II "Substrate Behavior." Admittedly, in a few cases, articles could be considered appropriate for either volume.

A gratifyingly significant number of contributions dealt with the behavior of biologically important polymers and related matters, sent in by authors having quite different investigational approaches. In addition, several discourses were concerned with molecular aggregates, e.g., micelles. All of these were incorporated into Volume III, "Macro- and Multimolecular Systems."

Whatever papers did not belong in Volumes I–III were combined and constitute Volume IV. Fortunately, in these remaining papers some elements of unity could be discerned; in fact, their entire content falls into the following categories: "Electron Transfer and Energy Conversion (photosynthesis, porphyrins, NAD^+, cytochromes); Cofactors (coenzymes, NAD^+, metal ions); Probes (cytokinin behavior, steroid hormone action, peptidyl transferase reactivity)."

Finally, early in this enterprise, we asked Derek Barton to compose a Foreword. Sir Derek complied graciously, and in every volume his personalized view on the nature of bioorganic chemistry appears.

E. E. VAN TAMELEN

Contents of Other Volumes

VOLUME III   Macro- and Multimolecular Systems

1

Direct Observation of Transient ES Complexes: Implications to Enzyme Mechanisms

D. S. Auld

INTRODUCTION

Determination of the mechanism of action of an enzyme requires a complete kinetic description of the individual steps involved in substrate binding and catalysis, the conformational changes that presumably connect these steps, and the nature of the intermediates involved. The direct visualization of enzyme-substrate (ES) complexes has not been a simple task, since the formation of the ES complex and its conversion to products are generally very rapid for natural substrates. The difficulties imposed by the high rate of enzymatic reactions has been reduced greatly by the now ready availability of stopped-flow and low-temperature equipment and methodology [1,2]. However, spectral systems that can directly detect the ES complex must be designed to be compatible with rapid mixing techniques. In addition, substrates must be synthesized that possess the appropriate probe properties and are readily turned over by the enzyme under investigation.

Resonance energy transfer between fluorescent enzyme tryptophanyl residues as intrinsic donors and an extrinsically placed acceptor, e.g., a dansyl group, in the substrate allows direct visualization of the enzyme–substrate complex [3]. The spectral overlap between the dansyl group absorption and tryptophan emission is excellent, and the dansyl emission

1

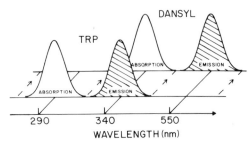

Fig. 1 Schematic representation of the spectral overlap relationships between the enzyme tryptophan and substrate dansyl groups, which constitute the energy donor–acceptor pair critical to observation of the ES complex.

spectrum is red shifted far enough not to overlap with its own absorption spectrum (Fig. 1), properties that make these an exceptionally good donor–acceptor pair. Quantitatively, the degree of energy transfer is sensitive to the distance between and orientation of the donor–acceptor pair and to the environment of the acceptor [4,5]. Differences in tryptophan to dansyl transfer efficiencies and/or dansyl quantum yields can characterize the ES species formed. Thus, if a set of reversible enzyme–substrate complexes and/or a covalent intermediate, EA, are formed in the course of an enzyme-catalyzed reaction it should be possible to determine the minimal number of significantly populated states, the rates of interconversion of molecules among these states, and the equilibrium constants determining the relative proportion of the populations [Eq. (1)].

$$E + S \rightleftharpoons (ES)_1 \rightleftharpoons (ES)_2 \rightleftharpoons (ES)_n \longrightarrow EA + P_1 \longrightarrow E + P_2 \quad (1)$$

This approach has been applied successfully to carboxypeptidase A and has allowed delineation of mechanisms of inhibition [6], enzymatic consequences of chemical modification [7], syncatalytic measurement of distances between the active-site metal atom and dansyl blocking group of peptide substrates [3,8], and differentiation of the mechanism of ester and peptide hydrolysis [7]. Most recently it has been applied to a number of other hydrolytic enzymes, such as yeast carboxypeptidase, chymotrypsin, and alkaline phosphatase [9]. The intermediates observed have allowed quantitative assessment of kinetic schemes for these enzymes.

CARBOXYPEPTIDASE A·DANSYL SUBSTRATE COMPLEXES: DIRECT OBSERVATION BY STOPPED-FLOW FLUORESCENCE

A series of oligopeptide and depsipeptide substrate pairs were synthesized for mechanistic studies of carboxypeptidase A [6,7,10–12]. They have the general form

$$R—(Gly)_n—X—\underset{\underset{COOH}{|}}{\overset{\overset{R'}{|}}{CH}}$$

where X is O or NH. For the peptide the C-terminal residue is an L-amino acid, while for the ester it is the corresponding β-substituted L-lactic acid. The exact ester analog for the peptide substrate containing the phenylalanyl residue (X = NH, R' = CH_2Ph) is therefore the L-phenyl lactate derivative (X = O, R' = CH_2Ph). Thus, each ester–peptide pair differs only by virtue of the susceptible bond. The N-terminal blocking group, R, is either the fluorescent dansyl group or the conventional benzoyl or carbobenzoxy groups. The length of the substrate varies with the number, n, of glycyl residues.

The oscilloscope tracing in Fig. 2 demonstrates the rapid enhancement of dansyl fluorescence following mixing of 1×10^{-4} M Dns-(Gly)$_3$-L-OPhe and 2.5×10^{-6} M carboxypeptidase A and reflects the extremely rapid equilibration of enzyme and substrate to form the ES complex. The decrease in the signal is a considerably slower process and, as hydrolysis reduces the ester concentration, reflects a concomitant diminution in the concentration of enzyme-bound ester. The complementary pattern is observed when quenching of enzyme tryptophan fluorescence by the dansyl group of the bound ester is measured (Fig. 2).

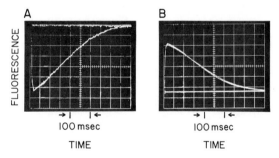

Fig. 2 Enzyme tryptophan (A) and substrate dansyl (B) fluorescence during the time course of hydrolysis of Dns-(Gly)$_3$-L-OPhe, 1×10^{-4} M, catalyzed by zinc carboxypeptidase, 2.5×10^{-6} M, in 1 M NaCl–0.03 M Tris, pH 7.5, 25°C. The fluorescence of either tryptophan (A) or dansyl (B) was measured as a function of time under stopped-flow conditions. Oscilloscope traces of duplicate reactions are shown in each case. Excitation was at 285 nm. Enzyme tryptophan fluorescence was measured by means of band-pass filter peaking at 360 nm, and dansyl emission was measured by a 430 nm cutoff filter. Scale sensitivities for (A) and (B) are 50 and 500 mV/div, respectively. The existence of the ES complex is signaled by either (A) the suppression of enzyme tryptophan fluorescence (quenching by the dansyl group) or (B) enhancement of the substrate dansyl group fluorescence (energy transfer from enzyme tryptophan).

QUANTITATIVE ANALYSIS OF ES COMPLEX
FORMATION AND BREAKDOWN

The rapid equilibrium of carboxypeptidase A and Dns-(Gly)$_3$-L-OPhe to form the ES complex (Fig. 2) and previous kinetic data for blocked oligopeptide substrates [6,10–12] indicate that the action of carboxypeptidase is entirely in accord with the classic Michaelis–Menten kinetic scheme (2),

$$E + S \xrightleftharpoons{K_s} ES \xrightarrow{k_2} P + E \tag{2}$$

where K_s is the dissociation constant of the ES complex and k_2 is the first-order rate constant for the breakdown of the complex into products, P.

The maximal fluorescence, F_{max}, is directly proportional to the maximal concentration of the enzyme–substrate complex, $[ES]_{max}$, formed at this substrate concentration (Fig. 3). The area under the oscilloscope tracing, A, is related to the concentration of the ES complex by Eq. (3), where C is a proportionality factor (Fig. 3).

$$A = C \int_0^\infty [ES]\, dt \tag{3}$$

The quantities of F_{max} and A are related to the equilibrium dissociation constant, K_s, and the rate constant for the rate-determining step, k_2 (Fig. 3), as shown in Eq. (4) in a manner similar to that suggested by Chance for absorbance changes in peroxidase [13]. Since the total substrate added, S_0, is converted completely to products, P, and $F_{max} = C[ES]_{max}$ (Fig. 3)

$$S_0 = \int_0^\infty dP = \int_0^\infty \frac{dP}{dt}\, dt = k_2 \int_0^\infty [ES]\, dt = \frac{k_2 A}{F_{max}} [ES]_{max} \tag{4}$$

Expressing $[ES]_{max}$ in terms of the total enzyme concentration, $[E]_T$ and K_s,

$$\frac{F_{max}[S]_0}{A} = k_2 [E]_T \frac{[S]_F}{K_s + [S]_F} \tag{5}$$

where $[S]_F$ is the free substrate concentration. A negligible amount of product is formed before attainment of F_{max}, due to the large difference between the time required for equilibration of the ES complex compared to that for hydrolysis. The concentration of free substrate, $[S]_F$, therefore equals $[S]_0 - [ES]_{max}$ and under the conditions employed, i.e., $[E]_T \ll [S]_0$, $[S]_F$ approximates $[S]_0$.

The term $(F_{max}/A)[S]_0$ is analogous to v, the steady-state reaction rate; k_2 is analogous to k_{cat} and K_s to K_m. The reciprocal of Eq. (5) is therefore

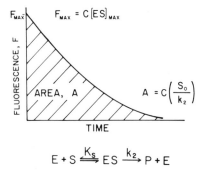

$$E + S \underset{}{\overset{K_s}{\rightleftharpoons}} ES \overset{k_2}{\longrightarrow} P + E$$

Fig. 3 Schematic form of the stopped-flow traces such as are obtained in Fig. 2. The maximal fluorescence, F_{max}, is proportional to the concentration of the ES complex present, while the area under the curve, A, is inversely proportional to the rate constant, k_2, for the rate-determining step.

equivalent to a Lineweaver–Burk plot [14] allowing determination of the parameters k_2 and K_s (Fig. 4; Table 1).

The values obtained for Dns-(Gly)$_3$-L-OPhe, $k_2 = 13{,}700$ min^{-1} and $K_s = 3.5 \times 10^{-5}$ M, are in excellent agreement with those obtained by steady-state pH-stat measurement of hydrogen ion production, $k_{cat} = 11{,}600$ min^{-1} and $K_m = 2.5 \times 10^{-5}$ M (Table 1).

The value of K_s can also be obtained directly from the equations relating the initial rapid change in tryptophan and dansyl fluorescence, F_{max}, to the

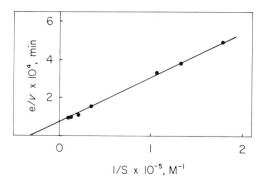

Fig. 4 Determination of the kinetic parameters, k_2 and K_s, for Dns-(Gly)$_3$-L-OPhe hydrolysis by stopped-flow fluorescence. Oscilloscope tracings of dansyl and tryptophan fluorescence were obtained at 2.5×10^{-6} M carboxypeptidase A and Dns-(Gly)$_3$-L-OPhe concentrations of from 6.0×10^{-6} to 1.0×10^{-4} M in 1 M NaCl–0.03 M Tris, pH 7.5, at 25°C. The rate, v, is obtained from the ratio of $F_{max}[S]_0/A$ for each substrate concentration. The kinetic parameters k_2 and K_s are obtained from the reciprocal of Eq. (5) by linear regression analysis.

TABLE 1

Hydrolysis of Dns-(Gly)$_3$-L-OPhe by Carboxypeptidase A^a

Method	k_{cat} (min^{-1})	K_m (mM)	K_s (mM)
E_T/v vs. $1/S_0$^{b}	11,600	0.025	—
$AE_T/F_{max}S_0$ vs. $1/S_F$^{c}	13,700	—	0.033
$1/T$ vs. $1/S_F$^{c}	—	—	0.030
$1/Q$ vs. $1/S_F$^{c}	—	—	0.025

^{a} All analyses performed at 25°C, in 1.0 M NaCl, and pH 7.5. Concentration of Tris buffer was 1×10^{-4} M for pH-stat assay and 3×10^{-2} M for stopped-flow assay.
^{b} Steady-state conditions are $[S]_0 = 0.04–0.4$ mM and $[E]_T = 0.56$ nM [7]. The initial rate determined by the pH-stat assay is v.
^{c} Stopped-flow fluorescence conditions are $[S]_0 = 0.006–0.10$ mM and $[E]_T = 2.5$ μM. The symbols are area, A; maximal fluorescence, F_{max}; tryptophan to dansyl transfer efficiency, T; quantum yield of dansyl group in ES complex, Q.

concentration of ES [Eqs. (6)–(8)]. The fractional quenching of enzyme tryptophan fluorescence, T, by bound substrate is

$$T = T_{max}[ES]/[E]_T \qquad (6)$$

where $[ES]/[E]_T$ is the fraction of enzyme complexed with substrate and T_{max} is the quenching observed when $[ES] = [E]_T$, i.e., when $[S]_0 > [E]_T$ and $[S]_0 \gg K_s$. For a given enzyme and substrate, T_{max} is a constant. A change in T as a function of substrate concentration, therefore, means that the ratio of $[ES]/[E]_T$ is changing. Substitution for $[ES]$ in terms of K_s, $[S]_F$, and $[E]_T$ leads to

$$T = \frac{T_{max}[S]_F}{K_s + [S]_F} \qquad (7)$$

A plot of $1/T$ vs. $1/[S]_F$ is linear and allows an independent determination of K_s (Fig. 5; Table 1). It can also be demonstrated that a similar relationship exists between the initial change in dansyl fluorescence of the ES complex and the $[S]_F$ as shown in Eq. (8), where Q_{max} is the dansyl quantum yield of

$$Q = \frac{Q_{max}[S]_F}{K_s + [S]_F} \qquad (8)$$

the enzyme saturated with substrate and Q is the apparent quantum yield for a particular concentration of substrate, S. Thus, a value of K_s can also be obtained from this relationship. The values derived for K_s by the three different ways of treating the stopped-flow data agree closely with the value

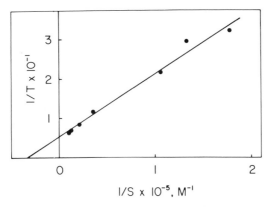

Fig. 5 Stopped-flow fluorescence determination of the dissociation constant for substrate binding, K_s, by measurement of the fractional quenching of carboxypeptidase tryptophan fluorescence by bound Dns-(Gly)$_3$-L-OPhe. The conditions are the same as in Fig. 4. The value of K_s is calculated from the reciprocal of Eq. (7) by linear regression analysis.

of K_m deduced by standard steady-state kinetic procedures, indicating that for this enzyme K_m is a good measure of the substrate dissociation constant, K_s (Table 1). Similar results have been obtained for peptide substrates of carboxypeptidase A [6,7]. The rate of equilibrium to form the ES complex is again much faster than the rate of breakdown of the complex to products, and K_m is therefore equivalent to K_s [Eq. (2)].

It should be noted, however, that at the temperature employed for these studies, 25°C, it is possible for the signal initially observed to reflect a distribution of ES complexes which have reached equilibrium within the mixing time of the instrument. Lowering the temperature in combination with rapid mixing techniques should allow the detection of such rapidly equilibrating species [2].

INHIBITION KINETICS

Enzyme-catalyzed reactions are usually characterized by the kinetic parameters K_m and V_{max}, which can be related mathematically to the steady-state concentration of the ES complex and its rate of breakdown [15]. Reversible inhibitors affect the apparent K_m, V_{max}, or both, just like other factors that influence the velocity of enzyme-catalyzed reactions. Hence, of necessity the assignment of inhibition modes is indirect. However, if the ES complex can be observed, the manner in which an inhibitor affects substrate binding and hydrolysis can be recognized directly.

NONCOMPETITIVE   COMPETITIVE

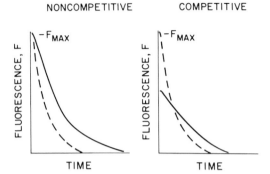

Fig. 6 Effect of inhibitors on the formation and breakdown of ES complexes. The uninhibited reaction is characterized by the symbol F_{max} on the ordinate and the dashed line. The solid lines are the traces expected for inhibitors that possess noncompetitive or competitive inhibition modes.

Since the area, A, reflects the catalytic rate and F_{max} reflects the binding strength of substrates (Fig. 3), modes of inhibition can be deduced directly from the effects of inhibitors on A and F_{max} (Fig. 6). Thus, a noncompetitive inhibitor retards hydrolysis of bound substrate and thereby increases A. However, F_{max} remains unchanged since substrate binding is unaffected. Conversely, a competitive inhibitor reduces F_{max}, owing to a reduction of substrate binding, but A remains constant since the rate-determining step is unchanged. This approach is also equally applicable to the study of the effect of activators on enzyme catalysis.

Inhibition constants are derived from the dependence of A and F_{max} on inhibitor concentrations, [I], at a single substrate concentration [6]. For noncompetitive inhibition, the inhibition constant, K_I, is calculated from the slope of the line A_I/A_C vs. [I] as in Eq. (9), where the subscripts C and I refer to control and inhibitor values, respectively.

$$A_I/A_C = 1 + [I]/K_I \qquad (9)$$

When this approach to inhibition kinetics is applied to carboxypeptidase A [6,7], identical K_I values, $1 \times 10^{-4}\ M$, are observed for noncompetitive inhibition of peptide hydrolysis by phenyl acetate both from stopped-flow fluorescence analysis of ES complex formation and from steady-state measurements, based on ninhydrin analysis of the rate of production of the C-terminal amino acid [6,7,10].

For competitive inhibition, K_I is derived from the relationship shown in Eq. (10). If the study is performed under conditions where $[S]_F \ll K_s$,

$$\frac{(F_{max})_C}{(F_{max})_I} = 1 + \frac{[I]/K_I}{1 + [S]_F/K_s} \qquad (10)$$

Eq. (10) can be simplified to a form similar to that for noncompetitive inhibition [9], i.e., Eq. (11). Stopped-flow fluorescence analysis for L-phenylalanine inhibition of the hydrolysis by carboxypeptidase A of Dns-Gly-L-Phe at pH 7.5 indicates competitive inhibition [6], and the K_I value obtained,

$$\frac{(F_{max})_C}{(F_{max})_I} = 1 + [I]/K_I \qquad (11)$$

$2 \times 10^{-3}\ M$, is in good agreement with results derived from steady-state spectrophotometric and fluorometric analysis [16–18]. Our recent studies of the action of serine proteases acting on N-dansylated peptide and ester substrates indicate these methods of analysis [Eqs. (3) through (11)] are applicable to enzymes with multiple intermediates. In addition, for such cases the effect of modifiers on individual steps of the reaction can be inspected (19).

Using this approach to inhibition kinetics, it has proved possible to resolve mixed inhibition observed for a number of inhibitors of carboxypeptidase into their noncompetitive and competitive components, from measurements at a single substrate concentration [6]. Further resolution of the mixed inhibition was obtained by examining the influence of pH on the inhibition constants of the component modes [6].

Stopped-flow fluorescence inhibitor studies have also demonstrated differences in the binding of exact ester–peptide analogs to carboxypeptidase A [9]. Thus, phenyl acetate does not decrease peptide binding but does retard its hydrolysis (Fig. 7), characteristic of *noncompetitive* inhibition (Fig. 6). In contrast, under the same conditions, this inhibitor reduces ester binding but does not affect hydrolysis, characteristic of *competitive* inhibition. In

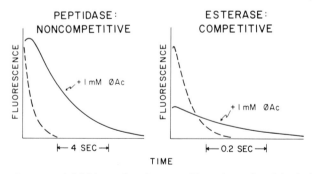

Fig. 7 Phenyl acetate inhibition of carboxypeptidase A-catalyzed hydrolysis of the peptide, Dns-(Gly)₃-L-Phe, $1 \times 10^{-4}\ M$, and the ester, Dns-(Gly)₃-L-OPhe, $4 \times 10^{-5}\ M$. Enzyme concentrations were 5×10^{-6} and $4 \times 10^{-5}\ M$ for ester and peptide hydrolysis, respectively. Assays were performed in the absence (dashed line) and presence (solid line) of $1 \times 10^{-3}\ M$ phenyl acetate, ϕAc, at 25°C and pH 6.5, 0.03 M Mes–1.0 M NaCl. Excitation was at 285 nm and dansyl emission, above 430 nm, was observed.

TABLE 2

Carboxylic Acid Inhibitors of the
Carboxypeptidase A-Catalyzed Hydrolysis of Esters and Peptides[a]

Substrate	Inhibitor	$K_I \times 10^4$ (M)	Type of inhibition
Dns-(Gly)$_3$-L-Phe	Phenyl acetate	3.3	Noncompetitive
Dns-(Gly)$_3$-L-OPhe	Phenyl acetate	3.2	Competitive
Bz-(Gly)$_2$-L-Phe	β-Phenyl propionate	1.2	Noncompetitive
Bz-(Gly)$_2$-L-OPhe	β-Phenyl propionate	1.2	Competitive
Cbz-(Gly)$_2$-L-Phe	Indole 3-acetate	1.7	Noncompetitive
Bz-(Gly)$_2$-L-OLeu	Indole 3-acetate	1.6	Competitive

[a] Assays performed at 25°C, pH 7.5, 1.0 M NaCl, 0.05 M Tris except for the phenyl acetate study, where the conditions were pH 6.5, 1.0 M NaCl, 0.03 M 2-(N-morpholino)-ethanesulfonic acid (Mes) buffer [7,10].

both cases, K_I values are identical (Table 2), suggesting identical enzyme–inhibitor complexes. However, since bound inhibitor allows the peptide but not the ester to bind to the enzyme, the binding of these substrates to the locus at which the inhibitor interacts with the enzyme must differ. Similar results have been observed under steady-state conditions with a number of such inhibitors acting on matched peptide and ester pairs (Table 2).

ENZYMATIC CONSEQUENCES OF CHEMICAL MODIFICATION

An enzyme that has been modified chemically often still exhibits a "residual activity." The question naturally arises as to whether this activity is due to the presence of unmodified native enzyme or to altered kinetic properties of the modified enzyme.

If the latter is the case, it is important to determine if loss in activity is due to weakened substrate binding or to reduced catalytic efficiency. Equilibrium binding techniques usually are precluded, since the residual activity is sufficient to catalyze the conversion of the substrate into products before the measurements can be completed. Again, examination of the enzyme in the presence of substrates under stopped-flow conditions can aid in resolving these questions.

The substitution of another metal for that present in the native state or the removal of any metal is the simplest chemical modification for a metallo-enzyme. Marked changes in activity are usually observed in either case [20].

Substitution of Cd for Zn first demonstrated a difference in the esterase and peptidase activities of carboxypeptidase A [21]. The activity of [(CPD)Cd] toward Bz-Gly-L-OPhe is increased, but that enzyme is virtually inactive toward Cbz-Gly-L-Phe.

Stopped-flow fluorescence studies of ES complexes provided a direct comparison of the peptide binding affinities of the zinc and cadmium enzymes and, simultaneously, an explanation for the different roles of metals in peptide and ester hydrolysis [7]. Cadmium carboxypeptidase binds the peptide Dns-(Gly)$_3$-L-Phe as readily as does [(CPD)Zn] but catalyzes its hydrolysis at a rate that is reduced considerably. Steady-state rate studies of oligopeptides are in agreement with this observation. For all peptides examined, the catalytic rate constants of the Cd enzyme are decreased markedly, but the association constants (K_m^{-1} values) of the Cd enzyme are identical to those of the Zn enzyme [7,10]. However, in marked contrast, for all esters examined, the catalytic rate constants of the Cd enzyme are nearly the same as those of the Zn enzyme, but the association constants are decreased greatly [7].

Removal of the metal atom from carboxypeptidase A drastically reduces its activity toward both peptide and ester substrates [22,23]. The mechanisms for the loss of activity are different, however [7]. Thus, the apoenzyme (CPD), although unable to catalyze the hydrolysis of peptide substrates, binds them to the same degree as does the zinc enzyme (Fig. 8). However, when zinc is removed from the native enzyme, binding of the exact ester analog decreases by orders of magnitude. Hence, the binding of peptides to metallocarboxypeptidase must differ from that of esters. Kinetic studies of the metalloenzymes are in agreement with these conclusions [7,10]. The results of the studies

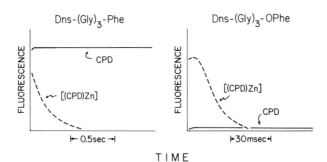

Fig. 8 Stopped-flow fluorescence measurements of Dns-(Gly)$_3$-L-Phe, 1×10^{-4} M, and Dns-(Gly)$_3$-L-OPhe, 1×10^{-4} M, both binding to zinc (dashed line) and apo- (solid line) carboxypeptidase A at pH 7.5 and 25°C in 0.03 M Tris–1.0 M NaCl. Concentration of the zinc and apoenzyme was 5×10^{-5} M for peptide and 2×10^{-5} M for ester hydrolysis. Enzyme tryptophans were excited at 285 nm and dansyl emission was measured by means of a 430 nm cutoff filter.

indicate that the metal atom interacts with the carboxyl group of esters during the binding step but with the carbonyl group of peptides during the catalytic step.

SYNCATALYTIC MEASUREMENT OF DISTANCES

N-Dansylated oligopeptide substrates have been used to report distances between the cobalt atom of carboxypeptidase and the substrate dansyl group within the enzyme active center while simultaneously signaling other aspects of active-center topography as the enzymatic reaction is in progress [3,8].

The cobalt absorption spectrum overlaps the fluorescence emission spectrum of the N-dansyl blocking group bound to the enzyme. The cobalt atom, the dansyl group, and tryptophanyl residues constitute two donor–acceptor pairs, with the dansyl group playing a dual role. Subsequent to excitation of tryptophan, energy is transferred in tandem to the dansyl group and from the dansyl group to the cobalt atom, constituting an energy-relay system, the operation of which is terminated by dissociation of the product after scission of the susceptible bond [3,8].

Energy transfer between tryptophanyl residues of the enzyme and the dansyl group of bound substrate allows observations of the formation and breakdown of the ES complexes for both the zinc and cobalt enzymes, irrespective of the degree of energy transfer between the dansyl group and the cobalt atom.

Energy transfer, T, from the dansyl group of the substrate in the ES complex to the cobalt atom of the enzyme is calculated from the relative fluorescence efficiency, F'_{Co}/F'_{Zn}, of the dansyl moiety of the substrate when bound to either of the two enzymes [Eq. (12)]. The ratio of donor–acceptor separation,

$$1 - T = F'_{Co}/F'_{Zn} \tag{12}$$

R, to the critical distance for 50% efficient energy transfer, R_0, calculated from the Förster equation [24,25] is

$$1 - T = 1/[1 + (R_0/R)^6] \tag{13}$$

and

$$R_0^6 = 8.78 \times 10^{-25}\kappa^2 QJ/n^4 \tag{14}$$

where the donor quantum yield in the absence of transfer, Q, and the overlap integral, J, are quantities determined experimentally; the index of refraction of the solvent, n, is 1.33; and a value of $\frac{2}{3}$ is employed for the random dipole orientation factor κ^2 [8].

Assignment of the probable value of κ^2 has consistently posed problems in the determination of R. The value of κ^2 can range from 0, when all vectors are mutually orthogonal, to 4, when all vectors are parallel. The random average for donor–acceptor orientation is $\frac{2}{3}$. The choice of cobalt as an energy acceptor has the great advantage that the nearly triply degenerate visible absorption transitions of cobalt [26] limit the possible range of κ^2 values from $\frac{1}{3}$ to $\frac{4}{3}$ with $\frac{2}{3}$ remaining the random average [8]. A metal–metal energy transfer system, such as has been recently demonstrated for terbium and cobalt in thermolysin, can reduce this range even further [27]. In this case, both the donor and acceptor energy levels possess degeneracies or near degeneracies that virtually eliminate the relative orientation problem, confining the value of κ^2 to about $\frac{2}{3}$ and hence justifying its use.

The distance between the cobalt atom, acting at the susceptible peptide bond, and the dansyl group of bound substrates, as determined experimentally by means of energy transfer [3,8], are within the limits of those measured on Corey–Pauling–Koltun (CPK) models of such peptides, assumed to be in an extended conformation (Table 3). Further, the increase in distance as a function of the chain length of the peptides is internally consistent for both the Phe and Trp sets of substrates, while the results obtained for corresponding members of the two sets of peptides agree well.

These dansyl–cobalt energy transfer measurements determine the radii of arcs about the cobalt atom along which the dansyl group might lie. The intersection of these arcs with the enzyme surface defines the regions surveyed by the dansyl peptide substrates. Substrate orientation within such contours remains to be determined.

TABLE 3

Dansyl–Cobalt Distances in Carboxypeptidase–Substrate Complexes[a]

Substrate	R (Å)[b]	R (Å), CPK models[c]
Dns-Gly-L-Phe	< 8	7
Dns-Gly-L-Trp	< 8	
Dns-Gly-Gly-L-Phe	11.1–11.3	10
Dns-Gly-Gly-L-Trp	10.8–12.3	
Dns-Gly-Gly-Gly-L-Phe	11.7–12.7	13
Dns-Gly-Gly-Gly-L-Trp	12.9–14.4	
Dns-Gly-Gly-Gly-Gly-L-Phe	14.1–14.7	16

[a] 1 M NaCl–0.02 M Tris, pH 7.5, 25°C [8].

[b] The range of experimental values for the distance, R, reflects that of possible bound substrate quantum yields.

[c] Measured on Corey–Pauling–Koltun molecular models from the center of the dansyl group to the cobalt atom assuming the peptide to be in an extended conformation and the metal to bind the oxygen of the C-terminal peptide bond.

This approach can be extended by placement of specific donor or acceptor groups at various strategic positions on the enzyme surface [3]. By this means it may be possible to measure catalysis-related distances and the rate of conformational movements of the enzyme. While the present approach already allows qualitative decisions in regard to the polarity of the active-center environment, the scope of the information can be increased further by rapid scanning techniques, as well as by polarization and lifetime measurements.

DETECTION OF MULTIPLE INTERMEDIATES

The effects of substrate structure, inhibitors, pH, and ionic strength on the steady-state parameters for hydrolytic enzymes have often been used to indicate the presence of an intermediate other than the Michaelis complex [28,29, and references therein]. The results of pre-steady-state studies using stopped-flow techniques are also consistent with such conclusions [28,29]. However, even when the stopped-flow technique was used, the ES complex was often inferred from the kinetics of the release of chromophoric products or substrate displacement of absorbing or fluorescing inhibitor complexes.

In many hydrolytic enzymes the existence of a covalent intermediate formed between the enzyme and the substrate has been inferred from the rapid release of a chromophoric product in a pre-steady-state reaction of such an enzyme with nonspecific substrates, e.g., p-nitrophenyl acetate or p-nitrophenyl phosphate [30–33]. After the initial rapid release the p-nitro-phenolate is released at a constant rate equal to the steady-state rate of hydrolysis. This has been interpreted to imply that there is a rapid formation of a covalent intermediate followed by its rate-determining breakdown to yield the first product.

Another indirect method involves formation of an enzyme–inhibitor complex with a distinctive spectrum. Proflavin, a competitive inhibitor of trypsin and α-chymotrypsin activity, has been used frequently [34,35]. For proflavin in water, the λ_{max}, 444 nm, shifts to 465 nm when it is bound to the enzyme. Upon addition of substrate the inhibitor is displaced. If substrate and inhibitor binding to the enzyme is always very rapid, slow equilibration of ES complexes results in a slow change in the concentration of free enzyme. The rapid equilibration of proflavin with the free enzyme thus allows indirect monitoring of the formation and equilibration of ES complexes.

While the results of all of the above methods can be interpreted in terms of the existence of more than one intermediate in the reaction, the experiments have not been designed to observe such intermediates directly.

Spectrophotometric identification of acyl derivatives of chymotrypsin

(α-CT) has been accomplished [36–40]. Thus, this enzyme reacts stoichiometrically with acylating agents such as cinnamoyl- and 3-(2-furyl)acryloyl-imidazole [39] and p'-nitrophenyl anthranilate [40]. The resulting derivatives are sufficiently stable to be studied spectrophotometrically by conventional techniques [39,40]. However, these derivatives are turned over so slowly the possibility exists that the pathway for hydrolysis of these acylating agents is not the same as that for substrates that are hydrolyzed rapidly.

Intermediates in the chymotrypsin-catalyzed hydrolysis of specific esters and peptides have been identified by absorbancy changes of the enzyme near 290 nm [41]. The reactions of α-CT with N-acetyl-L-phenylalanine methyl ester and N-furylacryloyl-L-tryptophanamide have been studied in this manner. Intermediates other than the Michaelis complex have been observed in both cases [41]. Only small absorbance changes are observed upon formation of the intermediates, the signal-to-noise ratio being often as low as 2/1, making it difficult to do extensive kinetic studies by this procedure.

For yeast carboxypeptidase and chymotrypsin acting on N-dansylated substrates, intermediates other than the Michaelis complexes are directly observed [9,42,43]. Four distinct steps are seen in the stopped-flow fluorescence studies of ester hydrolysis by these enzymes (Figs. 9 and 10): (1) A rapid increase in dansyl fluorescence occurs within the time resolution of the stopped-flow instrument. This step is considered to be the very rapid equilibration of enzyme with substrate to form the Michaelis complex. (2) A relatively slow, first-order increase in dansyl fluorescence occurs having rate constants in the range of 10 to $\sim$100 sec^{-1} depending on the enzyme, substrate, substrate concentration, pH, and temperature. This step signals the formation of a second ES complex, ES', and reflects, e.g., a conformational change in the enzyme or formation of a covalent enzyme substrate intermediate. (3) There is a time interval during which the fluorescence remains

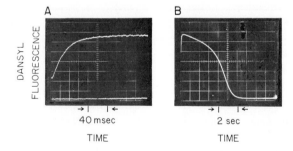

Fig. 9 Dansyl fluorescence during the time course of hydrolysis of Dns-(Gly)$_3$-L-OPhe, 2×10^{-5} M, catalyzed by yeast carboxypeptidase, 1×10^{-6} M, in 0.1 M NaCl–0.03 M acetate, pH 4.5, 20°C. Excitation was at 285 nm and dansyl emission was measured by a 430 nm cutoff filter. Scale sensitivities are 50 mV/div for (A) and (B).

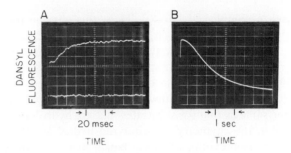

Fig. 10 Dansyl fluorescence during time course of hydrolysis of Dns-(Gly)$_2$-L-Phe-OMe, 1×10^{-4} M, catalyzed by α-chymotrypsin, 1×10^{-5} M, in 0.1 M KCl–0.03 M Mes, pH 6.5, 20°C. Excitation was at 285 nm and dansyl emission was measured using a 430 nm cutoff filter. Scale sensitivities are 20 mV/div for (A) and (B).

constant. Since in these experiments $[S]_0$ is usually far greater than $[E]_T$ the constant signal is believed to represent the maximum steady-state concentration of the ES complexes. (4) Finally, the fluorescence decreases as the dansyl intermediate is broken down to a dansyl product which does not bind to the enzyme. A completely analogous series of events occurs when the reaction is monitored by the quenching of tryptophan fluorescence.

These observations can be described minimally by the kinetic scheme (15),

$$ \text{E} + \text{S} \underset{}{\overset{K_s}{\rightleftharpoons}} \text{ES} \underset{k_{-2}}{\overset{k_2}{\rightleftharpoons}} \text{ES}' \xrightarrow{k_3} \text{E} + \text{P}_1 + \text{P}_2 \qquad (15) $$

where P_1 is liberated upon either the formation or the breakdown of the intermediate ES'.

TABLE 4

Hydrolysis of Dns-(Gly)$_3$-L-OPhe by Yeast Carboxypeptidase[a]

Method	K_m (μM)	K_s' (μM)	K_s (μM)	k_{cat} (min^{-1})
E_T/v vs. $1/S_0$[b]	6.5	—	—	630
$AE_T/F_{max}S_0$ vs. $1/S_F$[c]	—	3.4	—	615
$1/T$ vs. $1/S_F$[c]	—	3.7	36	—
$1/Q$ vs. $1/S_F$[c]	—	4.5	21	—

[a] All analyses performed at 20°C, in 0.1 M NaCl, and pH 5.5. Concentration of Mes buffer was 1×10^{-6} M for pH-stat assay and 3×10^{-2} M for stopped-flow assay [38].

[b] Steady-state conditions are $[S]_0 = 4$–130 μM and $[E]_T = 2$–20 nM.

[c] Stopped-flow fluorescence conditions are $[S]_0 = 0.5$–200 μM and $[E]_T = 0.3$ μM. Symbols defined in Table 1.

The different steps in the reaction can be quantitated by varying substrate concentration and measuring the initial (step 1) and intermediate (step 3) dansyl and tryptophan signals and the rate of interconversion of these two states. The dissociation constant for the Michaelis complex, K_s, can be obtained by measuring the F_{max} values for the first step [see Eq. (6)], and in a similar fashion a value of K_s' (defined by $[ES]_{tot} = [E]_{tot}[S]/(K_s' + [S])$) can be calculated for the second step. The latter values agree closely with the K_m value for the reaction obtained by steady-state kinetics (Table 4) [42].

Our present efforts are directed toward the identification of the chemical nature of intermediates and the study of these reactions at lower temperatures, under stopped-flow conditions, where it may be possible to visualize yet other steps along the pathway [44].

ACKNOWLEDGMENT

This work was supported by Grant-in-Aid GM-15003 from the National Institutes of Health of the Department of Health, Education and Welfare.

REFERENCES

1. B. Chance, *in* "Techniques of Chemistry" (A. Weissberger, ed.), 3rd ed., Vol. 6, "Investigation of Rates and Mechanisms of Reactions" (G. Hammes, ed.), Part II, p. 5. Wiley (Interscience), New York, 1974.
2. P. Douzou, *Method Biochem. Anal.* **22**, 401 (1974).
3. S. A. Latt, D. S. Auld, and B. L. Vallee, *Proc. Natl. Acad. Sci. U.S.A.* **67**, 1383 (1970).
4. R. F. Chen, *Arch. Biochem. Biophys.* **120**, 609 (1967).
5. R. H. Conrad and L. Brand, *Biochemistry* **7**, 777 (1968).
6. D. S. Auld, S. A. Latt, and B. L. Vallee, *Biochemistry* **11**, 4994 (1972).
7. D. S. Auld and B. Holmquist, *Biochemistry* **13**, 4355 (1974).
8. S. A. Latt, D. S. Auld, and B. L. Vallee, *Biochemistry* **11**, 3015 (1972).
9. D. S. Auld, *Fed. Proc., Fed. Am. Soc. Exp. Biol.* **35**, 703 (1976).
10. D. S. Auld and B. L. Vallee, *Biochemistry* **9**, 602 (1970).
11. D. S. Auld and B. L. Vallee, *Biochemistry* **9**, 4352 (1970).
12. D. S. Auld and B. L. Vallee, *Biochemistry* **10**, 2892 (1971).
13. B. Chance, *J. Biol. Chem.* **151**, 553 (1943).
14. H. Lineweaver and D. Burk, *J. Am. Chem. Soc.* **56**, 658 (1934).
15. M. Dixon and E. C. Webb, "Enzymes," 2nd ed., p. 316. Academic Press, New York, 1964.
16. E. Elkins-Kaufman and H. Neurath, *J. Biol. Chem.* **178**, 645 (1949).
17. J. R. Whitaker, F. Menger, and M. L. Bender, *Biochemistry* **5**, 386 (1966).
18. S. A. Latt, D. S. Auld, and B. L. Vallee, *Anal. Biochem.* **50**, 56 (1972).
19. R. R. Lobb and D. S. Auld, *Fed. Proc., Fed. Am. Soc. Exp. Biol.* **36**, 877 (1977).
20. B. L. Vallee and W. E. C. Wacker, *in* "The Proteins" (H. Neurath, ed.), 2nd ed., Vol. 5 (1970).
21. J. E. Coleman and B. L. Vallee, *J. Biol. Chem.* **235**, 390 (1960).

22. B. L. Vallee, J. F. Riordan, and J. E. Coleman, *Proc. Natl. Acad. Sci. U.S.A.* **49**, 109 (1963).
23. B. L. Vallee, J. F. Riordan, D. S. Auld, and S. A. Latt, *Philos. Trans. R. Soc. London, Ser. B* **257**, 215 (1970).
24. T. Förster, *Ann. Phys. (Leipzig)* **2**, 55 (1948).
25. T. Förster, *Mod. Quantum Chem.*, *Lect. Istanbul Summer Sch.*, *1964*, Part 3, p. 93. (1965).
26. C. J. Ballhausen, "Introduction to Ligand Field Theory." McGraw-Hill, New York, 1962.
27. W. DeW. Horrocks, B. Holmquist, and B. L. Vallee, *Proc. Natl. Acad. Sci. U.S.A.* **72**, 4764 (1975).
28. M. L. Bender, "Mechanism of Homogeneous Catalysis from Proton to Proteins," p. 487. Wiley (Interscience), New York, 1971.
29. K. J. Laidler and P. S. Bunting, "The Chemical Kinetics of Enzyme Action," 2nd ed., p. 312. Oxford Univ. Press (Clarendon), London and New York, 1973.
30. B. S. Hartley and B. A. Kilby, *Biochem. J.* **56**, 288 (1954).
31. H. Gutfreund and J. M. Sturtevant, *Proc. Natl. Acad. Sci. U.S.A.* **42**, 719 (1956).
32. F. J. Kezdy and M. L. Bender, *Biochemistry* **1**, 1097 (1962).
33. W. K. Fife, *Biochem. Biophys. Res. Commun.* **28**, 309 (1967).
34. H. Gutfreund and S. A. Bernhard, *Proc. Natl. Acad. Sci. U.S.A.* **53**, 1238 (1965).
35. K. G. Brandt, A. Himoe, and G. P. Hess, *J. Biol. Chem.* **242**, 3973 (1967).
36. M. L. Bender, G. R. Schonbaum, and B. Zerner, *J. Am. Chem. Soc.* **84**, 2540 (1962).
37. M. Caplow and W. P. Jencks, *Biochemistry* **1**, 883 (1962).
38. S. A. Bernhard, S. J. Lau, and H. Noller, *Biochemistry* **4**, 1108 (1965).
39. E. Charney and S. A. Bernhard, *J. Am. Chem. Soc.* **89**, 2726 (1967).
40. R. P. Haugland and L. Stryer, *Conform. Biopolym. Pap. Int. Symp.*, *1967*, Vol. 1, p. 321 (1967).
41. G. P. Hess, J. McConn, E. Ku, and G. McConkey, *Philos. Trans. R. Soc. London, Ser. B* **257**, 89 (1970).
42. J. Johansen and D. S. Auld (in preparation).
43. R. R. Lobb and D. S. Auld (in preparation).
44. D. S. Auld, *in* "Methods in Enzymology" (C. H. W. Hirs and S. N. Timasheff, eds.) (in preparation).

CHAPTER

2

Models of Hydrolytic Enzymes

Myron L. Bender and Makoto Komiyama

INTRODUCTION

The study of "enzyme models" has made important contributions to the understanding of enzymes. Enzyme models are simple organic molecules that contain one or more features found in enzymatic systems. For example, physical organic chemists have synthesized models in which a catalyst and an amide are present in the same molecule so that the two groups are forced to reside in close proximity. This is but one example of intramolecular catalysis, which has been admirably reviewed by Kirby and Fersht [1]. A recent example of intramolecular catalysis involves the cleavage of an amide bond by both the carboxylate ion and the phenolic group, which serves as a model for the enzyme carboxypeptidase [2].

Of course, there exist models of many other enzymatic systems [3]. These include redox systems [4,5], nonredox coenzyme systems, such as pyridoxal [6] and thiamine [7], a model for the enzyme glyoxylase [8], and many others. But by far the most extensive models have been made of hydrolytic enzymes, and this is what we cover in this chapter.

For example, we discuss carboxylic derivative reactions in apolar media. Enzymes exist in aqueous solution only, but there is a growing feeling that the active site of hydrolytic enzymes contains an apolar environment. Thus, we cover this area, which has never been reviewed before. In addition, we deal with complexes, primarily of the micellar [9], charge-transfer, and clathrate varieties, which can serve as models of hydrolytic enzymes. Furthermore, we discuss cycloamyloses as enzyme models. These were extensively reviewed by Griffiths and Bender in 1973 [10], but there have been considerable advances since then. Earlier it was pointed out that these cyclic oligomers of glucose could bind many benzenoid compounds and also catalyze their reactions (1) by interaction of one of the many hydroxyl groups in cyclo-amylose, (2) by binding of the guest molecule in a region of lower dielectric constant, or (3) by binding that leads to a restriction of conformation. Now there are many chemically modified cycloamyloses to consider.

Finally, we describe enzyme models that are based on polymers. The main premise of this area is that enzymes are polymeric systems and thus a properly constituted polymer should serve as an enzyme model. Polymers that contain electrostatic bonding have been described [11]. Also, anionic [12] and cationic [13] polymers have been described. The latter have been called "synzymes," since their reactivity has been purported to be of the same order of magnitude as that of an enzyme; thus, the term synthetic enzyme was used.

REACTIONS IN APOLAR MEDIA

The rates of nucleophilic reactions of carboxylic acid derivatives in media containing high concentrations of organic solvents are slow compared with the rates in pure water [14]. Jencks and Gilchrist [15] have shown, for example, that tetrahydrofuran and ethanol inhibit the reaction of methylamine with phenyl acetate in water. It is therefore tempting to assume that the reactive site of hydrolytic enzymes is a region of high water content, and this assumption usually has been made. On the basis of this assumption, most models of these enzymes have been studied in aqueous solution. It is possible, however, that the catalytic sites of the enzymes are apolar regions and that the transition states of the enzymatic reactions are neutral in character. A micro-scopic change in environment could conceivably lead to a large rate accelera-tion. On the surface of the protein, there are regions of low dielectric constant adjacent to regions of high dielectric constant [16]. Hydrolytic reactions obviously must involve water. Furthermore, the medium in which enzymes reside is aqueous. At least a portion of the active site must be aqueous, since a normal pK_a is found for the imidazole of the active site [17]. None of the transition states described in the most widely accepted mechanism for α-chymotrypsin involves creation of charge [18]. Thus, there appears to be conflicting evidence about the polarity of the active site.

In order to test the idea that a nucleophilic attack on an ester may occur readily in an apolar medium if the process proceeds by means of a neutral transition state and tetrahedral intermediate, the reaction of benzamidine with p-nitrophenyl acetate in chlorobenzene at 25°C was investigated [19]. It was found that benzamidine reacts with p-nitrophenyl acetate in chlorobenzene with a second-order rate constant of 3.45 M^{-1} sec^{-1}. On the other hand, n-butylamine, a nucleophile with a basicity similar to that of benzamidine, reacts with p-nitrophenyl acetate in chlorobenzene by means of a third-order process, the rate of which is little affected by the presence of large amounts of a tertiary amine. Benzamidine reacts at least 15,000 times faster than n-butylamine monomer. The reactivity of benzamidine with the ester in the apolar solvent was attributed to the bifunctional nature of the nucleophile, which cannot occur with n-butylamine.

The benzamidine model thus supports the suggestion that multifunctional catalysis by hydrolytic enzymes may occur in a cyclic fashion in apolar regions of the active sites. Since hydrolytic enzymes catalyze hydrolyses, water is obviously present in the regions of catalytic activity. This, however, does not mean that the regions are aqueous in nature. A single water molecule within a hydrophobic environment may be bound in a position suitable for reaction with an acyl-enzyme intermediate. This water molecule could react with the carbonyl carbon atom, simultaneously losing a proton which would be delivered (with the aid of intervening groups, such as an imidazole ring or other water molecules) to the carbonyl oxygen atom. In this way, hydrolysis could occur without charge generation. A similar mechanism could be envisioned for the formation of the acyl-enzyme involving the serine hydroxyl rather than a water molecule. Such a process was proposed by Bender and Kezdy earlier [20].

Further, the reactions of pyrrolidine with esters in acetonitrile and chlorobenzene are much more sensitive to substituents on the leaving group of the ester ($\rho = 4$–6) than to substituents on the acyl portion of the ester ($\rho = 1$–2). This is the reverse order of sensitivity found in reactions of esters with hydroxide ion in water. It was concluded from the values (and from the aminolysis rates of tertiary amine, pyrrolidinium ion, azide ion, and pyrrolidine-d) that collapse of a tetrahedral intermediate is rate determining [21]. Many previous studies had proposed that the formation of the tetrahedral intermediate was rate determining. A rate-determining collapse of a tetrahedral intermediate implies the possibility of its capture or detection. This has never been accomplished. Perhaps in such a system it might succeed. If in fact a hydrolytic enzyme works in the same manner, a tetrahedral intermediate could be observed not only in a model system, but also in an enzyme system [21].

In addition, tetra-n-hexylammonium benzoate hemihydrate (THAB) catalyzes the reaction between piperidine and p-nitrophenyl acetate in toluene. For example, 0.054 M THAB increases the rate more than 1200-fold. The acceleration arises from removal of a proton residing on the nitrogen atom of a tetrahedral intermediate. The intermediate can then collapse to products without first forming an N-protonated amide. In toluene, benzoate is a 10^3 times better proton acceptor than piperidine (corresponding to a 10^{10} reversal in basicity relative to that in water). Tetra-n-hexylammonium benzoate hemihydrate displays an even greater catalysis in the aminolysis of p-nitrophenyl acetate by imidazole. Thus, the half-life of the ester in toluene at 25°C with 0.0104 M imidazole and no THAB is 25 hr. Addition of 0.059 M THAB decreases the half-life to 7.5 sec. A plot of k_{obs} vs. [THAB] curves downward with the rate becoming independent of the [THAB] above 0.1 M. It was concluded that THAB induces a change in the rate-determining step from formation to decomposition of the tetrahedral intermediate. This phenomenon cannot be a micellar effect since it is not seen with octadecylammonium benzoate. The kinetic behavior of the reaction at high [THAB], where formation of the tetrahedral intermediate is rate limiting, has permitted the first determination of the rate constant for the addition of imidazole to an ester carbonyl in an apolar solvent, which may be related to the corresponding reaction in an enzyme. This rate is 29 times faster than the corresponding rate in water. Therefore, the slowness of ester aminolyses in apolar solvents, but possibly not in enzyme active sites, can be ascribed solely to an unfavorable partitioning of a tetrahedral intermediate to products [22].

There have been many proposals of the involvement of a carboxylate ion of an enzyme in catalytic processes. These might involve a carboxylate ion on the surface of the enzyme which would be hydrated or it might involve a "buried" carboxylate ion, which should be more nucleophilic because it would be stripped of its hydration shell. An examination was made of the aminolysis of the mono-o-nitrophenyl ester of oxalic acid by piperidine in toluene with the purpose of determining how a "buried" carboxylate at an active site might affect an enzyme-catalyzed reaction. The oxalate ester and piperidine formed an ion pair ($R_2NH_2^+ {}^-O_2CCO_2Ar$) even in extremely dilute toluene solutions. This conclusion was supported by the kinetic effects of acidic and basic additives and by a concentration "inversion" experiment; the oxalate ester was found to react more than three orders of magnitude faster than o-nitrophenyl acetate. The neighboring carboxylate in the ion pair apparently accelerates the decomposition of the tetrahedral intermediate by accepting a proton from the amine nitrogen. Immersing a carboxylate in an apolar portion of an active site need not necessarily lead to the rate enhancements observed in the model systems. This is because such a buried carboxylate is not only more reactive than normal; it is also more basic. If a buried carboxylate is not completely insulated from the external solvent, then the gain in catalytic activity could be negated by a corresponding decrease in carboxylate concentration. Only when the anionic charge is protected from hydrophilic regions at the active site (and this may require the presence of a bound substrate) is a large catalysis possible [23].

The hydrolysis of p-nitrophenyl acetate in acetonitrile is dependent on a term involving both imidazole and benzoate ion, indicating that in this apolar medium the "charge-relay" system is operative (even in an inter-molecular system) [24]. The aminolysis of aspirin shows saturation kinetics like an enzyme reaction [24a].

The above arguments are predicated on several simple experiments which indicate that the polarity of the solvent can have a large effect on the rate and equilibria of organic reactions. For example, chloride ion reacts with methyl iodide 2×10^6 times faster in acetonitrile than in methanol [25]. Further-more, the equilibrium heterolysis of trityl chloride in either increases 7×10^9-fold upon addition of $0.05\ M$ LiClO$_4$ [26]. If in fact the active site of an enzyme is partially or fully apolar in nature, large rate enhancements due solely to polarity might be seen. There is no question from X-ray data that the active sites of most enzymes consist of holes, clefts, or crevasses in the enzyme, which is a globular protein. Thus, it is conceivable that the active site of an enzyme is apolar even though the enzyme as a whole is water soluble. Then these studies become pertinent.

COMPLEX FORMATION

Complex formation between catalysts and substrates in solution can cause acceleration or deceleration of chemical reactions. The term "complex formation" includes the formation of micelles, charge-transfer complexes, and clathrates. Acceleration or deceleration is attributable to the proximity effect, the electrostatic effect, the medium effect, and others.

Micelles are aggregates of a large number of soap or detergent molecules loosely bound mainly through hydrophobic interaction. Micelles are usually formed in aqueous solution; the polar and nonpolar groups are found at the surface and in the interior of micelles, respectively. Micelles bind a variety of organic substrates through apolar interaction, resulting in catalyses of the substrates. Reactions catalyzed by micelles usually take place on the surface of the micelles, which is consistent with the fact that the majority of chemical reactions in biological systems occur on or in the vicinity of the boundaries between apolar and polar regions. Micellar catalyses exhibit enzyme-like features such as Michaelis–Menten type of kinetics, stereospecificity, and DL specificity. Thus, reactions in micelles can be used as models of enzymatic reactions [9,27–32]. Here, we cover only those micellar systems useful as models of hydrolytic enzymes, in which the surfactants have hydroxyl, imidazolyl, or carboxyl groups. Menger has discussed micellar effects from a physicochemical point of view in Chapter 7, Volume III, so the emphasis here is solely on model systems [32].

The effects of hydroxyl and of imidazolyl groups in surfactant molecules on the cationic micelle-catalyzed hydrolysis of p-nitrophenyl acetate were investigated by Moss and co-workers [33]. Hydroxyl groups in the surfactant molecules caused 12- to18-fold, and imidazolyl groups in the molecules caused 1200-fold, rate enhancements. But coexistence of the hydroxyl and imidazolyl groups in the surfactant molecules did not produce the anticipated rate enhancement; there was only 65%, whereas the cooperativity expected of the combined groups should have been much greater. In this cationic micellar catalysis, both the hydroxyl [34] and imidazolyl groups [35,36] are supposed to function as catalysts in their anionic states. The imidazolyl group, however, in some of the cationic surfactants functions in its neutral state in micellar systems [37,38]. The distance between imidazolyl group and cationic site in the surfactant, as well as other factors, may be important in deciding the nature of the catalytically active species.

N^{α}-Myristoyl-L-histidine (MirHis) forms a mixed micelle with cetyltri-methylammonium bromide (CTAB); this mixed micelle exhibits remarkable catalyses in p-nitrophenyl ester hydrolyses [39]. N^{α}-Myristoyl-L-histidine barely catalyzes the reaction, whereas CTAB retards the reaction. In this mixed micelle, CTAB provides the binding site for the substrate, and the imidazolyl residue in MirHis functions as the nucleophilic catalyst. The reaction was shown to proceed in two steps, with rapid acylation of the imidazolyl group of MirHis followed by rate-determining breakdown of this intermediate. The logarithm of the apparent binding constant increased linearly with the number of carbons in the acyl group of the p-nitrophenyl esters. The free-energy change (-630 cal/mole) for the transfer of each methylene group of the acyl chain from the bulk solution to the CTAB micellar phase supports apolar binding between the micelle and the substrate.

Surfactants were also used as models of the cooperativity of two functional groups. Sunamoto and co-workers [40] found that addition of N,N-dimethyl-N-hexadecyl-N-(4-imidazolium)methylammonium dichloride (Im-I) to the ester cleavage catalyzed by 10-hydroxyl-11-hydroxyimino[20]paracyclophane (Oxime-I) caused a rate acceleration of 22-fold over the catalysis by Oxime-I [41] and of 39-fold over catalysis by Im-I. Catalysis was attributed to the cooperativity of the imidazolyl group in Im-I and the oxime group in Oxime-I in the ternary complex composed of Im-I, Oxime-I, and the substrate.

Recently, the solubilization sites and orientation of the substrate within micelles were investigated by use of absorption and proton magnetic resonance (^{1}H nmr) spectroscopy [42,43]. These kinds of studies can make the micellar system a more useful model of enzymes.

Reversed micelles, surfactant aggregates in nonpolar solvents, are also good models of enzymes, since they have cavities with polar functional groups capable of binding substrates in specific orientations [44]. For example, such a large molecule as vitamin B_{12a} can be solubilized, together with some water molecules, in benzene by dodecylammonium propionate and sodium di(2-ethylhexyl) sulfosuccinate. In these reversed micelles, cobalamin is surrounded by some 300 surfactant molecules and is effectively shielded from the bulk apolar solvent [45]. The remarkable rate enhancements in reversed micelles have been ascribed to favorable substrate orientation in the interior of the reversed micelles, where bond breaking may be assisted by proton transfer [46–49].

In addition to micellar complexes, other complexes that can serve as enzyme models are charge-transfer complexes. Charge-transfer complexes are formed between electron acceptors (A) and electron donors (D). The charge-transfer force is basically due to the resonance between the nonbonded structure (D·A) and the bonded structure (D$^+$–A$^-$) [50]. Several chemical reactions proceed through the formation of charge-transfer complexes [51–53]. The acceleration of the solvolysis of 2,4,7-trinitrofluorenyl p-toluenesulfonate in acetic acid by electron donors such as phenanthrene, hexamethylbenzene, and naphthalene can be attributed to the stabilization of transition state by charge-transfer interaction [54–56]. However, charge-transfer interactions do not seem to be very important for binding substrates in enzymatic reactions [57], and thus charge-transfer complexes do not appear to be useful as enzyme models.

There have been few reports about the use of clathrates as enzymatic models, although they may become important in the future.

CYCLOAMYLOSES

Structure and Inclusion Complex Formation with Guest Compounds

The cycloamyloses, sometimes called Schardinger dextrins, cyclodextrins, or cycloglucans, are cyclic oligosaccharides consisting of α-1,4-linked D-glucopyranose rings. Cyclohexaamylose, cycloheptaamylose, and cyclooctaamylose (α-, β-, and γ-cyclodextrins) contain, respectively, six, seven, and eight glucose residues in a molecule. Furthermore, cyclononaamylose, which has nine glucose residues per molecule, was found by French et al. [58]. X-Ray crystallography [59–61] showed that cycloamyloses have a doughnut shape, with the glucose units in the C1 conformation. The inner diameters of the cavities in cyclohexaamylose, cycloheptaamylose, and cyclooctaamylose are approximately 4.5 [62], 7.0 [63], and 8.5 [63] Å, respectively. The primary hydroxyl groups at the C-6 atom of a glucose unit are arranged in one of the open ends of the cavity, while the secondary hydroxyl groups at the C-2 and C-3 atoms of a glucose unit are arranged in the other open end. The interior of the cavity contains a ring of C—H groups, a ring of glycosidic oxygens, and another ring of C—H groups. Therefore, the interior of cyclo-amylose torus has hydrophobic (apolar) character, as do the binding sites of most enzymes.

Cycloamyloses (hosts) form inclusion complexes with many kinds of mole-cules and ions (guests), either in the solid phase or in solution. The stoichi-ometry of guest compounds to host compounds in inclusion complexes is usually 1:1 in aqueous solution, with the exception of the cycloamylose–long-chain aliphatic acid complexes [64] and some others. X-Ray analyses for the cyclohexaamylose–potassium acetate complex [59], the cyclohexaamylose–iodine complex [65], and the cyclohexaamylose–p-iodoaniline complex [66] showed that the guest compounds are really included in the cavity of the cycloamylose. Formation of inclusion complexes was also detected by solu-bility methods [64,67–71], by absorption [72–75], by circular dichroism (CD) [76,77], and by ^{1}H nmr spectroscopy [78,79]. The solubility method is based on an increase in the solubility of the guest compound or a decrease in the solubility of cycloamylose due to inclusion complex formation. Absorption and CD spectroscopies measure the change in medium from aqueous to hydrophobic and the change in chirality of the cavity of the cycloamylose, respectively, while ^{1}H nmr spectroscopy makes use of the anisotropic shielding effect of an aromatic guest compound on the protons of cycloamylose.

The driving force for inclusion complex formation is still unidentified. The importance of hydrophobic interactions, however, seems to be most probable. The water molecules in the cavity of cycloamylose are enthalpy rich because of an inadequate complement of hydrogen bonds. Thus, inclusion complex formation is associated with release of high-energy water molecules, resulting in a favorable enthalpy change. Inclusion complex formations are usually unfavorable in organic solvents such as ether, carbon tetrachloride, and benzene [68]. Siegel and Breslow [80] found that inclusion complexes can be formed in polar nonaqueous media such as dimethyl sulfoxide and dimethylformamide, which have dielectric constants of 46.7 and 36.7, respectively. However, the dissociation constants of complexes in dimethyl sulfoxide are considerably larger (180 times for the cycloheptaamylose–m-tert-butylphenyl acetate complex and 80 times for the cycloheptaamylose–anisole complex) than of those in aqueous media. This result suggests the importance of hydrophobic interaction. Some other types of interaction forces, such as hydrogen bonding or ion pairing, are disrupted by highly polar solvents.

In addition to hydrophobic interaction, hydrogen bonding, van der Waals and London dispersion forces may all play roles in inclusion complex forma-tion [81]. The importance of hydrogen bonding is indicated by the fact that tert-butyl alcohol, which hydrogen bonds less easily than tert-butyl hydro-peroxide, does not form an inclusion complex with cycloamylose, while the latter does [82]. Release of conformational strain energy of cycloamylose can also function in inclusion complex formation [83–85], although it is less of a driving force in complexation [86].

One of the principal reasons for the utilization of the cycloamylose-catalyzed reactions as models of hydrolytic enzyme reactions is the formation of inclusion complexes between the catalyst and the substrate preceding the catalyses, which is comparable to the formation of a Michaelis–Menten complex in enzymatic reactions.

Manor and Saenger [83] proposed, on the basis of results of X-ray crystallography and potential energy calculations, that cycloamyloses can be even better models for enzymes than was hitherto assumed since there is a conformational change of the cycloamylose when substrates are included in it, much as enzymes exhibit "induced fit" in their interaction with substrates. The macrocyclic conformation of the cyclohexaamylose torus in solution before inclusion complex formation, assumed to be identical with that in cyclohexaamylose hexahydrate [83,87], is less symmetrical and of higher energy than the conformation of the cyclohexaamylose in inclusion complexes with iodine, 1-propanol, methanol, or potassium acetate. Thus, Manor and Saenger concluded that inclusion complex formation is associated with a conformational change of the cyclohexaamylose.

Catalysis by Cycloamyloses

It is well known that cycloamyloses can accelerate or decelerate chemical reactions, mostly because of inclusion complex formation of the cyclo-amylose with the substrate. Cycloamylose-catalyzed reactions exhibit many of the kinetic features shown by enzymatic reactions, including catalyst–substrate complex formation [69,73,80,88–97], competitive inhibition [73–88], saturation [73,80,88–96], stereospecific catalysis [73,92–94,98–100], and DL specificity [91,101]. Furthermore, a bell-shaped pH-rate constant profile, which is characteristic of many enzymatic reactions, was observed for the cycloamylose-catalyzed hydrolysis of mandelic acid esters [102].

It is useful to classify cycloamylose-catalyzed reactions into two categories: (1) covalent catalyses, in which cycloamyloses catalyze reactions via formation of covalent intermediates and (2) noncovalent catalyses, in which cyclo-amyloses provide their cavities as hydrophobic or sterically restricted reaction fields without formation of covalent intermediates.

COVALENT CATALYSIS

Covalent catalyses by cycloamyloses (C), used for models of hydrolytic enzymes, proceed as shown in Scheme 1, where S, S·C, and C–P_1 represent, respectively, the substrate, the inclusion complex between S and C, and the

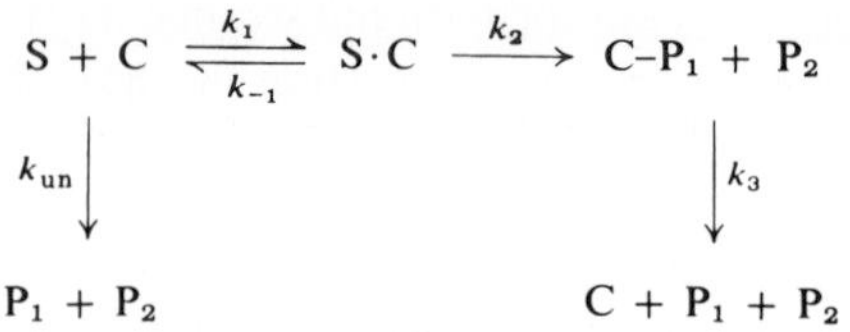

$$S + C \underset{k_{-1}}{\overset{k_1}{\rightleftharpoons}} S\cdot C \xrightarrow{k_2} C\text{--}P_1 + P_2$$

$$k_{un} \downarrow \qquad\qquad\qquad\qquad \downarrow k_3$$

$$P_1 + P_2 \qquad\qquad\qquad C + P_1 + P_2$$

Scheme 1

covalent intermediate, usually an acyl cycloamylose; P_1 and P_2 are final products, an acid and an alcohol from an ester; and k_1, k_{-1}, k_2, k_3, and k_{un} are rate constants. The formation of the $S\cdot C$ complex preceding the catalytic reactions has been shown by saturation phenomena. The reaction rate is not a linear function of cycloamylose concentration, but it approaches a maximum value asymptotically with an increase of the cycloamylose concentration. The formation of the $C\text{--}P_1$ intermediate has been proved both by spectroscopic detection of cycloamylose benzoate in the cycloamylose-catalyzed hydrolyses of phenyl benzoates [99,100] and by isolation of acyl cycloamyloses *in situ* [99,100,103]. Scheme 1 for the cycloamylose-catalyzed reactions is identical with the scheme proposed for many enzymatic reactions. Thus, the maximal rate constant (k_{cat}) and the binding (dissociation) constant, K_d, which is equal to k_{-1}/k_1, can be determined by a Lineweaver–Burk plot [Eq. (1)] [104] of the kinetic data under the condition that $[C]_0 \gg [S]_0$,

$$\frac{1}{k_{obs} - k_{un}} = \frac{K_d}{(k_{cat} - k_{un})[C]_0} + \frac{1}{k_{cat} - k_{un}} \tag{1}$$

where $[C]_0$ and $[S]_0$ are the initial concentrations of cycloamylose and substrate, respectively. When $[S]_0 \gg [C]_0$, $[C]_0$ in Eq. (1) is simply replaced by $[S]_0$. The values of k_{cat} and K_d, determined under the condition that $[C]_0 \gg [S]_0$, are identical to those obtained under the condition that $[S]_0 \gg [C]_0$, within experimental error [98].

Of many kinds of cycloamylose-catalyzed reactions, the cycloamylose-catalyzed hydrolyses of substituted phenyl acetates have been most extensively and precisely investigated [73,98–100]. The hydrolysis rates of meta- and para-substituted phenyl acetates catalyzed by cycloamyloses exhibit a large dependence on the positions and sizes of substituents [73]. Importantly, meta-substituted phenyl acetates show larger accelerations than the para analogs. The acceleration for phenyl acetate is midway between the meta- and para-substituted esters. Besides, meta/para specificity is larger for substituents of larger bulk. No linearity was found between the logarithm of the rate constant and the Hammett substituent constant in the presence of cycloamyloses, although the hydrolyses adhere to a linear Hammett relationship in the absence of cycloamylose. Thus, it is not electronic effects, but steric effects, that govern the rate of cycloamylose-catalyzed ester hydrolysis.

TABLE 1

Catalytic Rate Constants and Accelerations in the
Cyclohexaamylose-Catalyzed Hydrolyses of Phenyl Acetates [a,b]

Acetate	k_{cat} (10^{-2} sec^{-1})	$\dfrac{k_{cat}}{k_{un}}$	K_d (10^{-2} M)
Phenyl	2.19	27	2.2
m-Tolyl	6.58	95	1.7
p-Tolyl	0.22	3.3	1.1
m-tert-Butylphenyl	12.9	260	0.2
p-tert-Butylphenyl	0.067	1.1	0.65
m-Nitrophenyl	42.5	300	1.9
p-Nitrophenyl	2.43	3.4	1.2
m-Carboxyphenyl	5.55	68	10.5
p-Carboxyphenyl	0.67	5.3	15.0

[a] Reprinted with permission from R. L. VanEtten, J. F. Sebastian, G. A. Clowes, and M. L. Bender, *J. Am. Chem. Soc.* **89**, 3242 (1967). Copyright by the American Chemical Society.
[b] pH 10.6, Carbonate buffer, $I = 0.2$ M: 1.5% (v/v) acetonitrile–water.

Table 1 lists the k_{cat} and K_d values of the cycloamylose-substituted phenyl ester complexes, determined by Lineweaver–Burk plots of Eq. (1) [73]. The k_{cat} values for meta-substituted compounds are 5-, 236-, 88-, and 13-fold larger than those for para-substituted compounds, respectively, for methyl, *tert*-butyl, nitro, and carboxyl substitutions, whereas K_d values for meta-substituted compounds are 1.5, 0.3, 1.6, and 0.7 times those for para-substituted compounds, respectively, for corresponding substituents. Thus, the specificity for meta-substituted compounds in catalysis by cycloamyloses is derived from the larger value of k_{cat}, not from the smaller value of K_d. The sensitivity of k_{cat} to the stereochemistry of the substrate is quite similar to that found in enzymatic reactions.

The catalytic site of cycloamyloses in hydrolyses of esters involves the anion derived from a secondary hydroxyl group [100]. The pH–k_2 profile in the cyclohexaamylose-catalyzed hydrolysis of *m*-tolyl acetate corresponds to functional group of pK_a 12.1. Besides, hepta(6-O-mesyl) cycloheptaamylose, in which the primary hydroxyl groups are blocked, causes as large an acceleration of phenyl ester cleavage as native cycloheptaamylose does, while dodecamethylcyclohexaamylose, in which the primary hydroxyl groups and half of the secondary hydroxyl groups are blocked, causes a small inhibition of the hydrolysis.

The specificity for meta- and para-substituted phenyl acetates in cycloamylose-catalyzed hydrolyses can be explained by the geometry of the carbonyl carbon of the included ester around the secondary hydroxyl groups of the cycloamylose. Studies using Corey–Pauling–Koltun molecular models showed that the carbonyl carbon of the meta-substituted phenyl acetate included in cycloamylose is much closer to the secondary hydroxyl groups of the cyclo-amylose than that of the corresponding para-substituted ester [73,105]. The proximity between the catalytic site of the cycloamylose and the reactive site of the substrate in the catalyst–substrate complex is the origin of stereo-specificity in catalysis [106].

Another enzyme-like feature of cycloamylose-catalyzed ester hydrolysis is competitive inhibition by reaction products [73]. A variety of acid anions retard cycloamylose-catalyzed hydrolyses. Here, an anion containing a larger hydrophobic group is a more effective retardant, indicating that the retardation of hydrolysis takes place because the anion and substrate, in binding, compete for the cycloamylose cavity.

As described above, cycloamylose-catalyzed hydrolyses proceed quite similarly to enzymatic hydrolysis. Thus, it is interesting to compare the cycloamylose-catalyzed rate constants with the rate constants for the corre-sponding alkaline-catalyzed hydrolysis of the same substrate, as shown in Table 2. Here, the second-order rate constants for the cycloamylose reactions were obtained by division of k_2(lim) (the maximal rate constant) by the dissociation constant, K_d. The cycloamylose reactions were 10^3- to 10^4-fold superior to hydroxide ion reactions, while the chymotrypsin reactions are 10^4- to 10^6-fold superior to hydroxide ion reactions. Thus, with respect to the rate enhancement, cycloamylose seems to be as effective as chymotrypsin. However, there are two important differences between the cycloamylose-catalyzed hydrolysis and the chymotrypsin-catalyzed hydrolysis: (1) The rate constant for the cycloamylose reaction was determined at pH 13, its maximum,

TABLE 2

Second-Order Rate Constants for the Reactions of Substrates with Cycloamyloses and with Chymotrypsin[a]

Substrate		Catalyzed rate/ hydroxide ion rate
m-tert-Butylphenyl acetate	Cyclohexaamylose	1.8×10^3
m-tert-Butylphenyl acetate	Cycloheptaamylose	2.6×10^4
Acetyltryptophanamide	Chymotrypsin	4×10^4
Acetyltryptophan ethyl ester	Chymotrypsin	1×10^6
Acetyltyrosine ethyl ester	Chymotrypsin	3×10^4

[a] Reprinted with permission from R. L. VanEtten, G. A. Clowes, J. F. Sebastian, and M. L. Bender, *J. Am. Chem. Soc.* **89**, 3253 (1967). Copyright by the American Chemical Society.

while that for the chymotrypsin reaction was determined at pH 8, its maximum; and (2) the rate constant k_3 is much smaller than k_2 in the cycloamylose reactions, resulting in its inefficiency as a catalyst.

Besides phenyl ester hydrolyses, cycloamyloses catalyze the hydrolyses of penicillin derivatives to pencillin acids; thus, cycloamyloses can be a model of the enzyme penicillinase [89,90]. The hydrolysis rates of penicillin derivatives included in cycloheptaamylose are 21–89 times the alkaline hydrolysis rates. The mechanism of this catalysis is the same as that of the cycloamylose-catalyzed phenyl ester hydrolyses. Furthermore, cycloamyloses catalyze the hydrolyses of organophosphorus derivatives [88,92–94].

NONCOVALENT CATALYSIS

Noncovalent catalyses by cycloamyloses are attributable either (1) to microsolvent effects due to the hydrophobic character of the cycloamylose cavity or (2) to conformational effects due to the geometric requirements of the inclusion process. The microsolvent effect is associated with enzymatic solvent effects [107], while the importance of conformational restrictions, strains, and distortion in substrates on binding in enzymatic reactions is widely accepted [108–111].

Kinetically, the microsolvent effect in noncovalent catalyses by cycloamyloses is very different from that in covalent catalyses in that (1) the catalytic rate constant is not highly dependent on the structure of the substrate [95,112,113], (2) the Hammett relationship holds with a ρ value larger than that for the same reaction in H_2O, which is consistent with the smallness of the dielectric constant in the cycloamylose cavity compared to that in an aqueous medium [95], and (3) the catalytic rate constant does not depend at all on the ionization of the hydroxyl groups of the cycloamylose [95,114]. Besides, it is important to note that the activation parameters for the cycloamylose-catalyzed reactions due to the microsolvent effect are almost identical to those for the uncatalyzed reactions in a solvent of low dielectric constant [95]. For example, the activation energy (26.1 kcal/mole) and activation entropy (10 eu) for the cycloheptaamylose-catalyzed decarboxylation of 4-chlorophenyl cyanoacetate are equal to those (25.9 kcal/mole and 10.6 eu, respectively) for the uncatalyzed reaction in 2-propanol–H_2O (74.0 wt %) solution within experimental error, although they are much different from those (31.3 kcal/mole and 19 eu, respectively) for the uncatalyzed reaction in H_2O [95].

Some of the noncovalent catalyses by cycloamyloses have been ascribed to conformational effects rather than microsolvent effects. The decarboxylation of un-ionized benzoylacetic acids [115] and the migration of the trimethylacetyl group in 2-hydroxymethyl-4-nitrophenyl trimethylacetate [116] were accelerated by 7.6- and 7.4-fold, respectively, by cycloheptaamylose through the operation of conformational effect. A portion of the free energy gained from inclusion complex formation between the cycloamylose and the substrate can be used to impose a restriction of the substrate in a conformation favorable for reaction [116]. Interestingly, cyclohexaamylose decelerates the migration of the trimethylacetyl group in 2-hydroxymethyl-4-nitrophenyl trimethylacetate by 5-fold in spite of an acceleration of this reaction by cycloheptaamylose, indicating that the geometry of the fit of the reactant to the binding site governs the catalysis. Retardations due to conformational effects by cycloamyloses were also observed in the hydrolyses of mono-carboxyphenyl esters of 3-substituted glutaric acids [96] and the benzidine rearrangement of hydrazobenzene [97].

Finally, it is noteworthy that cycloamyloses can be a useful model for the strain and distortion effect in enzymatic reactions [109]. The acceleration of hydrolyses of aryl sulfates has been ascribed to the induction of strain in the substrates on the inclusion complex formation, as well as to a microsolvent effect [114].

Improvement of Cycloamylose Catalyses by Covalent Modification or by Noncovalent Addition

Cycloamylose-catalyzed reactions exhibit many of the features of enzymatic reactions. However, covalent catalyses by cycloamyloses have two defects as a perfect model of enzymes, as noted above. They are as follows: (1) Cyclo-amyloses effectively catalyze the reaction in the high pH region, whereas many hydrolytic enzymes are highly active in the neutral pH region, due to the involvement of imidazole of pK_a 7; and (2) the hydrolysis of acyl cyclo-amylose (rate constant, k_3) is much slower than that of the substrate by the cycloamylose (rate constant, k_2), especially in cycloamylose-catalyzed phenyl ester cleavage.

In order to obtain an even better model of enzymatic reactions, several attempts have been made to introduce other functional groups to cyclo-amylose-catalyzed reactions by either (1) selective chemical modification of the cycloamylose or (2) noncovalent binding of a compound with appropriate functional groups.

IMPROVEMENT BY COVALENT MODIFICATION

The first attempt by Cramer and Mackensen [117] at improvement of cycloamylose catalysis of ester hydrolysis by the introduction of an imida-zolyl group to cycloamylose resulted in only a slight rate enhancement over the hydrolysis of the combination of cycloamylose and imidazole. The small effect of their modification can be attributed to the fact that the imidazolyl groups preferentially substituted the primary hydroxyl groups at C-6 atoms of cycloamylose rather than the secondary hydroxyl groups at C-2 and C-3 atoms, which have been shown to be effective in catalysis. In this structure, the added imidazolyl groups cannot exhibit cooperativity with the secondary hydroxyl groups of cycloamylose. However, Iwakura and co-workers [118] succeeded in selective modification of one of the secondary hydroxyl groups of cyclohexaamylose by a histamine group. This cyclohexaamylose–histamine compound accelerates the p-nitrophenyl acetate hydrolysis 80 times more than cyclohexaamylose itself does and 6.3 times more than the mixture of cyclohexaamylose and histamine (Table 3). This modified cyclohexaamylose, which has a catalytic site containing imidazolyl and hydroxyl groups and a binding site (the hydrophobic cavity of cycloamylose) as in α-chymotrypsin, trypsin, elastase, and subtilisin, can be a better enzyme model than cyclohexaamylose itself, since it shows rate acceleration around the neutral pH region.

Introduction of the N-methylacetohydroxamic acid group [119] or the N-(N,N'-dimethylaminoethyl)acetohydroxamic acid group [120] into cycloamylose exhibited a larger rate enhancement than cycloamylose itself or acetohydroxamic acid derivatives. Besides, effective regeneration of the catalyst was observed at neutral pH in the hydrolysis of acetic anhydride by cycloamylose–N-(N,N'-dimethylaminoethyl)acetohydroxamic acid. Thus, cycloamylose–N-(N,N'-dimethylaminoethyl)acetohydroxamic acid is a true catalyst. However, it should be noted that the catalytic site in the acetohydroxamic acid-modified cycloamylose is not the hydroxyl group of cycloamylose but rather the acetohydroxamic acid group [121,122]. In these catalyses, the cycloamylose supplies the cavity as the binding site but does not function as the nucleophile.

Modified cycloamyloses containing appropriate ligand groups that bind metal ions can serve as models of metalloenzymes, since the metal ion and

TABLE 3

Pseudo First-Order Rate Constants for the Deacylation of p-Nitrophenyl Acetate in the Presence of Added Catalysts[a,b]

Catalyst	Rate constant $(10^{-4}\ sec^{-1})$
None	0.789
Cyclohexaamylose	1.27
Histamine	12.1
Cyclohexaamylose + histamine (1:1 mixture)	16.0
Cyclohexaamylose–histamine compound	101

[a] Reprinted with permission from Y. Iwakura, K. Uno, F. Toda, S. Onozuka, K. Hattori, and M. L. Bender, *J. Am. Chem. Soc.* **97**, 4432 (1975). Copyright by the American Chemical Society.

[b] pH 8.37, Tris-HCl buffer, $I = 0.2\ M$, 25°C.

cycloamylose function as the catalytic and binding sites, respectively [123]. Recently, polymers containing cycloamylose units as branches were synthesized [124]. Cycloamylose in the polymer is 3-fold more active than cycloamylose that is not incorporated in the polymer for hydrolyses of phenyl benzoates. This might be attributable to the cooperativity of two neighboring cycloamylose molecules.

IMPROVEMENT BY THE ADDITION OF NONCOVALENT
COMPOUNDS

Several kinds of amines were added to accelerate hydrolyses of acyl cycloamyloses, which are the stable intermediates in cycloamylose-catalyzed ester cleavages. 5-Nitrobenzimidazole exhibited only a 2- to 3-fold enhancement in the hydrolysis of either cycloamylose acetate or cycloamylose cinnamate [103]. However, Komiyama and Bender [125] found that the hydrolysis of cycloamylose cinnamate was enhanced 57-fold at pH 13.6 by the inclusion of 1,4-diazabicyclo[2.2.2]octane. Triethylamine, piperidine, and n-butylamine are also more effective in the acceleration of hydrolysis of cycloamylose cinnamate than 5-nitrobenzimidazole, while diisobutylamine hardly catalyzes the reaction at all.

The cycloamylose-catalyzed hydrolysis of phenyl esters was used as a probe of the "charge-relay" system in serine esterases [126]. The addition of 2-benzimidazoleacetic acid was found to accelerate the cyclohexaamylose-catalyzed hydrolysis of m-$tert$-butylphenyl acetate 12-fold, although neither benzimidazole nor 2-naphthaleneacetic acid exhibited measurable acceleration. Catalysis by the cycloamylose-2-benzimidazoleacetic acid system involves nucleophilic attack by the imidazole group, assisted by the carboxyl and alkoxide anions. The mechanism is apparently different from those shown by the "charge-relay" system in enzymatic reactions. However, the validity of the "charge-relay" system was supported by the general base-catalyzed hydrolysis of ethyl chloroacetate in the absence of cycloamyloses [127]. The catalytic rate constant for the reaction catalyzed by 2-benzimidazoleacetic acid is 11.4-fold larger than that for the reaction catalyzed by benzimidazole. However, the "charge-relay" system in hydrolytic enzymes was criticized by Rogers and Bruice [128], in whose model compound the introduction of the hydrogen-bonding carboxylate group enhanced the intramolecular general base-catalytic activity of the adjacent imidazolyl group by "only" 2.8-fold. The small acceleration observed by Rogers and Bruice probably results from the conjugated system of their model compound, in which both catalyst and substrate coexist. In such a system, hydrogen abstraction of the carboxyl anion from the imidazolyl group can result in a decrease in electrophilicity of the substrate because of electron transfer through the conjugated system.

Thus, cycloamyloses can bind many organic compounds and can accelerate their reactions either by covalent or by noncovalent catalysis.

Covalent catalysis usually involves a hydroxyl group of the cycloamylose, although other nucleophiles, such as the hydroxamic acid group, have been covalently attached to the cycloamylose. The rate-determining step in catalysis of ester hydrolysis is deacylation of the acyl cycloamylose, and this has been speeded up both by the covalent attachment of an imidazole or other group and by the noncovalent inclusion of imidazole derivatives and tertiary amines. Cycloamyloses may even serve as a model for the "charge-relay" system, which is seen in some enzymes.

Noncovalent catalyses by cycloamyloses are of two varieties. If a reaction is accelerated by a lower dielectric constant and if the appropriate molecule is bound by the cycloamylose, then acceleration can take place by a "microsolvent" effect. This was recognized in the decarboxylation reaction mentioned earlier. In addition, if the cycloamylose can bind the correct conformation of a molecule that is undergoing an intramolecular reaction but not bind the corresponding incorrect conformation, the cycloamylose can then effect a "conformational" acceleration of reaction. This has been described earlier in our discussion of a transesterification reaction.

Thus, cycloamyloses can show considerable stereospecificity, as enzymes do, and, although they show less spectacular rate enhancement, they can be considered model enzymes.

POLYMERS

Enzymes are polymeric systems composed of amino acids. There has therefore been a considerable amount of interest in various polymeric systems that can serve as enzyme models. These polymeric systems can be classified in different ways. One system is based on the kind of binding that occurs— electrostatic [11] or apolar [13]. Another scheme of classification depends on whether the polymer is anionic [12] or cationic [13]. A third scheme of classification distinguishes polymers that are water insoluble, usually based on some vinyl polymerization, from those that are water soluble.

The classification system used here is based on anionic polymers or cationic polymers, not because it is necessarily the best classification system, but because it leads to the least amount of ambiguity or overlap. However, there will still be some, because some polymers have both anionic and cationic groups in them. Furthermore, there is an obvious overlap that combines a polymer and a cycloamylose (see previous section). With these problems in mind, let us consider some polymeric systems.

In early investigations, the polymeric systems were intramolecular systems in which the reactive entity and the catalytic group were both covalently attached to the polymer. A recent example, involving copolymers of acrylamide with small proportions of reactive monomers (1) and catalytic monomers (2), was formulated with the knowledge that pyridines catalyze the hydrolysis of p-nitrophenyl esters [12].

$$CH_2{=}CHCONHCH_2COO{-}\langle\bigcirc\rangle{-}NO_2$$

(1)

$$CH_2{=}CHCONHCH_2{-}\langle\bigcirc\rangle N$$

(2)

An early example involved the hydrolysis of ester–acid copolymers, depending on the relative steric configurations of the carbon atoms to which the ester and carboxylate groups were attached. In the hydrolysis of copolymers of methacrylic acid containing about 1% of phenyl methacrylate, the reaction deviates sharply from first-order kinetics and behaves as if about 20% of the ester groups were 10 times as susceptible to neighboring carboxylate attack as the remaining ester groups [129,130].

A variant involves a poly(methacrylic acid) and poly(vinylpyridine betaine). Ladenheim and Morawetz studied bromine displacement from α-bromoacetamide and bromoacetate ion using these polymers. They found that the former polymer was particularly effective, especially when the degree of ionization of the carboxylate groups in the polymer increased [131].

Taking into account that each substituent in the reactive chain is spaced at a different distance from each of the catalytic groups attached to that chain, the catalytic efficiency of the polymeric system with respect to the corresponding intermolecular reaction was found to be 2.7 M. That is, the catalytic group in the polymeric system corresponds to an effective molarity of 2.7 M in the intermolecular system.

The catalytic group in this polymeric system is a neutral pyridine moiety. A neutral catalytic entity is seen in many cationic polymeric systems, such as those of Letsinger, Overberger, Kabanov, and Klotz. However, since the overall charge on these polymers is positive, these systems are discussed in detail later.

Anionic Polymers

A prime example of an anionic polymer is an ion-exchange resin. The ion-exchange resin Dowex-50 requires free ammonium ion for its catalytic activity. It catalyzes the hydrolysis of glycylglycine 112 times *faster* than does the equivalent concentration of hydrochloric acid, but it catalyzes the hydrolysis of *N*-acetylglycine 2 times *slower* than does the equivalent concentration of hydrochloric acid [132]. Similarly, a water-soluble poly(styrenesulfonic acid) hydrolyzes 2-amino-2-deoxy-β-D-glucopyranoside hydrochloride 30-fold and the diethylaminoethyl ether of starch hydrochloride 20-fold *faster* than the equivalent concentration of hydrochloric acid, but it hydrolyzes starch 3-fold *slower* than an equivalent concentration of hydrochloric acid [133]. These specific catalyses can be due to one factor only: the selective binding of the cationic substrates to the ion-exchange resin followed by catalysis by the high local concentration of oxonium ions. Binding must occur by electrostatic interaction in these cases.

In special circumstances, π molecular complexes can lead to binding of substrate to an ion-exchange resin. The hydrolysis of allyl acetate by a sulfonic acid resin is enhanced by partial exchange of the oxonium ions with silver ions [134]. Although the rate of hydrolysis of propyl acetate is monotonically decreased by increasing concentration of silver ion on the resin, the rate of hydrolysis of allyl acetate reaches a maximum as the concentration of silver ion on the resin is increased (at about 50%). The increased concentration of allyl acetate near the polymer surface thus produces a 2-fold specificity for its hydrolysis.

The polyanions, poly(ethenesulfonate) (3) and poly(β-sulfoethyl methacrylate) (4) accelerate the Hg^{2+}-catalyzed aquation of $Co(NH_3)_5Cl^{2+}$, presumably by an increase in electrostatic potential in the polymer domain

$$(-CH_2-CH-)_n \qquad \begin{array}{c} CH_3 \\ | \\ (-CH_2-C-)_n \\ | \\ C=O \\ | \\ OC_2H_4SO_3^{\ominus} \end{array}$$
$$\quad\;\; | \atop SO_3^{\ominus}$$

(3)   (4)

[12,135,136]. The reaction involves a compact inorganic system.

$$Co(NH_3)_5Cl^{2+} + Hg^{2+} + H_2O \longrightarrow Co(NH_3)_5H_2O^{3+} + HgCl^+ \qquad (2)$$

If apolar binding is important for catalysis by polymeric acids, the solvent can have an effect on catalytic activity. For example, in acetone–water mixtures, a polymeric acid is often a poorer catalyst per oxonium ion than hydrochloric acid [137,138], but, in water, the same polymeric acid becomes a much better catalyst than hydrochloric acid. This change can readily be explained on the basis that apolar binding of the substrate to the polymer is better in water solution than in a solution of an organic solvent.

Poly[triethyl(vinylbenzyl)ammonium hydroxide] is a 20-fold more effective catalyst than equivalent amounts of either sodium hydroxide or benzyltrimethylammonium ion for the transformation of glyoxal to glycolic acid [138a]. The efficiency of this polymeric base can be explained in similar electrostatic terms.

Thus, these rudimentary polymeric catalysts do show some of the attributes of binding, enhanced catalysis, and specificity exhibited by enzymes. Now let us look at some more sophisticated systems.

A classic reaction that has been investigated in the presence of anionic polyelectrolytes is

$$2[Co(NH_3)_5Br]^{2+} + Hg^{2+} + 2H_2O \longrightarrow 2[Co(NH_3)_5H_2O]^{3+} + HgBr_2 \quad (3)$$

which is very similar to Eq. (2). The kinetic salt effect on this reaction has been exhaustively investigated.

One group of investigators reported extremely large accelerations of this reaction in the presence of polyanions. A factor of 1.76×10^5 was reported for 5×10^{-5} equivalent/liter of poly(vinyl sulfonic acid); the factor was reported to be 2.47×10^5 by a similar concentration of poly(methacryloxysulfonic acid). However, other investigators found an acceleration of about $2-3 \times 10^{-5}$ equivalent/liter of poly(styrenesulfonic acid) and poly-(ethylenesulfonic acid), still larger than that by simple electrolytes (1.007 by NaCl at the same experimental conditions)[139]. The k_{2_0} value (the second-order rate at zero ionic strength) reported by the former group for the chloropentamine mercuric system is much lower (by a factor of 10^3) than that observed by other groups for the bromopentamine complex. With such a low k_{2_0} value, it is possible to obtain a value of $\sim 10^5$ as the acceleration factor (k_2/k_{2_0}), even though the k_2 values were correctly measured at finite ionic strength.

Serum albumin is a catalyst for the decomposition of p-nitrophenyl acetate [140]. Because it contains cationic and anionic groups and because many anions bind to it, it could be classified as a cationic polymer. However, the p-nitrophenyl acetate substrate, with which we are here concerned, is neutral and could bind to a polymer of either charge type. When human serum albumin reacts with p-nitrophenyl acetate, the ester is not completely hydrolyzed. Rather, the protein is acetylated.

Rapid acetylation of the protein accompanies and largely accounts for the easily observed rapid formation of p-nitrophenolate ion. One group is acetyl-ated much faster than all others. It appears to be located in a high-affinity binding site for small fatty acid anions and to have a pK_a of 8.7 and a limiting bimolecular rate of reaction with p-nitrophenyl acetate of $\sim 3 \times 10^4 \ M^{-1}$ sec^{-1} at alkaline pH values. Rapid reversible binding appears to be a major contributor to the high reaction velocity.

Cationic Polymers

Catalysis of hydrolysis by cationic polymers has been investigated in-tensively. This probably stems from the fact that nucleophiles that hydrolyze esters and related materials are primarily nitrogen bases and thus give, on protonation, cationic polymers. There are many varieties of these polymers. We will concentrate on three: (1) water-insoluble polymers, primarily based on some vinyl polymerization; (2) water-soluble polymers, primarily of the poly(ethylenimine) variety; and (3) polymers that have more than one functional group.

WATER-INSOLUBLE POLYMERS

The anionic polymeric catalysts show some of the attributes of binding, enhanced catalysis, and specificity exhibited by enzymes. Cationic catalysts in general are more sophisticated systems. For example, the rates of displace-ment of bromide ion, bromoacetate ion, and bromoacetamide by poly-(4-vinylpyridine), poly(methacrylic acid), and poly(vinylpyridine betaine) are sometimes faster than reactions with the corresponding monomeric com-pounds [131,141]. The reaction of 4-methylpyridine with these alkylating agents shows only a small positive salt effect, as does the reaction of bromo-acetamide with poly(4-vinylpyridine). But the reaction of partially protonated poly(4-vinylpyridine) with bromoacetate ion increases sharply with the degree of protonation of the polymer when α (the degree of protonation) = 0.1–0.5 and with decreasing ionic strength. These results indicate an electro-static interaction between bromoacetate ion and the positively charged polymer that facilitates reaction. The carboxylate anions in partially ionized poly(methacrylic acid) are 4–10 times more reactive toward bromoacetamide than those of simple dicarboxylic acids, their reactivity decreasing sharply with the increasing degree of ionization of the polymer. This rate enhance-ment results from hydrogen bonding of the substrate to the un-ionized carboxylic acid groups of the polymer while the carboxylate ions function as nucleophiles to displace bromide ion. Both bromoacetate ion and bromo-acetamide react with poly(4-vinylpyridine zwitterion) under conditions in which there is no reaction with the corresponding monomer. Since both the neutral and negatively charged substrates show the same effect, electrostatics are presumably not involved. But, in these reactions, as well as in the other described above, binding of the monomeric substrate to the polymeric reactant must be invoked in order to explain the rate enhancements.

The concept of substrate binding to a polymeric catalyst by electrostatic interaction was investigated by Letsinger [142]. A partially protonated poly(4-vinylpyridine) in ethanol–water solution was shown to serve as a particularly effective catalyst, relative to 4-picoline, nonprotonated polymer, or highly protonated polymer, for the solvolysis of a nitrophenyl ester substrate bearing a negative charge. The monomer is a better catalyst than the polymer toward the neutral substrate, perhaps because of the steric effect the latter exhibits. However, the polymer is a better catalyst than the monomer toward the anionic ester, at essentially all degrees of ionization. With the anionic ester, 3-nitro-4-acetoxybenzenesulfonate ion, the catalytic activity approaches a maximum when the polymer is partially protonated (very close to a bell-shaped curve), implying that both protonated pyridinium ions and neutral pyridine groups are necessary for this catalysis. The pyridine groups serve as nucleophilic catalysts. The pyridinium ions serve as electrostatic binding agents, increasing the local concentration of the anionic substrate in the region of the polymeric coil. Similar catalytic phenomena are seen in the solvolyses of 5-nitro-4-acetoxysalicylic acid and 3-nitro-4-acetoxybenzene-arsonic acid with poly(4-vinylpyridine). At its maximum, the polymer is a 9-fold better catalyst than the monomer toward 3-nitro-4-acetoxybenzene-sulfonate ion. The asymmetry of the rate constant–α profile is attributed to a conformational change as the acidity of the medium increases.

The behavior of poly[4(5)-vinylimidazole] with the anionic ester, 3-nitro-4-acetoxybenzenesulfonate, is similar [143,144]. Catalysis by the polymer exceeds that by the monomer 6-fold at maximal activity of the polymer. A copolymer of 4(5)-vinylimidazole and acrylic acid also exhibits a selective catalysis. The copolymer is a better catalyst than imidazole toward a positively charged p-nitrophenyl ester substrate when the imidazole groups in both the monomer and polymer are largely unprotonated. On the other hand, imidazole is a better catalyst than the copolymer toward a neutral ester. Finally, the copolymer is only one-twentieth to one-thirtieth as effective as is the monomer toward negatively charged ester substrates [145].

The solvolysis of other anionic substrates was also investigated. For example, the solvolysis of the anionic, long-chain substrate, 3-nitro-4-dodecanoyloxybenzoic acid (NDBA), catalyzed by poly[4(5)-vinylimidazole] (PVIm) was investigated as a function of temperature and pH in aqueous solutions containing various compositions of ethanol and water. Working on an assumption of catalyst–substrate complex formation, Overberger et al. found that the polymeric solvolytic reaction of NDBA follows the Michaelis–Menten mechanism in low ethanol composition. The maximum catalytic activity of the polymer was observed in about 30 vol % ethanol–water relative to that of its monomeric analog, imidazole. This was attributed to the extremely large value of the first-order rate constants, k_2, rather than a favorable Michaelis constant, K_m. The solvolytic reaction of NDBA catalyzed by PVIm in 43.7 vol % ethanol–water presented unusual kinetic features: a retardation behavior when [NDBA] > [PVIm] and an acceleration behavior when [PVIm] > [NDBA]. These results are probably related to the formation of the long-lived intermediate compound dodecanoyl poly[4(5)-vinylimidazole]. This formation also indicates that the deacylation step is overall rate determining in the solvolysis of NDBA catalyzed by PVIm. Apolar interactions between the substrate and the catalyst are probably superimposed on the electrostatic binding, and thus this compound undoubtedly binds better than the smaller compound mentioned above, especially in aqueous solution. This is not the case when the concentration of ethanol is increased [146].

Apolar influences on this solvolytic reaction were further investigated by varying the acyl chain length in the substrates (3-nitro-4-acyloxybenzoic acids) and the volume percent of water in the aqueous–alcohol solvent systems. The results demonstrate the importance of apolar bonding in the rate enhancements observed for reactions catalyzed by PVIm as compared to reactions catalyzed by imidazole (Im). In certain esterolysis reactions, deviations from pseudo first-order kinetics were observed. When the acyl group was seven and when its concentration was greater than that of PVIm in 15 and 20% 1-propanol–water systems, its kinetic behavior was indicative of "saturation" of the polymer by substrate. When [PVIm] > [S_2^-], [S_7^-], or [S_{12}^-] (S_n = number of carbons in the acyloxy group) was studied in 20% 1-propanol in water systems, an analogous saturation of the substrate by polymer was indicated by the kinetic behavior. The effect of varying the pH and temperature on esterolysis was also studied. An increase in r ($r = k_{PVIm}/k_{Im}$) was observed with decreasing pH. All of these observations can be explained on the basis of an apolar component to the binding [147].

Several variants of solvolysis by poly(vinylimidazole) involve copolymers. One example involves the hydrolysis of 3-nitro-4-acetoxybenzoic acid catalyzed by copolymers of 1-vinyl-2-methylimidazole with 1-vinylpyrrolidone (copolymer A) and with acrylamide (copolymer B) at 30°C. The rate of the catalytic hydrolysis was describable by Michaelis–Menten kinetics, showing substrate saturation phenomena at high substrate concentrations, as in enzymatic reactions. Copolymer A of low (< 20%) imidazole content showed the same kinetic pattern, independent of the copolymer composition, and gave K_m (dissociation constants of the polymer–substrate complex) of 9.3 mM and k_3 (first-order rate constant of the pseudo-intramolecular product formation) of 0.038 min^{-1}. Copolymer B gave K_m of 63 mM and k_3 of 0.11 min^{-1}. When, however, in copolymer A, which contained higher amounts of imidazole units, the kinetic pattern was not of the simple Michaelis–Menten type, the overall catalytic efficiency decreased, suggesting the presence of a catalytic site of a different nature. The substrate binding was attributed to an apolar interaction, since the electrostatic interaction between catalyst and substrate was found to be negligible. The kinetic characteristics of copolymer A were reasonably explained by assuming a loop formation of the polymer segment surrounding the substrate molecule [148].

Subsequently, hydrolyses of p-acetoxybenzoic acid catalyzed by copolymers of N-[p-4(5)-imidazolylbenzyl]acrylamide (PI) were studied. The rate of the catalytic hydrolysis could be described by Michaelis–Menten kinetics, showing leveling off of the rate at high substrate concentrations. On the other hand, the catalytic hydrolysis with a model compound of the catalytic unit, p-acetamidomethylphenyl-4(5)-imidazole, followed second-order kinetics. The PI–vinylpyrrolidone (PI–VP) copolymers had smaller dissociation constants of the catalyst–substrate complex and smaller intracomplex rates than PI–acrylamide (PI–AA) copolymers had, reflecting an increased hydrophobic property of the catalytic site in the PI–VP copolymers. The substrate binding with the polymer catalyst showed large negative entropy changes, supporting the supposition that the substrate binding is based on apolar forces. The entropies of activation of the intracomplex process were unusually large negative values (-58 to -60 eu). When the difference in reactivity of the imidazolyl groups due to different pK_a values was corrected according to Bruice and Schmir, the decrease in the free energy due to substrate binding was largely offset by the increase in the free energy of activation of the intracomplex process. The authors' interpretation of this result is that the apolar interaction of the polymer–substrate complex becomes destroyed in the transition state of the intracomplex reaction [149].

Polymers containing pyridine and imidazole are described above. Could polymers containing other nucleophiles also be effective catalysts for ester hydrolysis? Of particular interest in the five-membered nitrogen-containing heterocycles is 1,2,4-triazole. The pK_1 and pK_2 values of this compound are 4.4 units less than those of the respective values of imidazole. Consequently, at pH values in the region of pK_2, a strong catalytic effect by anionic triazole functions could be expected to occur in esterolytic reactions. However, monomeric 1,2,4-triazole as a catalyst was superior to the corresponding polymer in the pH region investigated [136–138]. Furthermore, there is a marked enhancement of the monomeric triazole-catalyzed reaction with increasing pH, indicative of the participation of anionic triazole groups [150]. Thus, it appears that polymers of pyridine and imidazole cannot be extended to the triazole family.

The hydrolysis of p-nitrophenyl acetate by polymers of incompletely alkylated poly(4-vinylpyridine) are of particular importance mechanistically and with respect to a direct comparison with the enzyme α-chymotrypsin. The alkylation was carried out by treatment of the polymer with benzyl chloride in a mixture of nitromethane and methanol. The molecular weight of the polymer was approximately 21,000. Kinetic measurements showed that the rate of hydrolysis of p-nitrophenyl acetate by the polymer (calculated per mole of free pyridine nuclei) was 2–2.5 orders of magnitude greater than the rate by the corresponding monomer, 4-ethylpyridine. Separate pre-steady-state and steady-state portions of the hydrolysis were seen, indicating discrete acylation and deacylation steps like those in enzymatic reactions. A comparison of the polymeric catalyst with α-chymotrypsin, shown in Table 4, indicates that in the rate-determining k_3 step, the polymer is only about 20-fold less efficient than the enzyme chymotrypsin [151].

This behavior implies the formation and decomposition of an acyl-polymer intermediate by analogy with acyl-enzyme formation in the enzyme-catalyzed reaction. The formation would be acylation, k_2, while the decomposition would be k_3. This analysis is borne out in the hydrolysis of 3-nitro-4-acyloxybenzoic acid substrates by poly[4(5)-vinylimidazole]. This reaction displayed a deviation (accelerative) from pseudo first-order kinetic behavior. This deviation was attributed to an appreciable buildup of a partially acylated polymer intermediate which is relatively long lived and is more apolar than the original catalyst. In all cases where accelerative kinetic behavior was observed, a knowledge of the relative rates of acylation and deacylation allows one to predict a buildup of acylated catalyst. The presence of this

TABLE 4

Hydrolysis of p-Nitrophenyl Acetate by Polymeric
Catalysts and Chymotrypsin[a,b]

Catalyst[c]	$K_m(M)$	k_2 (sec^{-1})	k_3 (sec^{-1})
PC-4	2.2×10^{-4}	1.6×10^{-3}	4.2×10^{-4}
PC-7	1.1×10^{-4}	4.4×10^{-3}	4.0×10^{-4}
PC-10	1.1×10^{-4}	4.4×10^{-3}	4.4×10^{-4}
PC-26	1.1×10^{-4}	4.4×10^{-3}	4.2×10^{-4}
PC-50	1.1×10^{-4}	4.4×10^{-3}	4.1×10^{-4}
α-Chymotrypsin	1.2×10^{-3}	4	8.0×10^{-3}

[a] From Y. E. Kirsh, V. A. Kabanov and V. A. Kargin, *Dokl. Akad. Nauk SSSR* **177**, 112 (1967).

[b] Catalysis conditions: for PC, 25°C, pH 8, [Tris–HCl] = 10^{-2} M, $I = 10^{-2}$ M; for α-chymotrypsin, 25°C, pH 8.6, [Tris–HCl] = 2×10^{-2} M, $I = 0.2$ M.

[c] PC-n = polymeric catalyst where n is percent benzylation.

acylated catalyst intermediate and the fact that it is more apolar than the original catalyst were demonstrated by double addition (of substrate) experiments and by destruction of the intermediate through the addition of hydroxylamine to the reaction solution [152].

The formation of an acyl polymer intermediate is also borne out in the hydrolysis of p-nitrophenyl acetate by poly(4-vinylpyridine) reacted with 2-(2'-chloroethyl)pyridine. In this reaction, like the reaction of the partially alkylated poly(vinylpyridine) discussed above, the activity of the polymeric catalyst exceeded the activity of monomeric models by several orders of magnitude. Product inhibition, like that of enzymatic reactions, was also noted. One can talk about the active site of the polymer somewhat like one can talk about the active site of an enzyme, except that although there are many active sites in the polymer there is only one in an enzyme molecule. The active site of the polymer is a nonalkylated 4-substituted pyridine ring, surrounded on both sides by fairly long sequences of 4-substituted pyridine rings that have been N-alkylated by 1'-[2'(2''-pyridyl)]ethyl chloride. The high activity of the polymeric catalyst is thought to be due to the formation of apolar active cavities in the macromolecular coil [153].

Thermodynamic parameters were also determined for these reactions. Of course, using different temperatures may have an effect both on the polymer and on the polymeric catalysis. The former was ruled out by viscosity studies. The experimental values of the free energy of formation of the Michaelis complex are about the same for all polymeric catalysts, and they differ little from the values for those enzymes (α-chymotrypsin, pepsin, trypsin) that bond the corresponding substrates through apolar interaction. The entropies and enthalpies of bonding of p-nitrophenyl acetate molecules by the polymeric catalysts PC are also about the same. Consideration of three-dimensional models reveals that, for polymeric catalysts containing isopropyl radicals, disposition of the substrate in the "active cavities" is possible only at a specific mutual orientation of the isopropyl groups at neighboring units of the macromolecule. In the other polymeric catalysts, the requirement of specific orientation of the alkyl groups is significantly more variable. On the other hand, the greatest number of contacts between the hydrocarbon portions of the "active cavity" and the molecule is attained in the p-nitrophenyl acetate–isopropylated poly(vinylpyridine) complex. Thus, the changes in the rate constants of the individual stages of p-nitrophenyl acetate hydrolysis in the presence of the different polymeric catalysts with a change in temperature (or in the nature of the medium) find satisfactory explanation in the hypothesis of the role of apolar interactions in the formation of substrate complexes with polymeric catalysts [154].

Some recent examples of esterolytic activity exhibited by polymers are described below. One example involves poly(N-alkylimidazoles) in an ethanol–

water solution. Both p-nitrophenyl acetate and 3-nitro-4-acetoxybenzoic acid were used as substrates. The polymeric materials were only 20% more effective than the corresponding monomeric catalyst—not a very impressive result [155]. When the chain lengths of both the substrates and the catalyst were increased, the polymer was 88-fold more effective than the corresponding monomer [156]. This indicates that an apolar interaction must be at least partially operative.

Selective polymeric catalysis at its best is seen in the reaction of the cationic polymeric catalyst poly(N-vinylimidazole) with the anionic polymeric substrate copoly(acrylic acid–2,4-dinitrophenyl p-vinylbenzoate) [157,158]. Comparison of the solvolysis of neutral and anionic monomeric esters, 2,4-dinitrophenyl p-isopropylbenzoate and p-nitrophenyl hydrogen terephthalate, by the cationic polymeric catalyst for both shows selectivity toward the anionic substrate, depending on the degree of ionization of the polymer due to electrostatic interactions between monomeric substrate and polymeric catalyst. Solvolysis of the polymeric substrate by the polymeric catalyst is twice as effective as solvolysis by the corresponding monomeric catalyst, although solvolysis of the monomeric substrate, 2,4-dinitrophenyl p-isopropylbenzoate, is 20-fold *less* effective with the polymer than with the monomeric catalyst. Thus, some specific interaction of the polymeric substrate and polymeric catalyst is indicated. The dependence of the catalytic rate constant for this reaction on the catalyst concentration substantiates this conclusion. Although the catalytic rate constant of the reaction of the polymeric substrate with the monomeric catalyst N-methylimidazole shows a linear (and low) dependence on the catalyst concentration, the catalytic rate constant of the reaction of the polymeric substrate with the polymeric catalyst poly(N-vinylimidazole) shows a typical saturation phenomenon. Since the calculated catalytic rate constant for the productive complex is greater than for the nonproductive complex, the complex formed between catalyst and substrate must be a productive one. The inverse phenomenon, in which the polymeric catalyst is saturated by the polymeric substrate, is also seen. Competitive inhibition of this reaction by poly(acrylic acid), a polymer containing the same binding groups as the substrate, also occurs. All these phenomena—specificity, saturation, and competitive inhibition—can be attributed to formation of a catalyst–substrate complex that increases the probability of encounter between the nucleophilic sites on the catalyst and the ester groups of the substrate. This system thus exhibits many of the characteristics of an enzymatic process, although its efficiency is still low.

A similar phenomenon, the interaction of a polymeric catalyst with a polymeric substrate, was observed in the poly(oxymethylene) ester of N-benzoyl-histidine–polyacrylic (or methacrylic) acid systems in the hydrolysis of neutral and positively charged substrates. It was shown that the catalytic efficiency of poly(oxyethylene) ester of N-benzoylhistidine in hydrolysis of p-nitrophenyl acetate, a neutral substrate, remains practically unchanged, while hydrolysis of a positively charged substrate, p-dimethylaminophenyl-acetate iodomethylate, in the presence of poly(oxyethylene) ester of N-benzoylhistidine is considerably accelerated on adding poly acid. It is assumed from kinetic results that a poly(oxyethylene) ester complex of N-benzoly-histidine–poly acid is formed. Further, the catalytic behavior of poly(oxy-ethylene) ester of N-benzoylhistidine was examined in hydrolysis of a polymer substrate—a copolymer of acrylic acid with p-nitrophenyl ester of p-vinyl-benzoic acid. Especially in the last case, which involves both a polymeric catalyst and a polymeric substrate, the enzymatic analogies mentioned above are seen [159].

WATER-SOLUBLE POLYMERIC CATALYSTS

Water-soluble synthetic polymers have been found to bind small molecules. Derivatives of highly branched water-soluble poly(ethylenimine) with a small proportion (10 residue mole %) of apolar side chains have an affinity for small molecules greater than anything extant, including serum albumin. Not only do they bind, but they also catalyze the cleavage of esters such as p-nitrophenyl acetate and p-nitrophenyl laurate. They cleave the latter compound more readily than the former compound, as would be expected from the greater apolar interaction of the latter compond with the polymer [160,161]. The cleavage of more esters by poly(ethylenimines) (PEI) of various molecular weights was investigated. Table 5 lists first-order rate constants, corrected for hydrolysis of ester in buffer alone. Propylamine served as a reference amine. In its presence k (in minutes^{-1}) for aminolysis decreased progressively from 0.98×10^{-2} to 0.51×10^{-2} to 0.05×10^{-2} as the length of the acyl group increased from 1 to 12 carbons. The sharp drop for nitrophenyl laurate may be the result of micelle formation, even at concentrations of 6×10^{-6} M. From Table 5, it is clear that the introduction of strong binding sites on the polymer leads to marked rate enhancements [162].

Furthermore, the synthetic polymer based on poly(ethylenimine) with pendent dodecyl groups and methyleneimidazolyl groups was prepared. Plots of ester concentration vs. time are consistent with a two-step pathway analogous to that of a hydrolytic enzyme such as chymotrypsin, in which an initial acylation burst is followed by a rate-determining deacylation reaction. Preceding all of this is undoubtedly a preequilibrium binding step for which much evidence has been given before. The rate of hydrolysis of p-nitrophenyl acetate by these polymers approaches that by the enzyme α-chymotrypsin as is seen in Table 6, and thus these polymers have been called synzymes (synthetic enzymes) [13] and are true catalysts, as opposed to earlier work, where the "catalyst" is not regenerated.

TABLE 5

First-Order Rate Constants for Amine Acylation by
p-Nitrophenyl Esters[a,b]

Amine	$k \times 10^2$ min$^{-1 c}$		
	p-Nitrophenyl acetate	p-Nitrophenyl caproate	p-Nitrophenyl laurate
Propyl	0.98	0.51	0.053
PEI-6[d]	3.60	1.47	0.11
PEI-18[d]	4.38	1.57	0.11
PEI-600[d]	4.60	1.80	0.17
L(10%)-PEI-6[e]	15.2	68.1	698

[a] Reprinted with permission from G. P. Royer and I. M. Klotz, *J. Am. Chem. Soc.* **91**, 5885 (1969). Copyright by the American Chemical Society.

[b] pH 9.0 in 0.02 M Tris, 25°C, 6.7% (v/v) acetonitrile.

[c] $k = k_a - k_0$, where k_a is the measured rate constant in the presence of amine and k_0 is that for the hydrolysis in Tris buffer alone; k_0 is 0.94 × 10^{-2} min^{-1} for the acetyl ester, 0.61 × 10^{-2} min^{-1} for the caproyl ester, and 0.023 × 10^{-2} min^{-1} for the lauroyl ester.

[d] The numeral following PEI (polyethylenimine) multiplied by 100 is the molecular weight of the polymer sample.

[e] This sample of PEI-6 has 10% of its nitrogens acylated with lauroyl (L) groups.

An acyl imidazole intermediate has been detected spectroscopically in this polymeric catalysis, and thus the postulate of acylation of the polymeric imidazole group followed by the hydrolysis of the acyl imidazole, leading to carboxylate ion and regenerated imidazole on the polymer, was proved [163].

Whereas rate enhancements of about 10^3-fold were found in the hydrolysis of p-nitrophenyl acetate by these polymers, rate enhancements of 10^{11}-fold [164] or 10^{12}-fold [165] (compared to unbound imidazole) were found in the hydrolysis of 2,4-dinitrophenyl sulfate, thus making the polymer 10^2 times more effective than the corresponding enzyme which hydrolyzes this ester (an aryl sulfatase). This polymeric catalysis is analogous to the corresponding enzymatic reaction in that it shows saturation of the polymer by the substrate and a fast acylation followed by a rate-determining deacylation.

The water solubility of poly(1-methyl-5-vinylimidazole) [poly(1-Me-5-VIm)], due to lack of intermolecular hydrogen bonding, has made it possible to achieve phenomenal rate enhancements of the order of enzymatic rates in the hydrolysis of nitrophenyl esters. The poly(1-Me-5-VIm)-catalyzed hydrolysis of a long-chain nitrophenyl ester exhibited saturation in excess catalyst and in excess substrate. Inhibition of the polymer-catalyzed hydrolysis of nitrophenyl ester substrates by analog inhibitors was also observed.

TABLE 6

Relative Effectiveness of Various Catalysts in
Cleavage of p-Nitrophenyl Esters[a,b]

Catalyst	"Catalytic constant," k $(M^{-1} min^{-1})$
Imidazole	10^c
α-Chymotrypsin	$10,000^d$
PEI-600-HA(25%)^e	420
PEI-600-HA(8%)-L(8%)-Im(6.6%)^f	3,100
PEI-600-D(10%)-Im(15%)^g	2,700

[a] From I. M. Klotz, G. P. Royer, and I. S. Scarpa, *Proc. Natl. Acad. Sci. U.S.A.* **68**, 263 (1971).

[b] The substrate used was p-nitrophenyl caproate, except for the first two reactions, in which p-nitrophenyl acetate was used at a pH near neutrality.

[c] Taken from T. C. Bruice and G. L. Schmir, *J. Am. Chem. Soc.* **79**, 1663 (1957).

[d] Taken from E. Katchalski, G. D. Fasman, E. Simons, and E. R. Blout, *Arch. Biochem. Biophys.* **88**, 361 (1960).

[e] In this derivative of PEI-600, 25% of the amine residues were aklylated with —$CH_2CON(OH)CH_3$ groups. This hydroxamate (HA) by itself has been studied by W. B. Gruhn and M. L. Bender, *J. Am. Chem. Soc.* **91**, 5883 (1969).

[f] This derivative of PEI-600 had also 8% of its residues alkylated with hydroxamate, 8% acylated by lauroyl (L) groups, and 6.6% acylated with

$$HC{=}C{-}CH_2{-}\overset{\displaystyle O}{\overset{\|}{C}}{-}$$

substituents.

[g] D, dodecyl groups.

Saturation apparently did not follow a simple Michaelis–Menten mechanism; however, the results were rationalized by analogy to certain enzymatic systems. Multisite enzymes have long been known to display kinetic patterns different from that exhibited by enzymes with only one active site, i.e., such phenomena as sigmoidal rate vs. [S] plots. These phenomena may arise entirely as a result of the multisite nature of the enzyme. Consequently, a synthetic macromolecular catalyst with multiple sites might also be expected to display such characteristics. The poly(1-Me-5-VIm)-catalyzed hydrolysis of a long-chain nitrophenyl ester is the first synthetic system in which such phenomena have been observed. The intermediacy of an apolar polymer–substrate complex for the polymer-catalyzed hydrolysis of the ester in water was given support by studies of the effect of temperature on the rate of hydrolysis. Activation parameters were determined for catalysis by 1,5-dimethylimidazole and by poly(1-Me-5-VIm). A comparison of results for the polymeric and monomeric systems showed that the rate enhancement exhibited by the polymer was due entirely to a favorable entropy term, presumably due to apolar binding [166].

The decomposition of aspirin in aqueous solutions of cationic poly-electrolytes was studied over a considerable pH range (1.5–11.0). Poly-(ethylenimine), a weak polyelectrolyte containing free amino groups which act as nucleophilic reagents on the substrate, and the strong polyelectrolyte poly(vinylbenzyltrimethylammonium chloride), without such reactive groups in its molecule, were employed. Poly(vinylbenzyltrimethylammonium chloride) was found to modify only slightly the rate of hydrolysis of aspirin in the pH-independent region (5–9) but increases by a factor of 9 the rate of bimolecular saponification of the ester in alkaline solutions. This acceleration can be explained satisfactorily with an electrostatic model which predicts an en-hanced local concentration of the substrate anion and OH$^-$ near the chains due to the large charge density of the polymeric chains. On the other hand, in poly(ethylenimine) solutions, the rate of decomposition of aspirin is sub-stantially increased, passing through a maximum at pH 7.8, where the rate constant is 1275 times greater than in the absence of polyelectrolyte. A further increase in pH causes a decrease in rate constant until the value corresponding to solutions without polyelectrolyte is reached at pH $\sim$ 11. The explanation of this behavior is given in terms of two competing effects. When the pH increases, the fraction of amino groups that are free also increases, thus enhancing the possibility of nucleophilic attack on the substrate; on the other hand, the concomitant decrease of charged groups on the macro ion reduces the local concentration of the charged substrate near the polymeric chain [167].

However, there appears to be some question as to whether polyelectrolytes can be considered to be true catalysts since they perturb the position of equilibrium considerably, which a true catalyst cannot do [168]. However, the same authors describe the hydrolysis of nitrophenyl esters catalyzed by polyelectrolytes [169]. Thus, this question appears to be open at this time.

POLYMERIC CATALYSTS WITH MORE THAN ONE
FUNCTIONAL GROUP

Polymeric catalysts can have more than one group at the "active site" of the polymer. We mentioned before that polymers containing electrostatic groups or apolar groups for binding also contain nucleophilic groups for catalysis. Here, we wish to cover polymers that contain two or more catalytic groups, exclusive of the groups that are used for binding.

One interesting example of this is the poly[4(5)-vinylimidazole]- and poly[5(6)-vinylbenzimidazole]-catalyzed ester hydrolysis. The esters involved were p-nitrophenyl acetate and 3-nitro-4-acetoxybenzoic acid. One would expect that, with the anionic ester, the rate would be dependent on the degree of ionization of the polymer, and it is. However, with the neutral substrate, p-nitrophenyl acetate, at high pH (above 8.0), the catalytic rate of the polymers was higher than that of the corresponding monomers, and the rate enhancement was pH dependent [143,144]. This can be explained only by a cooperative interaction between neutral imidazolyl or benzimidazolyl groups of the polymer and corresponding *anionic* groups of the polymers. The two neutral benzimidazolyl functions on the polymer also show cooperative interaction at intermediate pH [170].

Selective catalysis is seen in ester hydrolysis catalyzed by a copolymer of 4(5)-vinylimidazole and acrylic acid, which serves as a model for the enzyme acetylcholinesterase. The copolymer provides a binding site (anionic groups) and a catalytic site (imidazolyl groups) and thus might be classified as an anionic polymer. But the specificity is so large in this instance, approximately 20-fold (at pH 9) higher for the hydrolysis of the cationic ester, 3-acetoxy-N-trimethylanilium iodide, than for the hydrolysis of the anionic ester, 4-acetoxy-3-nitrobenzoic acid, that it seems appropriate to include this polymeric catalysis here [145].

The active sites of hydrolytic enzymes are normally comprised of several catalytic species. In order to ascertain if cooperative interactions between pendent imidazolyl and hydroxyl functions could occur in synthetic polymers, esterolytic reactions catalyzed by copolymers of 4(5)-vinylimidazole with vinyl alcohol and with p-vinylphenol were investigated. A cooperative interaction of imidazolyl and hydroxyl groups underlies most of the mechanisms of α-chymotrypsin-catalyzed hydrolyses. These copolymers are only slightly more active than poly[4(5)-vinylimidazole] in esterolytic reactions [150].

The hydrolyses of p-nitrophenyl acetate and of acetoxy-N,N,N-trimethylanilium iodide catalyzed by copolymers of 4(5)-vinylimidazole and γ-butyrolactone were also investigated. The catalytic activity of the copolymers in these catalyses was found to be more than twice as high as that of a random terpolymer containing imidazolyl, carboxylate, and hydroxyl moieties [171].

The hydrolysis of these two esters by copolymers of 4(5)-vinylimidazole and acrylic acid was also investigated. The most effective interaction of the positively charged ester appeared when the copolymer contained less than 30% poly(acrylic acid). The kinetic results, involving rates and Brönsted plots, and pH dependencies, indicate a cooperative interaction of the imidazolyl and carboxylate moieties [172].

In an effort to further investigate bifunctional catalysis in polymers, the solvolytic reactions of the neutral esters p-nitrophenyl acetate and p-nitrophenyl heptanoate catalyzed by poly[4(5)-vinylimidazole] and by imidazole were investigated in solutions containing various compositions of ethanol and water. Temperature and pH were postulated as the perturbing variable in order to elucidate the roles of apolar polymer–substrate interactions and bifunctional catalysis in the synthetic polymeric catalyses. Further, the solution properties of poly[4(5)-vinylimidazole] were examined in various ethanol–water compositions and at various pH values. The conformation of the polymer chain of poly[4(5)-vinylimidazole] in an ethanol–water mixture is dramatically affected by the ethanol–water composition and by the degree of neutralization of the pendent imidazole groups. The polymeric–catalytic enhancement for the solvolyses of neutral substrates in low and high ethanol compositions at pH ~ 8 was attributed to increased bifunctional catalysis with the shrinkage of the macromolecules in solution. The solvolysis of p-nitrophenyl heptanoate at low ethanol concentration by poly[4(5)-vinylimidazole] appeared to involve an increased accumulation of the substrate in the polymer domain because of apolar polymer–substrate interactions [173].

Heretofore, bifunctional catalysis in polymeric systems has involved two imidazolyl groups or one imidazolyl group and one carboxylate group. p-Nitrophenyl acetate can also be hydrolyzed by the cooperative interaction of carboxyl residues in polycarboxyl amino acids, such as poly-L-glutamic acid (I), poly-L-aspartic acid (II), their copolymer (III), and their mixtures. These reactions showed bell-shaped profiles with pH or temperature. Optimum pH values at 40°C were 5.2 for I, 5.55 for II, and 5.44 for III, and three peaks of activity were shown at 5.2, 5.51, and 5.65 for the mixture. These activities appeared in the initial stage of random coil formation, just after the disappearance of helical sense. The optima corresponded to the peaks of pH–rigidity curves of the solutions. Poly(acrylic acid) and poly(D-glutamic acid) had no activity toward p-nitrophenyl acetate [174].

It has been a favorite pastime to attach oxime groups to polymers since they have been so successful in enhancing the rate of cycloamylose catalyses. One recent report involves a copolymer of 4-vinyl-N-(phenacyloxime)-pyridinium bromide and vinylpyridine. In the region of pH 7.5–9, the copolymer hydrolyzed p-nitrophenyl acetate faster than the corresponding monomeric components [175].

Copolymers containing both imidazole and mercaptan groups were investigated as possible synthetic models for the proteolytic enzymes ficin and papain. However, kinetic data from the study indicated that the catalytic activity of the copolymers was composed mainly of additive hydrolytic effects of separate imidazolyl and mercaptan groups. Cooperative effects seemed to be present in the hydrolysis of 3-nitro-4-acetoxybenzoic acid by poly(vinylmercaptan), probably because of the interaction of each sulfhydryl group and its corresponding anion [176].

As mentioned previously, polymers containing cycloamyloses have been prepared [124]. Since catalysis occurs via the cycloamylose and since the polymer serves primarily as a support, these reactions are not discussed here.

CONCLUSIONS

Enzyme models have progressed a long way in the last decade. There has been considerable interest in synthesizing a relatively small organic molecule that in some way simulates enzyme action. Of these, the two most successful approaches have involved the cycloamyloses and the polymers. The former shows remarkable specificity and the latter can show considerable speed.

Morawetz, in 1969, stated, "In spite of the many interesting results that I have tried to survey, . . . I do not believe that a single new fact of interest to the enzyme chemist has been uncovered" [29]. We believe he was being unduly pessimistic, for in the intervening years, particularly with the advent of water-soluble polymers, catalysts have been prepared which give rates of the order of enzymatic reactions. These certainly can say something to the enzyme chemist, namely, that enzymes do not have the exclusive market on high catalytic rates but that polymers can have them also. Of course, not all polymeric catalyses are fast. Some are not much faster than those of the corresponding monomer. A complete listing of polymeric rates has been given by the Fendlers in their new book [9].

The cycloamyloses have not shown the speed of polymers, but they have surpassed the polymers in their specificity. As noted above, this is shown by the stereochemistry of binding.

In the years that follow, considerable progress can be expected in this field until we have enzyme models that will show both speed and specificity of enzymes. Thus, the billions of years of poor experiments by nature should be equaled in decades by clever experiments by chemists and biochemists. Of course, the poor experiments were in the evolutionary scheme, and thus had an important constraint that the chemist does not have. Evolution was limited to mainly the approximately 20 amino acids of a protein (plus some selected metal ions or small organic molecules). The chemist or biochemist has no such constraints and is thus in a better position to make a synthetic enzyme, which can catalyze any reaction.

ACKNOWLEDGMENT

This research was supported by grant MPS 72-05013 of the National Science Foundation.

REFERENCES

1. A. J. Kirby and A. R. Fersht, *Prog. Bioorg. Chem.* **1**, 1–82 (1971).
2. R. Breslow and D. E. McClure, *J. Am. Chem. Soc.* **98**, 258 (1976).
3. M. L. Bender, "Mechanisms of Homogeneous Catalysis from Protons to Proteins." Wiley (Interscience), New York, 1971.
4. B. Vennesland and F. H. Westheimer, *in* "The Mechanism of Enzyme Action" (W. D. McElroy and B. Glass, eds.), p. 357. Johns Hopkins Press, Baltimore, Maryland, 1954.
5. G. A. Hamilton, *Prog. Bioorg. Chem.* **1**, 83 (1971).
6. D. E. Metzler, M. Ikawa, and E. E. Snell, *J. Am. Chem. Soc.* **76**, 648 (1954).
7. R. Breslow, *in* "The Mechanism of Action of Water Soluble Vitamins" (A. V. S. de Reuck and M. O'Connor, eds.), p. 65. Churchill, London, 1961.
8. V. Franzen, *Chem. Ber.* **88**, 1361 (1955).
9. J. H. Fendler and E. J. Fendler, "Catalysis in Micellar and Macromolecular Systems." Academic Press, New York, 1975.
10. D. W. Griffiths and M. L. Bender, *Adv. Catal.* **23**, 209 (1973).
11. R. L. Letsinger and T. J. Savereide, *J. Am. Chem. Soc.* **84**, 114 (1962).
12. H. Morawetz, *Acc. Chem. Res.* **3**, 354 (1970).
13. I. M. Klotz, G. P. Royer, and I. S. Scarpa, *Proc. Natl. Acad. Sci. U.S.A.* **68**, 263 (1971).
14. M. Gordon, J. G. Miller, and A. R. Day, *J. Am. Chem. Soc.* **70**, 1946 (1948).
15. W. P. Jencks and M. Gilchrist, *J. Am. Chem. Soc.* **88**, 104 (1966).
16. D. E. Koshland, Jr. and K. E. Neet, *Annu. Rev. Biochem.* **37**, 377 (1968).
17. A. Yapel and R. Lumry, *J. Am. Chem. Soc.* **86**, 4499 (1964).
18. K. Tanizawa and M. L. Bender, *J. Biol. Chem.* **249**, 2130 (1974).
19. F. M. Menger, *J. Am. Chem. Soc.* **88**, 3081 (1966).
20. M. L. Bender and F. J. Kezdy, *J. Am. Chem. Soc.* **86**, 3704 (1964).
21. F. M. Menger and J. H. Smith, *Tetrahedron Lett.* p. 4163 (1970); *J. Am. Chem. Soc.* **94**, 3824 (1972).
22. F. M. Menger and A. C. Vitale, *J. Am. Chem. Soc.* **95**, 4931 (1973).
23. F. J. Menger, S. Wrenn, and H. S. Rhee, *Bioorg. Chem.* **4**, 194 (1975).
24. G. Wallerberg, J. Boger, and P. Haake, *J. Am. Chem. Soc.* **93**, 4938 (1971).
24a. T. Komives, A. F. Marton, and F. Dutka, *Chem. Ind.* (London) p. 567 (1975).
25. R. Alexander, E. C. F. Ko, A. J. Parker, and T. J. Broxton, *J. Am. Chem. Soc.* **90**, 5049 (1968).
26. Y. Pocker and R. F. Buchholz, *J. Am. Chem. Soc.* **92**, 2075 (1970).
27. R. B. Dunlap and E. H. Cordes, *J. Am. Chem. Soc.* **90**, 4395 (1968).
28. E. H. Cordes and R. B. Dunlap, *Acc. Chem. Res.* **2**, 329 (1969).
29. H. Morawetz, *Adv. Catal. Relat. Subj.* **20**, 341 (1969).
30. E. J. Fendler and J. H. Fendler, *Adv. Phys. Org. Chem.* **8**, 271 (1970).
31. T. C. Bruice, *in* "The Enzymes" (P. D. Boyer, ed.), 3rd ed., Vol. 2, p. 217. Academic Press, New York, 1970.

32. F. M. Menger, in "Bioorganic Chemistry" (E. E. van Tamelen, ed.), Vol. 3, Chapter 7, Academic Press, New York (1977).
33. R. A. Moss, R. C. Nahas, S. Ramaswami, and W. J. Sanders, *Tetrahedron Lett.* p. 3379 (1975).
34. K. Martinek, A. V. Levashov, and I. V. Berezin, *Tetrahedron Lett.* p. 1275 (1975).
35. K. Martinek, A. P. Osipov, A. K. Yatsimirski, V. A. Dadali, and I. V. Berezin, *Tetrahedron Lett.* p. 1279 (1975).
36. K. Martinek, A. P. Osipov, A. K. Yatsimirski, and I. V. Berezin, *Tetrahedron* **31**, 709 (1975).
37. W. Tagaki, M. Chigira, T. Amada, and Y. Yano, *Chem. Commun.* p. 219 (1972).
38. J. M. Brown, C. A. Bunton, and S. Diaz, *Chem. Commun.* p. 971 (1974).
39. C. Gitler and A. Ochoa-Solano, *J. Am. Chem. Soc.* **90**, 5004 (1968).
40. J. Sunamoto, H. Okamoto, H. Kondo, and Y. Murakami, *Tetrahedron Lett.* p. 2761 (1975).
41. Y. Murakami, J. Sunamoto, and K. Kano, *Bull. Chem. Soc. Jpn.* **47**, 1238 (1974).
42. J. H. Fendler, E. J. Fendler, G. A. Infante, P.-S. Shih, and L. K. Patterson, *J. Am. Chem. Soc.* **97**, 89 (1975).
43. J. H. Fendler, *Acc. Chem. Res.* **9**, 153 (1976).
44. E. J. Fendler, S. A. Chang, J. H. Fendler, R. T. Medary, O. A. El Seoud, and V. A. Woods, *in* "Reaction Kinetics in Micelles" (E. H. Cordes, ed.), p. 127. Plenum, New York, 1973.
45. J. H. Fendler, F. Nome, and H. C. Van Woert, *J. Am. Chem. Soc.* **96**, 6745 (1974).
46. J. H. Fendler, E. J. Fendler, R. T. Medary, and V. A. Woods, *J. Am. Chem. Soc.* **94**, 7288 (1972).
47. C. J. O'Connor, E. J. Fendler, and J. H. Fendler, *J. Am. Chem. Soc.* **95**, 600 (1973).
48. J. H. Fendler, E. J. Fendler, and S. A. Chang, *J. Am. Chem. Soc.* **95**, 3273 (1973).
49. C. J. O'Connor, E. J. Fendler, and J. H. Fendler, *J. Am. Chem. Soc.* **96**, 370 (1974).
50. R. S. Mulliken and W. B. Person, "Molecular Complexes." Wiley (Interscience), New York, 1969.
51. R. Foster and T. J. Thomson, *Trans. Faraday Soc.* **58**, 860 (1962).
52. I. Isenberg and S. L. Baird, Jr., *J. Am. Chem. Soc.* **84**, 3803 (1962).
53. W. Liptay, G. Briegleb, and K. Schindler, *Z. Elektrochem.* **66**, 331 (1962).
54. A. K. Colter and S. S. Wang, *J. Am. Chem. Soc.* **85**, 114 (1963).
55. A. K. Colter, S. S. Wang, G. H. Megerle, and P. S. Ossip, *J. Am. Chem. Soc.* **86**, 3106 (1964).
56. A. K. Colter and S. H. Hui, *J. Org. Chem.* **33**, 1935 (1968).
57. J. B. Jones and C. Niemann, *Biochemistry* **2**, 498 (1963).
58. D. French, A. O. Pulley, J. A. Effenberger, M. A. Rougvie, and M. Abdullah, *Arch. Biochem. Biophys.* **111**, 153 (1965).
59. A. Hybl, R. E. Rundle, and D. E. Williams, *J. Am. Chem. Soc.* **87**, 2779 (1965).
60. J. A. Hamilton, L. K. Steinrauf, and R. L. VanEtten, *Acta Crystallogr., Sect. B* **24**, 1560 (1968).
61. K. Takeo and T. Kuge, *Agric. Biol. Chem.* **33**, 1174 (1969).
62. W. J. James, D. French, and R. E. Rundle, *Acta Crystallogr.* **12**, 385 (1959).
63. J. A. Thoma and L. Stewart, *Starch: Chem. Technol.* **1**, 209 (1965).
64. H. Schlenk and D. M. Sand, *J. Am. Chem. Soc.* **83**, 2312 (1961).
65. R. K. McMullan, W. Saenger, J. Fayos, and D. Mootz, *Carbohydr. Res.* **31**, 211 (1973).
66. W. Saenger, K. Beyer, and P. C. Manor, *Acta Crystallogr., Sect. B* **32**, 120 (1976).
67. J. Cohen and J. L. Lach, *J. Pharm. Sci.* **52**, 132 (1963).

68. J. L. Lach and T. F. Chin, *J. Pharm. Sci.* **53**, 69 (1964).
69. J. L. Lach and T. F. Chin, *J. Pharm. Sci.* **53**, 924 (1964).
70. W. A. Pauli and J. L. Lach, *J. Pharm. Sci.* **54**, 1745 (1965).
71. J. L. Lach and W. A. Pauli, *J. Pharm. Sci.* **55**, 32 (1966).
72. F. Cramer, *Chem. Ber.* **84**, 851 (1951).
73. R. L. VanEtten, J. F. Sebastian, G. A. Clowes, and M. L. Bender, *J. Am. Chem. Soc.* **89**, 3242 (1967).
74. F. Cramer and H. Hettler, *Naturwissenschaften* **54**, 625 (1967).
75. J. L. Hoffman and R. M. Bock, *Biochemistry* **9**, 3542 (1970).
76. C. Formoso, *Biochem. Biophys. Res. Commun.* **50**, 999 (1973).
77. C. Formoso, *Biopolymers* **13**, 909 (1974).
78. P. V. Demarco and A. L. Thakkar, *Chem. Commun.* p. 2 (1970).
79. A. L. Thakkar and P. V. Demarco, *J. Pharm. Sci.* **60**, 652 (1971).
80. B. Siegel and R. Breslow, *J. Am. Chem. Soc.* **97**, 6869 (1975).
81. C. VanHooidonk and J. C. A. E. Breebaart-Hansen, *Recl. Trav. Chim. Pays-Bas* **90**, 680 (1971).
82. Y. Matsui, H. Naruse, K. Mochida, and Y. Date, *Bull. Chem. Soc. Jpn.* **43**, 1909 (1970).
83. P. C. Manor and W. Saenger, *J. Am. Chem. Soc.* **96**, 3630 (1974).
84. W. Saenger, R. K. McMullan, J. Fayos, and D. Mootz, *Acta Crystallogr., Sect. B* **30**, 2019 (1974).
85. W. Saenger and M. Noltemeyer, *Chem. Ber.* **109**, 503 (1976).
86. R. J. Bergeron and M. P. Meeley, *Bioorg. Chem.* **5**, 197 (1976).
87. P. C. Manor and W. Saenger, *Nature (London)* **237**, 392 (1972).
88. N. Hennrich and F. Cramer, *J. Am. Chem. Soc.* **87**, 1121 (1965).
89. D. E. Tutt and M. A. Schwartz, *Chem. Commun.* p. 113 (1970).
90. D. E. Tutt and M. A. Schwartz, *J. Am. Chem. Soc.* **93**, 767 (1971).
91. K. Flohr, R. M. Paton, and E. T. Kaiser, *Chem. Commun.* p. 1621 (1971).
92. C. VanHooidonk and J. C. A. E. Breebaart-Hansen, *Recl. Trav. Chim. Pays-Bas* **89**, 289 (1970).
93. C. VanHooidonk and C. C. Groos, *Recl. Trav. Chim. Pays-Bas* **89**, 845 (1970).
94. H. J. Brass and M. L. Bender, *J. Am. Chem. Soc.* **95**, 5391 (1973).
95. T. S. Straub and M. L. Bender, *J. Am. Chem. Soc.* **94**, 8875 (1972).
96. D. L. VanderJagt, F. L. Killian, and M. L. Bender, *J. Am. Chem. Soc.* **92**, 1016 (1970).
97. Y. Matsui, K. Mochida, O. Fukumoto, and Y. Date, *Bull. Chem. Soc. Jpn.* **48**, 3645 (1975).
98. M. L. Bender, R. L. VanEtten, G. A. Clowes, and J. F. Sebastian, *J. Amer. Chem. Soc.* **88**, 2318 (1966).
99. M. L. Bender, R. L. VanEtten, and G. A. Clowes, *J. Am. Chem. Soc.* **88**, 2319 (1966).
100. R. L. VanEtten, G. A. Clowes, J. F. Sebastian, and M. L. Bender, *J. Am. Chem. Soc.* **89**, 3253 (1967).
101. F. Cramer and W. Dietsche, *Chem. Ber.* **92**, 1739 (1959).
102. F. Cramer, *Chem. Ber.* **86**, 1576 (1953).
103. Y. Kurono, V. Stamoudis, and M. L. Bender, *Bioorg. Chem.* **5**, 393 (1976).
104. H. Lineweaver and D. Burk, *J. Am. Chem. Soc.* **56**, 658 (1934).
105. M. L. Bender, *Trans. N.Y. Acad. Sci.* [2] **29**, 301 (1967).
106. M. L. Bender and F. J. Kezdy, *Annu. Rev. Biochem.* **34**, 49 (1965).
107. J. Crosby, R. Stone, and G. E. Lienhard, *J. Am. Chem. Soc.* **92**, 2891 (1970).

108. T. C. Bruice and S. J. Benkovic, "Bioorganic Mechanisms." Benjamin, New York, 1966.
109. W. P. Jencks, "Catalysis in Chemistry and Enzymology." McGraw-Hill, New York, 1969.
110. M. L. Bender, F. J. Kezdy, and C. R. Gunter, *J. Am. Chem. Soc.* **86**, 3714 (1964).
111. D. R. Storm and D. E. Koshland, Jr., *Proc. Nat. Acad. Sci. U.S.A.* **66**, 445 (1970).
112. F. Cramer and W. Kampe, *Tetrahedron Lett.* p. 353 (1962).
113. F. Cramer and W. Kampe, *J. Am. Chem. Soc.* **87**, 1115 (1965).
114. W. I. Congdon and M. L. Bender, *Bioorg. Chem.* **1**, 424 (1971).
115. T. S. Straub and M. L. Bender, *J. Am. Chem. Soc.* **94**, 8881 (1972).
116. D. W. Griffiths and M. L. Bender, *J. Am. Chem. Soc.* **95**, 1679 (1973).
117. F. Cramer and G. Mackensen, *Angew. Chem.* **78**, 641 (1966).
118. Y. Iwakura, K. Uno, F. Toda, S. Onozuka, K. Hattori, and M. L. Bender, *J. Am. Chem. Soc.* **97**, 4432 (1975).
119. W. B. Gruhn and M. L. Bender, *Bioorg. Chem.* **3**, 324 (1974).
120. Y. Kitaura and M. L. Bender, *Bioorg. Chem.* **4**, 237 (1975).
121. W. B. Gruhn and M. L. Bender, *J. Am. Chem. Soc.* **91**, 5883 (1969).
122. W. B. Gruhn and M. L. Bender, *Bioorg. Chem.* **4**, 219 (1975).
123. R. Breslow and L. E. Overman, *J. Am. Chem. Soc.* **92**, 1075 (1970).
124. M. Furue, A. Harada, and S. Nozakura, *J. Polym. Sci., Polym. Lett. Ed.* **13**, 357 (1975).
125. M. Komiyama and M. L. Bender, *Proc. Natl. Acad. Sci. U.S.A.* **73**, 2969 (1976).
126. M. Komiyama, E. J. Breaux, and M. L. Bender, *Bioorg. Chem.* (in press).
127. M. Komiyama and M. L. Bender, *Bioorg. Chem.* (in press).
128. G. A. Rogers and T. C. Bruice, *J. Am. Chem. Soc.* **96**, 2473 (1974).
129. E. Gaetjens and H. Morawetz, *J. Am. Chem. Soc.* **82**, 5328 (1960).
130. H. Morawetz and E. Gaetjens, *J. Polym. Sci.* **32**, 536 (1958).
131. H. Ladenheim and H. Morawetz, *J. Am. Chem. Soc.* **81**, 4860 (1959).
132. J. R. Whitaker and F. E. Deatherage, *J. Am. Chem. Soc.* **77**, 5298 (1955).
133. T. J. Painter and W. T. J. Morgan, *Chem. Ind. (London)* p. 437 (1961).
134. S. Affrossman and J. P. Murray, *J. Chem. Soc. B* p. 1015 (1966).
135. H. Morawetz and B. Vogel, *J. Am. Chem. Soc.* **91**, 563 (1969).
136. H. Morawetz and G. Gordimer, *J. Am. Chem. Soc.* **92**, 7532 (1970).
137. S. Yoshikawa and O. K. Kim, *Bull. Chem. Soc. Jpn.* **39**, 1515 (1966).
138. S. A. Bernhard and L. P. Hammett, *J. Am. Chem. Soc.* **75**, 1798 (1953).
138a. C. L. Arcus and B. A. Jackson, *Chem. Ind. (London)* p. 2022 (1964).
139. N. Ise, *Fortschr. Hochpolym.-Forsch.* **7**, 536 (1971).
140. G. E. Means and M. L. Bender, *Biochemistry* **14**, 4989 (1975).
141. H. Ladenheim, E. M. Loebl, and H. Morawetz, *J. Am. Chem. Soc.* **81** 20 (1959).
142. R. L. Letsinger and T. J. Savereide, *J. Am. Chem. Soc.* **84**, 114 and 3122 (1962).
143. C. G. Overberger, T. St. Pierre, N. Vorchheimer, and S. Yaroslavsky, *J. Am. Chem. Soc.* **85**, 3513 (1963).
144. C. G. Overberger, T. St. Pierre, N. Vorchheimer, J. Lee, and S. Yaroslavsky, *J. Am. Chem. Soc.* **87**, 296 (1965).
145. C. G. Overberger, R. Sitaramaiah, T. St. Pierre, and S. Yaroslavsky, *J. Am. Chem. Soc.* **87**, 3270 (1965).
146. C. G. Overberger, M. Marimoto, I. Cho, and J. C. Salamone, *J. Am. Chem. Soc.* **93**, 3228 (1971).
147. C. G. Overberger, R. C. Glowaky, and P. H. Vandewyer, *J. Am. Chem. Soc.* **95**, 6008 (1973).

148. T. Kunitake, F. Shimada, and C. Aso, *J. Am. Chem. Soc.* **91**, 2716 (1969).
149. T. Kunitake and S. Shinkai, *J. Am. Chem. Soc.* **93**, 4256 (1971).
150. C. G. Overberger and J. C. Salamone, *Acc. Chem. Res.* **2**, 217 (1969).
151. Y. E. Kirsh, V. A. Kabanov, and V. A. Kargin, *Dokl. Akad. Nauk SSSR* **177**, 112 (1967).
152. C. G. Overberger and R. C. Glowaky, *J. Am. Chem. Soc.* **95**, 6014 (1973).
153. Yu. E. Kirsh, V. A. Kabanov, and V. A. Kargin, *Polym. Sci. USSR (Engl. Transl.)* **10**, 407 (1968).
154. S. K. Pluzhnov, Yu. E. Kirsh, V. A. Kabanov, and V. A. Kargin, *Dokl. Akad. Nauk SSSR* **185**, 843 (1969).
155. C. G. Overberger and T. W. Smith, *Macromolecules* **8**, 401 (1975).
156. C. G. Overberger and T. W. Smith, *Macromolecules* **8**, 407 (1975).
157. R. L. Letsinger and I. Klaus, *J. Am. Chem. Soc.* **86**, 3884 (1964).
158. R. L. Letsinger and I. Klaus, *J. Am. Chem. Soc.* **87**, 3380 (1965).
159. I. N. Topchieva, A. B. Solov'eva and V. A. Kabanov, *Polym. Sci. USSR (Engl. Transl.)* **15**, 2136 (1973).
160. I. M. Klotz and V. H. Stryker, *J. Am. Chem. Soc.* **90**, 2717 (1968).
161. I. M. Klotz, G. P. Royer, and A. R. Sloniewsky, *Biochemistry* **8**, 4752 (1969).
162. G. P. Royer and I. M. Klotz, *J. Am. Chem. Soc.* **91**, 5885 (1969).
163. T. W. Johnson and I. M. Klotz, *Macromolecules* **6**, 788 (1973).
164. H. C. Kiefer, W. I. Congdon, and I. M. Klotz, *Fed. Proc., Am. Soc. Exp. Biol.* (abstr.) 1423 (1972).
165. H. C. Kiefer, W. I. Congdon, I. S. Scarpa, and I. M. Klotz, *Proc. Natl. Acad. Sci. U.S.A.* **69**, 2155 (1972).
166. C. G. Overberger and T. W. Smith, *Macromolecules* **8**, 416 (1975).
167. R. Fernandez-Prini and E. Baumgartner, *J. Am. Chem. Soc.* **96**, 4489 (1974).
168. N. Ise and T. Okubo, *Nature (London)* **242**, 605 (1973).
169. N. Ise, T. Okubo, H. Kitano, and S. Kunugi, *J. Am. Chem. Soc.* **97**, 2882 (1975).
170. C. G. Overberger, T. St. Pierre, and S. Yaroslavsky, *J. Am. Chem. Soc.* **87**, 4310 (1965).
171. T. Shimidzu, A. Furuta, T. Watanabe, and S. Kato, *Makromol. Chem.* **175**, 119 (1974).
172. T. Shimidzu, A. Furuta, and Y. Nakamoto, *Macromolecules* **7**, 160 (1974).
173. C. G. Overberger and M. Morimoto, *J. Am. Chem. Soc.* **93**, 3222 (1971).
174. J. Noguchi, S. Tokura, T. Komai, K. Kokazi, and T. Azuma, *J. Biochem. (Tokyo)* **69**, 1033 (1971).
175. Yu. E. Kirsh and V. A. Kabanov, *Dokl. Akad. Nauk SSSR* **193**, 889 (1973).
176. C. G. Overberger, T. J. Pacansky, J. Lee, T. St. Pierre, and S. Yaroslavsky, *J. Polym. Sci., Polym. Symp.* **46**, 209 (1974).

CHAPTER

3

Aldol-Type Reactions and
2-Keto-4-hydroxyglutarate Aldolase

Eugene E. Dekker

INTRODUCTION*

An important and common method for lengthening or shortening the carbon chains of organic molecules, both chemically and biochemically, is the aldol-type reaction. The reaction can be represented generally as shown in Fig. 1. In 1887, Fischer and Tafel [1] first described an alkali-catalyzed condensation of two trioses to form a hexose. Some years later (in 1934), an enzyme in muscle which catalyzed a similar condensation of triose phosphates (Fig. 2) was described by Myerhof and Lohmann [2]. The pioneering and perceptive studies of B. L. Horecker, W. J. Rutter, I. A. Rose, A. H. Mehler, R. Barker, their associates, and many other investigators have made this

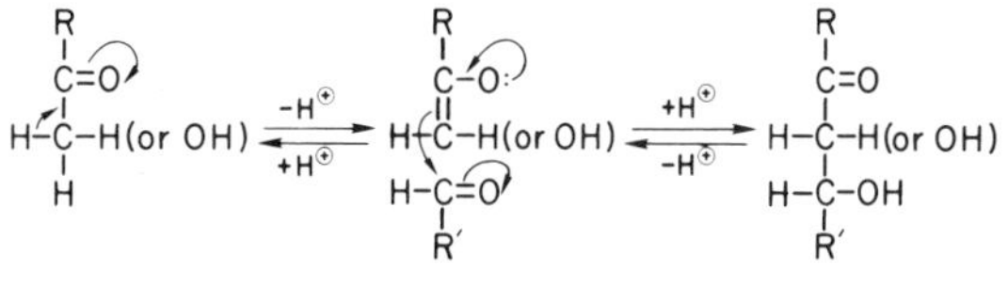

Fig. 1 Aldol-type reactions.

* Abbreviations used are FDP, fructose 1,6-diphosphate; DHAP, dihydroxyacetone phosphate; G3P, glyceraldehyde 3-phosphate; KHG, 2-keto-4-hydroxyglutarate; TNM, tetranitromethane; DTNB, 5,5′-dithiobis(2-nitrobenzoic acid).

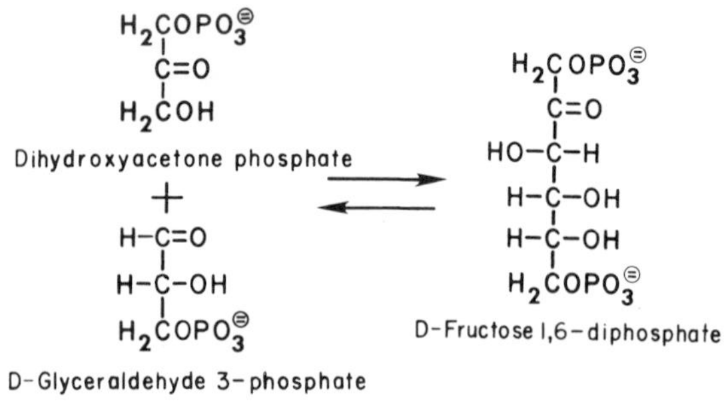

$$H_2COPO_3^{\ominus}$$
$$C=O$$
$$H_2COH$$
Dihydroxyacetone phosphate

$$+$$

$$H-C=O$$
$$H-C-OH$$
$$H_2COPO_3^{\ominus}$$
D-Glyceraldehyde 3-phosphate

$$\xrightarrow{\hspace{1cm}} \atop \xleftarrow{\hspace{1cm}}$$

$$H_2COPO_3^{\ominus}$$
$$C=O$$
$$HO-C-H$$
$$H-C-OH$$
$$H-C-OH$$
$$H_2COPO_3^{\ominus}$$
D-Fructose 1,6-diphosphate

Fig. 2   Reaction catalyzed by fructose 1,6-diphosphate aldolase.

reaction, as catalyzed by fructose 1,6-diphosphate aldolase, the best characterized of the enzyme-catalyzed aldol-type reactions. It is therefore the reference for all comparative aldolase studies. FDP-Aldolase appears to be ubiquitous in all biological systems. Early studies [3] with muscle FDP-aldolase showed that (a) it stereospecifically catalyzes the formation of only FDP from DHAP plus G3P, (b) it is completely specific for DHAP whereas a variety of aldehydes can replace G3P, (c) phosphorylated aldehydes, like G3P, are utilized much more effectively than their nonphosphorylated counterparts, and (d) because the carbanion resonance form of DHAP is involved as an intermediate, it stereospecifically catalyzes the exchange of one atom of carbon-bound hydrogen of this substrate with the solvent. Subsequent studies with FDP-aldolases from a wide variety of sources led Rutter to propose [4] that two classes (I and II) of FDP-aldolases exist, each having distinct catalytic elements and molecular characteristics. Class I FDP-aldolases ("lysine class") are those whose properties resemble the mammalian muscle enzyme, and those resembling yeast FDP-aldolase are designated as Class II ("metal-ion class"). Some general properties of Class I and Class II FDP-aldolases are listed in Table 1.

Similarity with these two classes of FDP-aldolases can be found in enzymes catalyzing β-keto decarboxylations. Considerable evidence has been presented [5–10] indicating that oxaloacetate decarboxylase* has an absolute divalent metal ion requirement; the mechanism of action of this enzyme has been suggested as involving a metal ion complex with the substrate. Acetoacetate decarboxylase, however, is not a metalloprotein. Westheimer and his associates have shown definitively [12–16] that this β-decarboxylase functions via a Schiff base mechanism involving an active-site lysyl residue in the pro-

---

* It should be noted that oxaloacetate decarboxylase from codfish muscle has recently been shown to be identical with pyruvate kinase [11]. The two activities appear to take place at the same or overlapping sites on the enzyme.

TABLE 1

Properties of Class I and Class II FDP-Aldolases

Class I	Class II
1. Found in animals, plants, protozoans, green algae	1. Found in bacteria, yeast, fungi, blue-green algae
2. MW $\sim 160,000$	2. MW $\sim 70,000$
3. Tetrameric subunit structure	3. Dimeric subunit structure
4. Activity not inhibited by metal chelating agents	4. Metalloproteins; activity inhibited by chelating agents
5. Activity not stimulated by K^+	5. Activity stimulated severalfold by K^+
6. Functional C-terminal (tyrosine) residues	6. Functional —SH groups
7. Schiff base mechanism, involving active-site lysyl residues	7. Mechanism unknown; possibly involves a metal chelate intermediate in which the metal ion serves as an electrophile
a. Enzyme–substrate complex stabilized and activity destroyed by reduction with BH_4^- plus either FDP or DHAP (not with G3P); between 3 and 4 moles DHAP bound per 160,000 MW	a. Activity not affected by BH_4^- in the presence of any substrate
b. ^{18}O exchanges between FDP (or DHAP) and the medium	b. No ^{18}O exchange occurs between substrates and the medium
c. Due to enamine formation from the imine, hydrogen of DHAP exchanges stereospecifically with the solvent	c. Proton exchange occurs, possibly as a consequence of keto–enol isomerization effected by the electrophilic metal ion
d. Reversibly inhibited by cyanide in the presence of DHAP (not with G3P)	d. Effect of cyanide unknown
8. pH profiles for exchange and for overall reaction are broad and congruent	8. pH profiles for exchange and for overall reaction are sharp and displaced from each other

tein molecule (i.e. analogous to Class I FDP-aldolases). Therefore, although it must still be conclusively established, the possibility remains that two fundamental catalytic sites may be utilized in most enzymes which facilitate aldol-type and β-decarboxylation reactions in biological systems. Furthermore, the mechanistic similarity between these enzymes led Rutter [4] to advance the phylogenetic concept that aldolases may structurally resemble a preexisting enzyme such as a β-decarboxylase.

Since the time that Myerhof and Lohmann discovered FDP-aldolase [2], a wide variety of different enzymes that catalyze aldol-type reactions has been

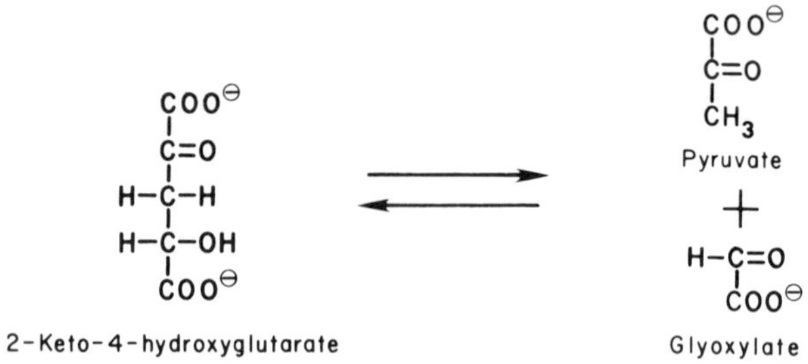

$$
\begin{array}{ccc}
& & \begin{array}{l} COO^{\ominus} \\ | \\ C=O \\ | \\ CH_3 \\ \text{Pyruvate} \end{array} \\
\begin{array}{l} COO^{\ominus} \\ | \\ C=O \\ | \\ H-C-H \\ | \\ H-C-OH \\ | \\ COO^{\ominus} \end{array} & \xrightleftharpoons{\qquad} & + \\
\text{2-Keto-4-hydroxyglutarate} & & \begin{array}{l} H-C=O \\ | \\ COO^{\ominus} \\ \text{Glyoxylate} \end{array}
\end{array}
$$

Fig. 3   Reaction catalyzed by 2-keto-4-hydroxyglutarate aldolase.

reported. One of these aldolases which seems to be especially interesting is KHG-aldolase, an enzyme that catalyzes a terminal step in the mammalian catabolism of L-hydroxyproline. Figure 3 shows the primary reaction catalyzed by this enzyme. Much work remains to be done, but KHG-aldolase (and the reaction it catalyzes) has already been found to have many novel and atypical properties when compared with other aldolases. This chapter reviews information currently available about KHG-aldolases and compares them with FDP-aldolase.

So far, the presence of KHG-aldolase activity has been reported in extracts of liver, kidney, and certain bacteria [17,18]. The involvement of this aldolase in mammalian catabolism of L-hydroxyproline was firmly established about 12 years ago. Before then, the most definitive report on hydroxyproline metabolism in mammals was that of Wolf et al. [19] who found that, after intraperitoneal injection of [2-^{14}C]DL-hydroxyproline into rats, alanine was the most highly labeled amino acid in liver proteins. The enzymatic oxidation of L-hydroxyproline to γ-hydroxyglutamate* via Δ^1-pyrroline-3-hydroxy-5-carboxylate was shown by Adams and Goldstone [22,23]. Our finding [24] that liver preparations catalyze the conversion of γ-hydroxy-glutamate to glyoxylate plus alanine, together with the report by Kuratomi and Fukunaga [25] that extracts of rat liver acetone powders catalyze a reversible condensation of glyoxylate with pyruvate yielding KHG, allowed for complete elucidation of the individual enzymatic steps whereby mammals convert L-hydroxyproline to glyoxylate plus alanine [26–31]. This pathway is outlined in Fig. 4; the required enzymes are identified by number. Although the reaction catalyzed by KHG-aldolase in this pathway is reversible, all information currently available appears to indicate that the primary role of this aldolase in mammals is to catalyze KHG degradation rather than its synthesis. KHG-Aldolase may also play a secondary role in L-homoserine

---

* Large amounts of threo-γ-hydroxy-L-glutamic acid are present as the free amino acid in certain plants [20,21]. Its metabolic significance in the plant kingdom is unknown.

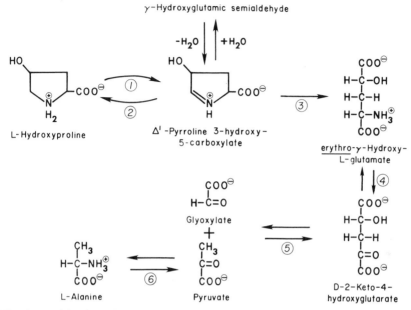

Fig. 4 Participation of 2-keto-4-hydroxyglutarate aldolase in mammalian catabolism of L-hydroxyproline. (1), Proline oxidase; (2), Δ^1-pyrroline-5-carboxylate reductase; (3), Δ^1-pyrroline dehydrogenase; (4), glutamate-aspartate transaminase; (5), 2-keto-4-hydroxyglutarate aldolase; (6), glutamate (glutamine)-alanine transaminase.

metabolism. The initial observation [32] that liver extracts catalyze a condensation of pyruvate with formaldehyde has been established as due to KHG-aldolase activity [33]. The reaction catalyzed by KHG-aldolase in mammalian L-homoserine metabolism is shown in Fig. 5 [34]. The primary role of KHG-aldolase in a bacterium like *Escherichia coli* has not been established.

KHG-Aldolase was initially obtained in partially purified form (about 100-fold purified) from rat liver extracts [17,30]; Rosso and Adams [35] succeeded in purifying the enzyme much more extensively (400- to 600-fold over an initial acetone powder extract) from the same source. The preparation of essentially homogeneous KHG-aldolase (over 1000-fold purified) from extracts of bovine liver was subsequently reported by Kobes and Dekker [36,37]. The discussion here focuses on results obtained with bovine liver KHG-aldolase since most of the published work has utilized the enzyme from this source and, to the extent that the enzyme from rat liver has been studied, no differences have been noted from the bovine liver aldolase. So far, KHG-aldolase has been studied from only two bacterial sources. A 10-fold

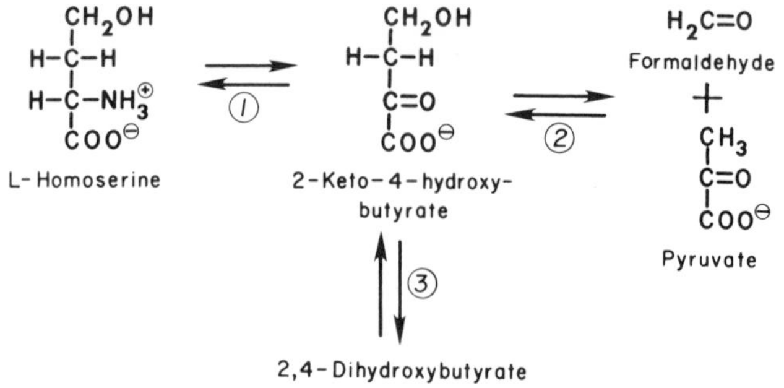

$$\underset{\text{L- Homoserine}}{\overset{\displaystyle CH_2OH}{\underset{\displaystyle COO^{\ominus}}{\overset{\displaystyle |}{\underset{\displaystyle |}{H-C-H}}\atop H-C-NH_3^{\oplus}}}} \underset{(1)}{\overset{\longrightarrow}{\longleftarrow}} \underset{\substack{\text{2-Keto-4-hydroxy-}\\\text{butyrate}}}{\overset{\displaystyle CH_2OH}{\underset{\displaystyle COO^{\ominus}}{\overset{\displaystyle |}{\underset{\displaystyle |}{H-C-H}}\atop C=O}}} \underset{(2)}{\overset{\longrightarrow}{\longleftarrow}} \begin{array}{l} H_2C=O \\ \text{Formaldehyde} \\ + \\ \underset{\text{Pyruvate}}{\overset{\displaystyle CH_3}{\underset{\displaystyle COO^{\ominus}}{\overset{|}{C=O}\atop |}}} \end{array}$$

$$\Big\updownarrow (3)$$

2,4-Dihydroxybutyrate

Fig. 5 Participation of 2-keto-4-hydroxyglutarate aldolase in L-homoserine metabolism. (1), Glutamate-aspartate or glutamate-alanine transaminase; (2), 2-keto-4-hydroxyglutarate aldolase; (3), lactate dehydrogenase [34].

purification of KHG-aldolase from a soil bacterium grown on α-keto-glutarate as carbon source was reported in a short communication [18]. For the enzyme from this source, only the questions of stereospecificity towards the two KHG isomers and of inactivation by $NaBH_4$ in the presence of substrates were addressed; in these two respects, no marked differences from liver aldolase were observed. The other bacterium from which KHG-aldolase has been obtained in homogeneous form (several thousandfold purified) is *E. coli* K-12 [38,39]. The *E. coli* aldolase differs uniquely in a number of respects from the bovine liver enzyme; KHG-aldolases from these two sources are compared and contrasted.

BOVINE LIVER KHG-ALDOLASE

KHG-Aldolase from bovine liver has an average molecular weight of 120,000, as determined by elution from a calibrated column of Sephadex G-200 [40], by sucrose density gradient centrifugation [41], and by standardized polyacrylamide gel electrophoretic procedures [42]. Some data have been obtained which would appear to suggest that, under certain experimental conditions that have not yet been well explored, the enzyme may undergo a process of association–dissociation. Preparations of the enzyme have been obtained [36,33] which, when subjected to polyacrylamide gel electrophoresis, showed two (and only two) protein bands, both of which were enzymatically active. When the sedimentation patterns of these same preparations were examined in the ultracentrifuge, two protein peaks of approximately equal concentration were observed. Since these two peaks overlapped somewhat, only estimated molecular weights of 113,000 and 208,000 for the two components could be obtained. Whether the separate proteins detected by these two techniques are, indeed, the same forms of KHG-aldolase and what their structural or catalytic interrelations may be await their individual isolation and examination. The enzyme has a sharp pH optimum of 8.8 and is neither stimulated by divalent metal ions nor inhibited by a number of metal chelating agents [36]. The equilibrium constant for the reaction at 37°C is 11 mM in Tris–HCl buffer (pH 8.4) and 1.32 mM in Krebs original Ringer phosphate buffer (pH 7.3); this difference is most likely due to Tris forming a complex with glyoxylate shifting the equilibrium in the direction of KHG cleavage.

Surprisingly, the bovine liver aldolase catalyzes the cleavage and the formation of both optical isomers* of KHG at essentially the same rate and to the same extent; K_m values for DL-, L-, and D-KHG are 1.0×10^{-4}, 7.1×10^{-5}, and 1.4×10^{-4} M, respectively, whereas the relative V_{max} values are 100, 105, and 84, respectively. This unusual nonstereospecific character of the mammalian aldolase is interesting. Since both of the isomers of KHG formed by enzymatic condensation of glyoxylate with pyruvate are, in turn, converted by glutamate-aspartate transaminase to the erythro and threo isomers of γ-hydroxy-L-glutamate, a pathway is available whereby either diastereoisomer of γ-hydroxy-L-glutamate can be utilized or formed by mammals (see Fig. 4). Since FDP-aldolase and many other aldolases have strict optical isomer requirements, it is postulated that such enzymes must catalyze either a stereospecific formation of an enzyme-bound carbanion or a stereospecific polarization of the aldehyde substrate so that only one stereo-isomer is cleaved or formed. Since stereospecific handling of one of the three methyl-group hydrogen atoms of pyruvate does not occur (as noted later), the nonstereospecific character of bovine liver KHG-aldolase suggests that an asymmetric polarization of glyoxylate does not occur in this enzyme-catalyzed reaction. The nonstereospecific nature of mammalian KHG-aldolases must be examined further in studies that include the separate enantiomers of various substrate analogs.

The enzyme is essentially specific for the cleavage of KHG; only 2-keto-4,5-dihydroxyvalerate, 2-keto-4-hydroxy-4-methylglutarate, 5-keto-4-deoxy-glucarate, 2-keto-3-deoxy-6-phosphogluconate, and 2-keto-4-hydroxybuty-rate are cleaved at 33, 8, 3, 2, and 1 %, respectively, the rate of KHG-cleavage

---

* The convention is followed that L-malic acid is formed by oxidative decarboxylation (with alkaline H_2O_2) of L-KHG, and D-KHG correspondingly yields D-malic acid [17,29]. The isomeric form of malate is established with malate dehydrogenase, which utilizes only the L-isomer as substrate.

[43]. All of the compounds that are cleaved enzymatically have the following structure on four carbon atoms at one end of the molecule:

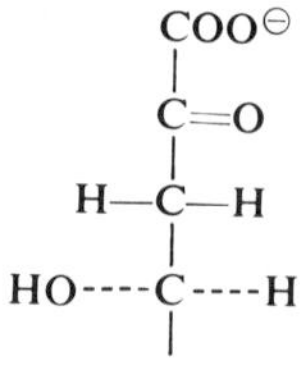

$$
\begin{array}{c}
COO^{\ominus} \\
| \\
C\!=\!O \\
| \\
H\!-\!C\!-\!H \\
| \\
HO\text{----}C\text{----}H \\
|
\end{array}
$$

The specificity of azomethine (Schiff base) formation with liver KHG-aldolase was tested by treating the enzyme with a variety of substrate analogs in the presence of borohydride; any loss in enzymatic activity was regarded as a measure of Schiff base binding. Of some 40 compounds tested, inactivation occurs (in order of decreasing effectiveness) with 2-keto-4-hydroxy-4-methylglutarate, 2-ketoglutarate, 2-keto-4-hydroxybutyrate, 2-keto-3-deoxy-6-phosphogluconate, fructose 1,6-diphosphate, 2-keto-4,5-dihydroxyvalerate, 2-keto-3-deoxygluconate, and 5-keto-4-deoxyglucarate (among KHG analogs); only with bromopyruvate and 2-ketobutyrate (among pyruvate analogs); and also with glyoxal, formaldehyde, acetaldehyde, and glycolaldehyde (glyoxylate analogs). In this regard, therefore, a high degree of specificity is shown for pyruvate but not for glyoxylate [43]. Furthermore, the process of azomethine formation can be clearly dissociated from aldol-type carbon–carbon cleavage; this is especially evident with 2-ketoglutarate and 2-keto-3-deoxygluconate, which substrate analogs are effectively bound as Schiff base intermediates but are not cleaved.

If considered according to Rutter's classification of FDP-aldolases, bovine liver KHG-aldolase is a Class I ("lysine class") aldolase. When either KHG or pyruvate is present, KHG-aldolase is completely inactivated by borohydride. When, however, this aldolase is incubated with glyoxylate in the presence of $NaBH_4$, enzymatic activity is also completely and irreversibly lost [44,45]. Azomethine formation between the ε-amino group of a lysyl residue (or residues) in the enzyme molecule and [^{14}C]glyoxylate or [^{14}C]pyruvate was established by reduction with $NaBH_4$ followed by acid hydrolysis of the stabilized adducts; N^6-(carboxymethyl)lysine and N^6-(1'-carboxyethyl)lysine, respectively, were isolated and identified in each case [45]. Furthermore, concomitant addition of these two ^{14}C substrates to the enzyme in the presence of $NaBH_4$ does not result in the incorporation of an additive amount of radioactivity, and the presence of one substrate unlabeled and the other labeled with ^{14}C significantly dilutes the total amount of radioactivity bound to the enzyme. Also, a prior incubation of KHG-aldolase with either unlabeled glyoxylate or pyruvate in the presence of $NaBH_4$ completely prevents the subsequent binding of the radioactive cosubstrate. Identical results are obtained when labeled or unlabeled KHG is used as one of the two competing substrates [45]. KHG-Aldolase, therefore, has the novel ability not only of binding as a Schiff base the two substrates (i.e., KHG and pyruvate) one would expect mechanistically, but also of forming an "abortive" azomethine with glyoxylate. The significance of Schiff base formation with glyoxylate is not entirely clear; the most tenable proposal at present is that glyoxylate is bound nonspecifically by virtue of being an analog of pyruvate and its binding by the enzyme may have no physiological significance. The stoichiometry of binding (moles of substrate/mole of enzyme), which is the same for KHG, pyruvate, and glyoxylate (or formaldehyde [46]), is 1:1, and all three substrates appear to compete for either the same site or binding sites that are mutually exclusive when one is occupied. This binding ratio is unusually low and is still regarded as a minimum value rather than as an absolute number of binding sites.

As would be expected for an aldolase that functions via a Schiff base mechanism, liver KHG-aldolase catalyzes a proton exchange reaction with the methyl hydrogen atoms of pyruvate [47]. This exchange reaction is nonstereospecific in that all three methyl hydrogen atoms are eventually exchanged with the medium. If KHG is incubated with the aldolase in tritiated water, tritium is incorporated into both KHG and pyruvate but not into glyoxylate. The rate of pyruvate tritiation in ^{3}HOH is about six times slower than the rate of KHG synthesis; for KHG-aldolase from liver, therefore, formation of the carbanion could be partially rate limiting in the overall reaction.

In studies with acetoacetate decarboxylase and FDP-aldolase, Westheimer [14] and Cash and Wilson [48], respectively, observed inactivation of enzymatic activity by cyanide only in the presence of those substrates capable of forming carbanions; this inactivation was readily reversed by dilution or dialysis. Inactivation of KHG-aldolase by cyanide differs markedly in three ways with the aforementioned Class I enzymes. For KHG-aldolase, (a) very low levels of cyanide cause inactivation levels approaching 100%, (b) the inhibition is *irreversible* and occurs stoichiometrically between enzyme–substrate–cyanide, and (c) the inactivation occurs only with aldehydic substrates or substrate analogs that are not able to form carbanions. No loss of KHG-aldolase activity is observed when the enzyme is incubated with KHG or pyruvate in the presence of cyanide, but with glyoxylate (or other aldehydes, notably formaldehyde, glyoxal, or glycolaldehyde) a very rapid and irreversible loss of activity occurs [46,49]. Other studies [46] had shown that formaldehyde selectively forms a Schiff base complex with apparently the active-site lysyl residue of KHG-aldolase (a novel interaction in itself in that the transfer of one-carbon moieties is usually enzyme mediated via tetrahydrofolate). Working with either [^{14}C]cyanide or [^{14}C]formaldehyde (by inference, any other aldehydes noted above should react similarly), we showed that cyanide inactivation is due, in this instance, to cyanide addition

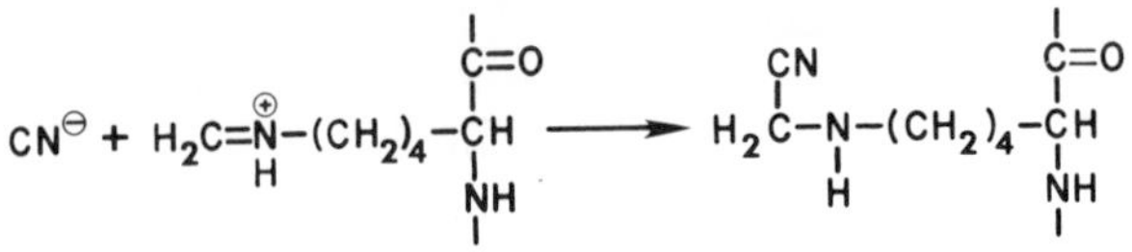

Fig. 6 Aminonitrile formation between cyanide and KHG-aldolase–formaldehyde aldimine.

to the Schiff base aldimine with stable formation of an aminonitrile (Fig. 6). The evidence for aminonitrile formation is 2-fold: (a) The mole per mole stoichiometry of the radioactive complex formed with the enzyme is 1:1:1 (KHG-aldolase to formaldehyde to cyanide), and (b) after acid hydrolysis of the enzyme–[^{14}C]formaldehyde–cyanide adduct, N^6-(carboxymethyl)lysine can be isolated and identified [46]. Now that the existence of enzyme-bound aminonitriles has been established, a basis is provided for using cyanide as well as borohydride to detect Schiff base intermediates. No meaningful evidence is available yet to explain why KHG-aldolase is irreversibly inactivated by cyanide only in the presence of aldehydic substrates (not with pyruvate or KHG). It would seem that with this aldolase some unusual aldehyde (Schiff base)–enzyme complex is formed which is different from the typical carbanion (Schiff base)–enzyme complex. When initial enzymatic rates of KHG cleavage are measured in the presence of cyanide, mixed competitive–noncompetitive inhibition is observed. In this case, the inhibition is freely reversible by dilution and does not lead to irreversible inactivation; the K_i for cyanide is 5.7×10^{-4} M [50]. The competitive aspect of such inhibition is more than likely due to an ionic interaction of cyanide at a carboxylate substrate binding site on the enzyme.

What do kinetic studies show regarding the irreversible inactivation of liver KHG-aldolase that occurs when enzyme is incubated with an aldehydic substrate in the presence of cyanide? The process is biphasic and can, under certain conditions, involve a direct interaction between KHG-aldolase and cyanide [50]. The data are consistent with the occurrence of three competing reactions: (a) irreversible addition of cyanide to the enzyme–substrate Schiff base intermediate with aminonitrile formation, (b) reversible cyano-hydrin formation between cyanide and the aldehydic substrate, and (c) an interaction of cyanide with the aldolase which is not substrate dependent. Approximately 0.4 mole of cyanide is associated with a mole (120,000 daltons) of enzyme when KHG-aldolase is incubated with [^{14}C]cyanide followed by exhaustive dialysis. Whereas native aldolase, not treated with cyanide, has ten titratable sulfhydryl groups, approximately one less such group reacts when the enzyme is incubated with cyanide (in the absence of an aldehydic substrate). As confirmed by circular dichroism spectra, all indications are that the binding of cyanide by KHG-aldolase results in a conformational change of the enzyme [50].

A most striking and novel property of this aldolase is that it is also an effective β-decarboxylase. In Tris–HCl buffer (pH 8.1), liver KHG-aldolase catalyzes the loss of carbon dioxide from oxaloacetate with formation of pyruvate at 50% the rate of KHG cleavage [43,49]. Since the β-decarboxylase *and* aldolase activities of the enzyme are both destroyed by a prior incubation of the enzyme with either pyruvate or glyoxylate plus $NaBH_4$ as well as by incubation of the enzyme with glyoxylate and cyanide [43], the same active site must catalyze the two processes (i.e., decarboxylation and aldol cleavage or condensation). Decarboxylation of acetoacetate is not catalyzed by KHG-aldolase.

For liver KHG-aldolase, sulfhydryl and histidyl residues (in addition to lysyl residues) appear to be involved in the catalytic activity of the enzyme. We have found that treatment of this enzyme with a 42-fold molar excess of TNM at pH 8.0 and 20°C rapidly and completely destroys both the aldolase and oxaloacetate β-decarboxylase activities of the enzyme [51]. Tetranitro-methane inactivation alters the same site that catalyzes both enzymatic reactions (aldolase and β-decarboxylase) since (a) the kinetics of inactivation are identical for both catalytic activities when the enzyme is treated with either a 19- or a 42-fold molar excess of the reagent, (b) the level of inactivation obtained with various concentrations of TNM is essentially the same for both activities, and (c) the protective effects of α-ketoglutarate against TNM inactivation are the same for aldolase and decarboxylase activities. Various substrates and competitive inhibitors protect both enzymatic activities. Loss of aldolase and β-decarboxylase activities occurs concomitant with the modification of free sulfhydryl groups; four such groups are oxidized in the completely inactivated enzyme. Spectral and amino acid analyses establish that the reaction is limited solely to cysteinyl residues. The apparent Michaelis constants for substrates of KHG-aldolase are not altered by modification of the enzyme with TNM, but the TNM-inactivated aldolase does not form an azomethine with either pyruvate or glyoxylate [52]. Loss of both aldolase and β-decarboxylase activities after reaction with TNM, therefore, is most likely due to impaired Schiff base binding of substrates. A comparable inactivation of FDP-aldolase by TNM is also associated with the oxidation of sulfhydryl groups in the molecule [53].

Liver KHG-aldolase has a total of eight to ten sulfhydryl groups in the molecule [54]. In the absence of sodium dodecyl sulfate, only four cysteinyl residues are accessible to titration with DTNB. These four sulfhydryl groups are of two types in the disulfide exchange reaction with DTNB; one —SH group (type A) reacts rapidly, whereas three additional thiols (type B) titrate at approximately 0.1 the rate of the type A —SH residue. Either pyruvate or glyoxylate protects one of the three type B —SH residues from reaction with DTNB. Reaction of KHG-aldolase with DTNB or *p*-mercuribenzoate results in a progressive loss of aldolase activity which is not proportional to the number of —SH groups modified; complete loss of enzymatic activity is never observed even when all eight to ten sulfhydryl groups in the molecule are titrated. Since the presence of either pyruvate or glyoxylate abolishes the reactivity of one thiol group toward DTNB and since each substrate also substantially protects the enzyme against inactivation by this reagent, all indications are that KHG-aldolase exists in an altered conformational state in the presence of its cosubstrates. A direct binding of glyoxylate or pyruvate by an —SH residue in the molecule is not envisioned. These effects of substrates on KHG-aldolase reinforce the observation noted earlier wherein cyanide, possibly interacting with a binding site, causes a similar conformational change in the protein molecule which affects the reactivity of a sulfhydryl group toward DTNB. Such results, involving DTNB or cyanide with —SH groups of KHG-aldolase, most likely differ from those obtained with TNM because of the nature of the product formed and/or the change in protein structure accompanying modification. Reaction of the aldolase with TNM produces cysteic acid and other unidentified products; in addition, whereas inactivation of KHG-aldolase by DTNB (or p-mercuribenzoate) is almost completely reversed by the addition of excess 2-mercaptoethanol, thiols are completely ineffective in reactivating the TNM-modified enzyme.

Preliminary data [55] also suggest a possible involvement of histidyl residues in liver KHG-aldolase activity. Photooxidation of the enzyme at pH 8.0 in the presence of methylene blue results in a rapid, pseudo first-order, and simultaneous loss of both aldolase and β-decarboxylase activities. The entire effect of photooxidation is reflected in a lowered V_{max} value (no change in K_m). Photooxidized aldolase does not form Schiff base intermediates with glyoxylate or pyruvate. Amino acid analyses of the photoinactivated enzyme show extensive loss of both histidyl and cysteinyl residues. Analysis of data according to the method of Ray and Koshland [56] yields identical apparent first-order rate constants (1.67×10^{-1} min^{-1}) for the loss of activity and for the destruction of histidyl residues; the corresponding rate constant for the destruction of cysteinyl residues is 8.64×10^{-2} min^{-1}. Further studies remain to be done with photooxidized KHG-aldolase to determine, if possible, how the loss of enzymatic activity relates to the precise number and kind of aminoacyl residues modified.

Escherichia coli KHG-ALDOLASE

KHG-Aldolase appears to be a constitutive enzyme in *E. coli* K-12; homogeneous preparations of the enzyme can be obtained from extracts of cells grown either on a medium of nutrient broth or on a glycerol, casein hydrolyzate, salts medium [38,39]. Purity of the enzyme is confirmed by polyacrylamide gel electrophoresis at three different pH values as well as by antibody precipitin tests. Antibody to the pure *E. coli* enzyme does not cross-react with KHG-aldolase from bovine liver.

KHG-Aldolase from *E. coli* has an average molecular weight of 63,000, approximately half that of the liver aldolase. The enzyme has a pH optimum of 8.6 and is neither stimulated by divalent metal ions nor inhibited by various metal chelating agents. KHG-Aldolase from *E. coli* and from liver are similar in the following four respects. First, they are alike in being highly specific toward KHG as a substrate for aldol-type cleavage. For the bacterial aldolase, of a large number of compounds tested only 2-keto-4-hydroxybutyrate is cleaved 8% relative to the rate of cleavage of DL-KHG at pH 8.1 and 25°C. Second, and in contrast to most bacterial FDP-aldolases which are "metal-ion type" (Class II), KHG-aldolase from *E. coli* is a Class I ("lysine type") aldolase. Like liver KHG-aldolase, the bacterial enzyme forms Schiff base intermediates with KHG and pyruvate, but again not only with these substrates (as required mechanistically) but also with glyoxylate. Competition studies, carried out as with the liver aldolase, indicate that in this case, too, all three substrates are apparently bound at the same active-site lysyl residue [38]. In the presence of borohydride, 1 mole of either pyruvate or glyoxylate is bound per mole (63,000 daltons) of enzyme. As determined by borohydride inactivation, Schiff base formation with the bacterial aldolase is highly specific for pyruvate but a number of analogs (including formaldehyde, glyoxal, acetaldehyde, and glycolaldehyde) replace glyoxylate. Third, *E. coli* KHG-aldolase is irreversibly inactivated by cyanide *only* in the presence of glyoxylate (*not* with KHG or pyruvate) and also in this instance, where an aminonitrile is formed, 1 mole of glyoxylate is stably bound per mole of enzyme. A final similarity is that, like liver KHG-aldolase, the bacterial aldolase catalyzes a typical hydrogen exchange half-reaction with pyruvate [57]. A rapid first-order exchange of all three methyl hydrogens of pyruvate occurs in the absence of glyoxylate. For the bacterial aldolase, this exchange reaction is not rate limiting for aldol condensation or cleavage. When 2-ketobutyrate is used in place of pyruvate, the exchange reaction appears to be stereoselective in that, although both methylene hydrogens participate in an exchange process with the medium, the rate of exchange for one carbon-bound hydrogen is six times faster than for the other.

Escherichia coli KHG-aldolase, however, in addition to being a smaller molecule, has a number of interesting properties that contrast sharply with the liver aldolase. Whereas bovine liver KHG-aldolase is essentially non-stereospecific toward the enantiomers of KHG, the bacterial aldolase preferentially utilizes the L-isomer as substrate. The respective K_m values for L-, D-, and DL-KHG are 2.3×10^{-3}, 25×10^{-3}, and 4.2×10^{-3} M, whereas the relative V_{max} values are 100, 19, and 84, respectively. *Escherichia coli*

KHG-aldolase, therefore, is stereoselective in the sense that, although both isomers are substrates, the enzyme exhibits stereopreference for L-KHG. This characteristic is due to a much more facile handling by the enzyme of the carbonyl carbon of glyoxylate oriented for si* face rather than for re face attack [58]. In contrast, with bovine liver KHG-aldolase either the si or the re face of the carbonyl carbon atom of glyoxylate can attack bound enol-pyruvate with equal facility [59].

The E. coli enzyme has a total of five titratable sulfhydryl groups, two of which are more reactive than the other three; the number titrated is not changed by addition of either KHG, pyruvate, glyoxylate, or pyruvate in combination with glyoxylate. When all five —SH groups are reacted with p-mercuribenzoate, aldolase activity toward KHG is reduced approximately 45%, whereas only an 11% loss is observed in the tritium exchange activity. Furthermore, no significant loss in either KHG cleavage activity or [^{3}H]-pyruvate detritiation activity is seen when all histidyl residues in the E. coli aldolase are reacted with ethoxyformic anhydride. Present indications are, therefore, that for KHG-aldolase from E. coli neither a histidyl nor a cysteinyl residue functions as the active-site amphoteric proton acceptor–donor group. Participation of these two aminoacyl residues in some aspect of the catalytic properties of bovine liver KHG-aldolase is more evident (as noted before), whereas one or the other of these same two residues has also been implicated in the proton donation and accepting steps for a number of other enzymes, including FDP-aldolase [60–65], transaldolase [66], glutamate-aspartate transaminase [67–69], and 2-keto-3-deoxy-6-phosphogluconate aldolase [70,71].

Another clear and interesting difference between KHG-aldolase from bovine liver and that from E. coli is in their relative β-decarboxylase activities. Whereas the liver enzyme catalyzes the β-decarboxylation of oxaloacetate at 50% the rate of KHG cleavage, the pure bacterial aldolase is a better β-decarboxylase toward oxaloacetate than it is an aldolase toward KHG [38]. Such results, with pure KHG-aldolase from two different sources, both of which have very significant decarboxylase activity, would seem to be the first indication of positive support for the phylogenetic concept of Rutter [4] that Class I ("lysine type") aldolases may structurally resemble a previously existing enzyme such as a β-decarboxylase.

In view of the many atypical properties this aldolase has when compared with FDP-aldolase and also the striking differences that exist between liver and bacterial KHG-aldolase, a need to establish the subunit structure of KHG-aldolases is obvious. Latest studies in our laboratory have elucidated the general substructure of E. coli KHG-aldolase [72]. As noted earlier, the

---

* The si face of the aldehyde carbon of glyoxylate is that described by the anticlockwise orientation of ligands substituted by carbonyl oxygen, carboxyl group, and hydrogen.

native protein has an average molecular weight of 63,000. After dissociation of the molecule with sodium dodecyl sulfate and subsequent polyacrylamide gel electrophoresis, one protein band with a molecular weight of 21,000 is seen. When the molecule is cross-linked with dimethyl suberimidate [73] (or glutaraldehyde), three protein bands with molecular weights of 21,000, 43,000, and 63,000 are detected after electrophoresis in sodium dodecyl sulfate. Amino acid analyses of the protein molecule allow for calculation of a minimum molecular weight of 20,600 and show a normal spectrum of amino acids which includes 3 tryptophanyl, 3 histidyl, 9 tyrosyl, 18 arginyl, and 36 lysyl residues per 63,000 daltons. Tryptic peptide maps show 18–20 ninhydrin-positive (total) peptides, whereas peptides containing 1 tryptophanyl, 4 histidyl plus tyrosinyl, and 6 arginyl residues are seen on maps analyzed by specific methods. In addition, when dissociated native enzyme and dissociated ^{14}C-maleylated KHG-aldolase are mixed and subsequently allowed to reassociate, four hybrid species with expected relative levels of radioactivity and aldolase activity are observed by polyacrylamide gel electrophoresis. All data in hand, therefore, indicate that KHG-aldolase of *E. coli* is a trimeric molecule consisting of three identical or nearly identical subunits. These latest findings underline the caution expressed earlier about the number of binding sites in KHG-aldolases and stress the need for re-examining the stoichiometry of substrate binding by as wide a variety of experimental methods and with as many different substrates (and/or analogs) as is possible.

SUMMARY

Since hydroxyproline constitutes 13% of collagen, and collagenlike extra-cellular protein has been estimated to account for about 40% of body protein, hydroxyproline represents a major constituent of total animal protein. It is apparent that elucidation of the pathway, on the enzyme level, whereby this abundant amino acid is catabolized by mammals has uncovered a most interesting and novel aldolase, i.e., 2-keto-4-hydroxyglutarate aldolase. Bovine liver KHG-aldolase differs from other aldolases in a number of respects. For example, (a) it is completely nonstereospecific in that it catalyzes the cleavage or the formation of both enantiomers of KHG at equal rates, (b) it forms an "abortive" Schiff base intermediate with glyoxylate or other low molecular weight aldehydes, (c) it is *irreversibly* inactivated by cyanide only in the presence of aldehydic substrates (such as glyoxylate and form-aldehyde) with stable formation of an aminonitrile, and (d) it is highly effective as a β-decarboxylase toward oxaloacetate. KHG-Aldolase has also been obtained in pure form from *E. coli* K-12. The bacterial aldolase is similar

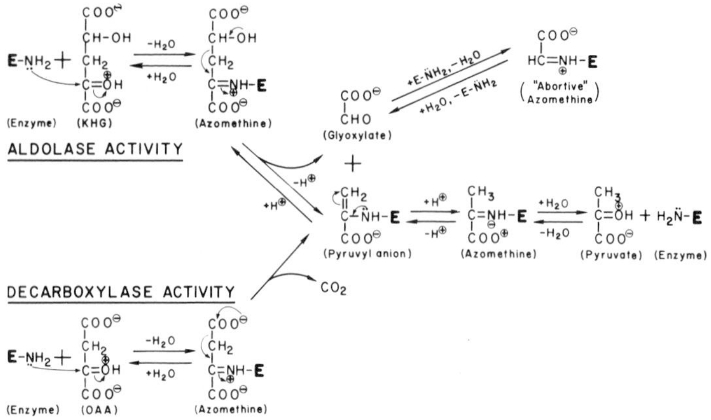

Fig. 7 Mechanistic aspects of 2-keto-4-hydroxyglutarate aldolase (OAA, oxaloacetic acid).

TABLE 2

Properties of 2-Keto-4-hydroxyglutarate Aldolase

Bovine liver	*Escherichia coli*
1. Schiff base mechanism, involving an active-site lysyl residue (or residues)	
2. Form an "abortive" Schiff base intermediate with glyoxylate or other low molecular weight aldehydes	
3. Irreversibly inactivated by cyanide only in the presence of aldehydic substrates (like glyoxylate and formaldehyde)	
4. Highly specific toward KHG as substrate for aldol cleavage; also highly specific for azomethine formation with pyruvate but not with glyoxylate	
5. Catalyze a hydrogen exchange half-reaction with all three methyl hydrogens of pyruvate	
6. Activity not inhibited by metal chelating agents nor stimulated by K^+	
7. MW $\sim 120,000$	7. MW $\sim 63,000$
8. Nonstereospecific, catalyzing the cleavage or formation of D- and L-KHG equally well	8. Stereoselective toward L-KHG
9. Histidyl and/or sulfhydryl residues required for maintaining full catalytic activity	9. Neither histidyl nor sulfhydryl residues seem to function as the amphoteric proton carrier in the molecule
10. About 50% as effective as a β-decarboxylase toward oxaloacetate than as an aldolase toward KHG	10. A better decarboxylase toward oxaloacetate than as an aldolase toward KHG
11. Subunit structure not yet elucidated	11. Trimeric subunit structure with three identical or nearly identical subunits

in some ways to liver KHG-aldolase, but it is also uniquely different in a number of other respects. Mechanistic similarities of liver and *E. coli* KHG-aldolase are presented in Fig. 7; common and different properties of KHG-aldolase from these two sources are summarized in Table 2. Although it would be desirable to have this enzyme purified from still other biological sources, having KHG-aldolase already in pure form from liver and *E. coli* provides a highly interesting system for studies that correlate enzymatic properties, molecular structure, and function. Further insights on enzyme-catalyzed aldol-type reactions may also be provided.

ACKNOWLEDGMENTS

Experimental work described in this paper was supported in part by Grant AM-03718 from the National Institute of Arthritis, Metabolism, and Digestive Diseases, U.S. Public Health Service. It is also a pleasure to acknowledge my many associates who contributed so importantly in various phases of this work; included are U. Maitra, R. D. Kobes, R. S. Lane, H. Nishihara, B. A. Hansen, A. Shapley, S. R. Grady, and J. Wang.

REFERENCES

1. E. Fischer and J. Tafel, *Ber. Dtsch. Chem. Ges.* **20**, 2566 (1887).
2. O. Meyerhof and K. Lohmann, *Biochem. Z.* **271**, 89 (1934).
3. See B. L. Horecker, *J. Cell. Comp. Physiol.* **54**, Suppl. 1, 89 (1959).
4. W. J. Rutter, *Fed. Proc., Fed. Am. Soc. Exp. Biol.* **23**, 1248 (1964).
5. A. Kornberg, S. Ochoa, and A. H. Mehler, *J. Biol. Chem.* **174**, 159 (1948).
6. D. Herbert, *Symp. Soc. Exp. Biol.* **5**, 52 (1951).
7. R. Steinberg and F. H. Westheimer, *J. Am. Chem. Soc.* **73**, 429 (1951).
8. A. Schmitt, I. Bottke, and B. Siebert, *Hoppe-Seyler's Z. Physiol. Chem.* **347**, 18 (1966).
9. G. W. Kosicki and F. H. Westheimer, *Biochemistry* **7**, 4303 (1968).
10. G. W. Kosicki, *Biochemistry* **7**, 4310 (1968).
11. D. J. Creighton and I. A. Rose, *J. Biol. Chem.* **251**, 69 (1976).
12. G. Hamilton and F. H. Westheimer, *J. Am. Chem. Soc.* **81**, 6332 (1959).
13. I. Fridovich and F. H. Westheimer, *J. Am. Chem. Soc.* **84**, 3208 (1962).
14. F. H. Westheimer, *Proc. Chem. Soc., London* p. 253 (1963).
15. B. Zerner, S. M. Coutts, F. Lederer, H. H. Waters, and F. H. Westheimer, *Biochemistry* **5**, 813 (1966).
16. R. A. Laursen and F. H. Westheimer, *J. Am. Chem. Soc.* **88**, 3426 (1966).
17. U. Maitra and E. E. Dekker, *J. Biol. Chem.* **239**, 1485 (1964).
18. L. D. Aronson, R. G. Rosso, and E. Adams, *Biochim. Biophys. Acta* **132**, 200 (1967).
19. G. Wolf, W. W. Heck, and J. C. Leak, *J. Biol. Chem.* **223**, 95 (1956).
20. A. I. Virtanen and P. K. Hietala, *Acta Chem. Scand.* **9**, 175 (1955).
21. E. E. Dekker, *Biochem. Prep.* **9**, 69 (1962).
22. E. Adams and A. Goldstone, *J. Biol. Chem.* **235**, 3492 (1960).
23. E. Adams and A. Goldstone, *J. Biol. Chem.* **235**, 3504 (1960).
24. E. E. Dekker, *Biochim. Biophys. Acta* **40**, 174 (1960).
25. K. Kuratomi and K. Fukunaga, *Biochim. Biophys. Acta* **43**, 562 (1960).

26. U. Maitra and E. E. Dekker, *Biochim. Biophys. Acta* **51**, 416 (1961).
27. E. E. Dekker and U. Maitra, *J. Biol. Chem.* **237**, 2218 (1962).
28. U. Maitra and E. E. Dekker, *J. Biol. Chem.* **238**, 3660 (1963).
29. A. Goldstone and E. Adams, *J. Biol. Chem.* **237**, 3476 (1962).
30. K. Kuratomi and K. Fukunaga, *Biochim. Biophys. Acta* **78**, 617 (1963).
31. K. Kuratomi, K. Fukunaga, and Y. Kobayashi, *Biochim. Biophys. Acta* **78**, 629 (1963).
32. H. Hift and H. R. Mahler, *J. Biol. Chem.* **198**, 901 (1952).
33. R. S. Lane, A. Shapley, and E. E. Dekker, *Biochemistry* **10**, 1353 (1971).
34. R. S. Lane and E. E. Dekker, *Biochemistry* **8**, 2958 (1969).
35. R. G. Rosso and E. Adams, *J. Biol. Chem.* **242**, 5524 (1967).
36. R. D. Kobes and E. E. Dekker, *J. Biol. Chem.* **244**, 1919 (1969).
37. E. E. Dekker, R. D. Kobes, and S. R. Grady, *in* "Methods in Enzymology" (S. P. Colowick and N. O. Kaplan, eds.), Vol. 42, p. 280. Academic Press, New York, 1975.
38. H. Nishihara and E. E. Dekker, *J. Biol. Chem.* **247**, 5079 (1972).
39. E. E. Dekker, H. Nishihara, and S. R. Grady, *in* "Methods in Enzymology" (S. P. Colowick and N. O. Kaplan, eds.), Vol. 42, p. 285. Academic Press, New York, 1975.
40. P. Andrews, *Biochem. J.* **96**, 595 (1965).
41. R. G. Martin and B. N. Ames, *J. Biol. Chem.* **236**, 1372 (1961).
42. J. L. Hedrick and A. J. Smith, *Arch. Biochem. Biophys.* **126**, 155 (1968).
43. R. D. Kobes and E. E. Dekker, *Biochim. Biophys. Acta* **250**, 238 (1971).
44. R. D. Kobes and E. E. Dekker, *Biochem. Biophys. Res. Commun.* **25**, 329 (1966).
45. R. D. Kobes and E. E. Dekker, *Biochemistry* **10**, 388 (1971).
46. B. A. Hansen, R. S. Lane, and E. E. Dekker, *J. Biol. Chem.* **249**, 4891 (1974).
47. S. R. Grady, Ph.D. Thesis, The University of Michigan, Ann Arbor (1973).
48. D. J. Cash and I. B. Wilson, *J. Biol. Chem.* **241**, 4290 (1966).
49. R. D. Kobes and E. E. Dekker, *Biochem. Biophys. Res. Commun.* **27**, 607 (1967).
50. B. A. Hansen and E. E. Dekker, *Biochemistry* **15**, 2912 (1976).
51. R. S. Lane and E. E. Dekker, *Biochem. Biophys. Res. Commun.* **36**, 973 (1969).
52. R. S. Lane and E. E. Dekker, *Biochemistry* **11**, 3295 (1972).
53. J. F. Riordan and P. Christen, *Biochemistry* **7**, 1525 (1968).
54. R. S. Lane, B. A. Hansen, and E. E. Dekker, *Biochim. Biophys. Acta* (in press).
55. B. A. Hansen, Ph.D. Thesis, The University of Michigan, Ann Arbor (1971).
56. W. J. Ray, Jr. and D. E. Koshland, Jr., *J. Biol. Chem.* **236**, 1973 (1961).
57. S. R. Grady and D. J. Dunham, *Fed. Proc., Fed. Am. Soc. Exp. Biol.* **32**, 667 (abstr.) (1973).
58. H. P. Meloche, C. T. Monti, and E. E. Dekker, *Biochem. Biophys. Res. Commun.* **65**, 1033 (1975).
59. H. P. Meloche and L. Mehler, *J. Biol. Chem.* **248**, 6333 (1973).
60. C. Y. Lai and P. Hoffee, *Fed. Proc., Fed. Am. Soc. Exp. Biol.* **25**, 408 (1966).
61. P. Hoffee, C. Y. Lai, E. L. Pugh, and B. L. Horecker, *Proc. Natl. Acad. Sci. U.S.A.* **57**, 107 (1967).
62. L. C. Davis, L. W. Brox, R. W. Gracy, G. Ribereau-Gayon, and B. L. Horecker, *Arch. Biochem. Biophys.* **140**, 215 (1970).
63. L. C. Davis, G. Ribereau-Gayon, and B. L. Horecker, *Proc. Natl. Acad. Sci. U.S.A.* **68**, 416 (1971).
64. F. C. Hartman, B. Suh, and R. Barker, *Fed. Proc., Fed. Am. Soc. Exp. Biol.* **32**, 473 (abstr.) (1973).

65. B. L. Horecker, O. Tsolas, and C. Y. Lai, in "The Enzymes" (P. D. Boyer, ed.), 3rd ed., Vol. 7, p. 213. Academic Press, New York, 1972.
66. K. Brand, O. Tsolas, and B. L. Horecker, *Arch. Biochem. Biophys.* **130**, 521 (1969).
67. J. G. Farrelly and J. E. Churchich, *Biochim. Biophys. Acta* **167**, 280 (1968).
68. D. L. Peterson and M. Martinez-Carrion, *J. Biol. Chem.* **245**, 806 (1970).
69. M. Martinez-Carrion, C. Turano, F. Riva, and P. Fasella, *J. Biol. Chem.* **242**, 1426 (1967).
70. H. P. Meloche, *Fed. Proc., Fed. Am. Soc. Exp. Biol.* **29**, 462 (1970).
71. H. P. Meloche, *Biochemistry* **9**, 5050 (1970).
72. E. E. Dekker and J. K. Wang, *Fed. Proc., Fed. Am. Soc. Exp. Biol.* **35**, 1521 (1976).
73. G. E. Davies and G. R. Stark, *Proc. Natl. Acad. Sci. U.S.A.* **66**, 651 (1970).

CHAPTER

4

The Hydroxylation of Alkanes

N. C. Deno, Elizabeth J. Jedziniak, Lauren A. Messer,
and Edward S. Tomezsko

INTRODUCTION

The conversion of simple linear alkanes to terminal primary alcohols is one of the most intriguing of enzyme reactions. In these reactions, the concept of functionality in organic chemistry has been reversed. Not only has the end of the alkane chain become the functional group, but a host of the usual functional groups have been found to be inert. These include alcohol, ether, carboxylic acid, ester, amide, chloro, cyano, nitro, and sulfonate [1] and even sometimes alkene [2–7].

The precedent for such hydroxylations has been much enlarged by the recent discovery that alkanes and unactivated CH bonds can be hydroxylated via aminium radicals (R_3N^+) [8–10]. A feature of these reactions is that the alcohol is inert to further oxidation [8–10]. These hydroxylations were first reported for intramolecular cases exemplified by Eqs. (1)–(3) [8]. They were soon extended to intermolecular cases [9,10], of which Eqs. (4) and (5) are examples.

The Fe(II) is a catalyst. In the intermolecular examples in CF_3COOH, the yields were nearly constant from 0.01 to 0.5 mole of Fe(II) per mole of reactant [10]. In the intramolecular examples, 2.5 moles of Fe(II) were used per mole of amine oxide [8]. It is not known if such large excesses were required.

In intramolecular reactions with $C_4H_9N(CH_3)_2O$, 3-hydroxylation accompanied 4-hydroxylation [8]. Labeling experiments showed that the 3-OH product arose from 1,2 shifts of hydrogen. Since such shifts are characteristic

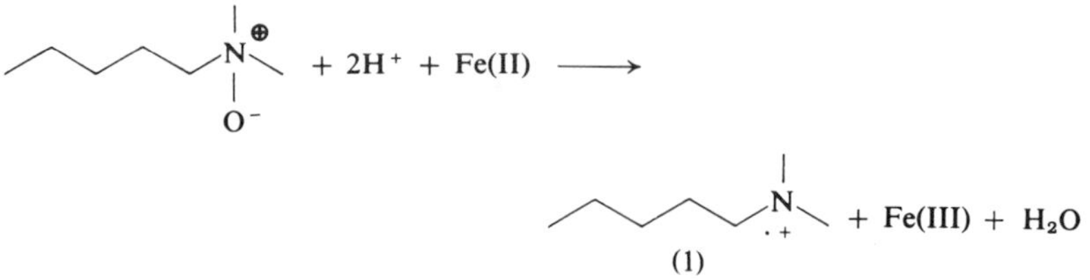

$$+ 2H^+ + Fe(II) \longrightarrow$$

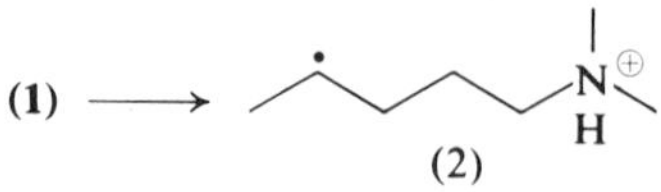

$$+ Fe(III) + H_2O \quad (1)$$

(1)

$$(1) \longrightarrow \qquad\qquad (2)$$

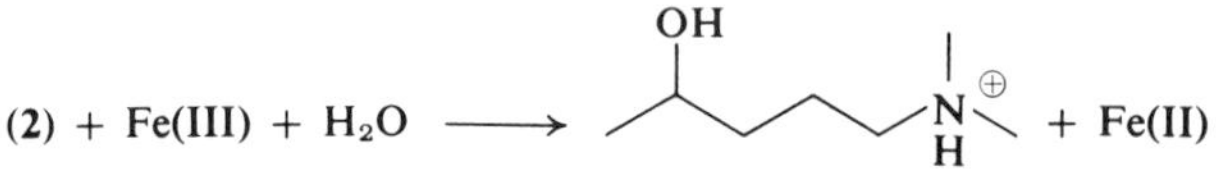

(2)

$$(2) + Fe(III) + H_2O \longrightarrow \qquad\qquad + Fe(II) \quad (3)$$

$$+ Fe(II) + R_3NO + CF_3COOH \longrightarrow$$

$$\xrightarrow{H_2O} \qquad\qquad (4)$$

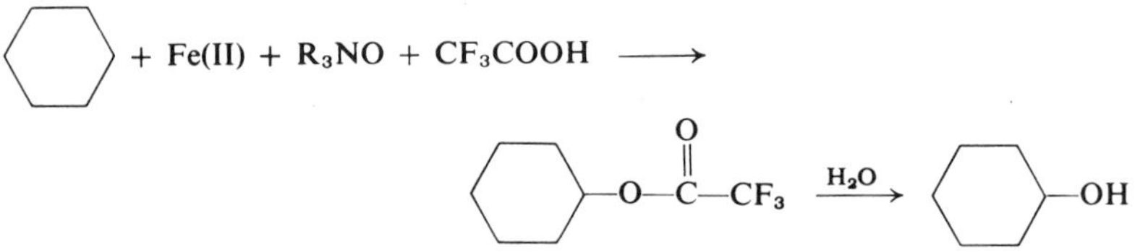

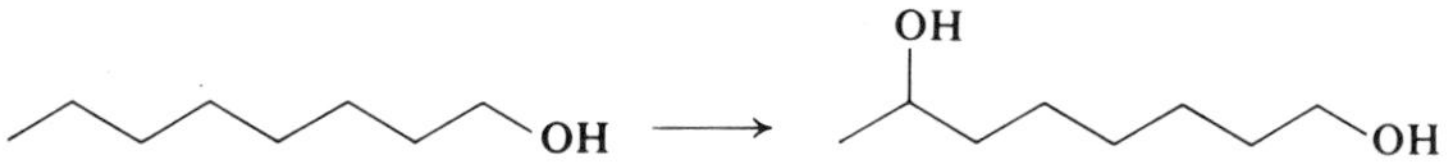

$$\longrightarrow \qquad\qquad (5)$$

70% selectivity for hydroxylation
at C-7

of alkyl cations, Eq. (3) was proposed to proceed in two steps through an intermediate alkyl cation [8]. However, rearrangement was not complete and comparable amounts of 3-OH and 4-OH formed. In the intermolecular cases in CF_3COOH, the tendency to rearrange is not fully known [9].

Closely related to these hydroxylations are chlorinations via aminium radicals [11–13]. This is a two-step chain reaction, and the propagation steps are Eqs. (6) and (7). Equation (6) is identical to the hydrogen abstraction step [Eq. (2)] in hydroxylation. Since this is the step that dictates selectivity and the position substituted, evidence from both hydroxylations and chlorinations will be regarded as indicating the selectivities of aminium radicals.

$$R_3N^{+} + R'H \longrightarrow R_3NH^+ + R'\cdot \quad (6)$$

$$R'\cdot + R_3NCl^+ \longrightarrow R'Cl + R_3N^{+} \quad (7)$$

In addition to the aminium radical hydroxylations, two other types of reagents have been proposed as models for alkane hydroxylations. One type centers around Fe(II) + H_2O_2, in which some form of oxy radical abstracts H from the alkane [14–18]. The second type consists of peroxyacids and variants [18–24] which act in some form of the concerted process shown in Eq. (8). The peroxyacid model has become more interesting with the discovery that primary and secondary alcohols are inert to trifluoroperoxyacetic acid. Thus linear alkanes (decane), cycloalkanes (cyclohexane and cyclododecane), and linear alkyl chains (1-octanol and octanoic acid) can be hydroxylated in high yield at 25°C by trifluoroperoxyacetic acid even in the presence of 5–10% water [10]. The alcohol forms first as shown in Eq. (8).

$$
\begin{array}{ccc}
\text{R—H} & & \text{R} \\
\text{H—O} \diagdown & \text{O} & \text{HO} \diagup \\
\diagdown & & \diagdown \text{OH} \\
\text{O—C} & \longrightarrow & \text{O=C} \\
\diagdown \text{R} & & \diagdown \text{R}
\end{array}
\qquad (8)
$$

The plan of this review is to single out the features of enzymatic hydroxylation of alkanes that have been established either from microbiological studies [1–7,25–34] or from mammalian enzyme studies [35]. These are compared with the simple chemical models to see the degree to which the models duplicate the features of the enzymes.

A critical point of logic is the following. The enzymes that hydroxylate alkanes are effective on linear alkanes of widely varying chain length. The hydroxylase from *Pseudomonas oleovorans* showed rates of reaction with C_6–C_{12} linear alkanes that varied by a factor of only 2 [36]. Other systems prefer C_{10}–C_{20} linear alkanes [25–33]. This is interpreted to mean that only the end of the alkane chain fits into the enzyme. To be otherwise would require that the active site accommodate chains ranging from C_6 to C_{20} and beyond, an extraordinary latitude in view of the severe spatial restraints ascribed to enzymes.

If only the end of the alkane chain fits into the enzyme, it follows that the enzyme must be capable of oxidizing all manner of derivatives of fatty acids at the end remote from the functional group. This has been directly demonstrated, most extensively by studies on ω- and ω-1-hydroxylation by a yeast, *Torulopsis gropengiesseri* [1]. The terminal group on the long chain included all the common functional groups of organic chemistry.

Extending this line of thought, it means that any features or characteristics demonstrated for the remote ω- and ω-1-hydroxylations of functional series can be regarded as equally established for the hydroxylation of linear alkanes and vice versa. In contrast hydroxylations that occur at a functional group or even three to four carbons away from a functional group are regarded as outside the scope of this review. In doing this it is recognized that many enzymologists will regard hydroxylation at or adjacent to functional groups as closely related to the remote hydroxylations, Both types are accomplished by mammalian liver homogenates [16,35,37], and cytochrome *P*-450 is generally implicated [37].

Our preference for maintaining a distinction between remote and adjacent hydroxylation is influenced by precedents from organic chemistry. Hydroxylation at or adjacent to a functional group is common in organic chemistry, and many model systems could be put forward. Remote hydroxylation is rare and is limited at present to hydroxylations by amine oxides plus Fe(II) and to hydroxylations by trifluoroperoxyacetic acid.

TERMINAL HYDROXYLATION

The extensive studies on C_6–C_{20} linear alkanes show that such alkanes are exclusively attacked at the ω (terminal) and ω-1 (penultimate) carbons [1,25–33,36,38–41]. The selectivity is so general as to justify the name ω-hydroxylases for such enzymes. The evidence for ω attack is usually that fatty acids are produced which have the same chain length and carbon number as the alkane substrate [28]. Less common is to isolate esters derived from the 1-alkanol [42] or even the 1-alkanols and 2-alkanols themselves [4,43].

It is evident that such selectivities must be entirely of steric origin since the CH_2 groups in long chains are so nearly equivalent electronically that electronic effects could not direct attack to the ω-1 methylene. This view is supported by the preferential hydroxylation at the ω position. These are primary hydrogen, and it is well known that electronic effects strongly promote loss of secondary hydrogen in preference to primary hydrogen in abstraction of either H· by radicals or H$^-$ by alkyl cations. The overcoming and reversal of these electronic preferences must be due to steric effects (R$^-$ might prefer to abstract H$^+$ from primary positions, but the extreme basicities required for this seem ruled out for biological systems).

If only the end of the chain interacts with the enzyme, any functional group at the other end should have little effect. Such compounds should show the same ω- and ω-1-hydroxylations as found with alkanes. Accordingly, fatty acids [1,28,44] and all manner of their derivatives [1] are hydroxylated exclusively at the ω and ω-1 positions. The ω-hydroxylases not only are widespread in bacteria and yeasts, but are also generally found in mammalian liver [37,38]. Such mammalian systems ω-hydroxylate alkanes [45], fatty acids [37,38,46–49], α,α-dimethyl fatty acids [50–52], and prostaglandin [53].

Any model system that hopes to duplicate such spatial effects must have the capacity to accommodate limitless structure of the peptide type. The aminium radicals seem uniquely fitted for such a role. All the common functional groups are inert except C=C [9–13,54,55], and thus peptide and other structure could be introduced. Also there are three R groups in R_3N^+ from which structure could be attached.

Already there is evidence that even the simple steric effect of large R groups is sufficient to generate some ω-1 selectivity. Chlorination of pentane with N-chloro-2,2,6,6-tetramethylpiperidine gave 1-Cl-, 2-Cl-, and 3-Cl-pentane in relative yields of 9, 87, and 4% [56]. Further spatial selectivities can be expected when peptide structure is introduced into the R_3N^+ reagent.

Of the free radicals that attack alkanes, the capacity for structure is much reduced in RO· and is zero in chlorine and bromine atoms. The Fe(II) + H_2O_2 systems have been regarded as HO· generators [17], but recent work on the hydroxylation of cyclohexanol strongly suggests two independent mechanisms [57]. One involves HO·, which attacks remote from the alcohol group like Cl· [58] and R_3N^+ [9,12,13], and the other involves an Fe–O system that gives attack at or close to the functional group. The latter is possibly capable of structure, although its attachment to the alcohol function coupled with attack at or near that function makes it a better candidate as a model for nonremote hydroxylations.

HYDROXYLASE ACTIVITY

There are sufficient examples in which the alcohol is isolated, notably from species of *Candida* [4,43] and *Torulopsis* [1] yeasts and the bacterium *Pseudomonas oleovorans* [36,39–41], to presume that the enzyme converts RH to ROH without further oxidation. Only the aminium radical hydroxylations duplicate this hydroxylase action [9,10].

It has been reported that the enzymes responsible for hydroxylating methyl groups in lanosterol effect further hydroxylation at the same site [38,59,60]. These hydroxylations occur adjacent to OH and C=C and are outside the scope of this review, although they would raise a problem if they were shown to be closely related to ω-hydroxylases.

INERTNESS OF FUNCTIONAL GROUPS

Linear alkyl series containing terminal CH_2OH, $CHOHCH_3$, OR, ketone, COOH, amide, ester, CN, NO_2, Cl, Br, sulfonate, and sulfonamide can be ω- and ω-1-hydroxylated by yeasts without affecting the functional groups [1]. The inertness of most of these groups toward aminium radicals has already been demonstrated [9–13].

It is easy to dismiss the inertness of the functional groups toward ω-hydroxylases by postulating specific spatial requirements for the enzyme.

The fact that the steric requirements of $—(CH_2)_nCH_3$ are closely approximated by $—(CH_2)_nCH_2OH$, $—(CH_2)_nOCH_3$, and $—(CH_2)_nCH_2Cl$ strains such a rationalization. Now that this inertness has been completely duplicated in the aminium radical chlorinations and hydroxylations by nonspatial effects, more serious consideration must be given to explaining the inertness in terms of the enormous polar effects demonstrated for aminium radicals.

Although the ω-hydroxylases can sometimes hydroxylate without affecting terminal C=C groups [2–7], this is not always the case. A purified enzyme from *Pseudomonas oleovorans* can ω-hydroxylate and epoxidize terminal alkene at comparable rates [39–41]. The aminium radicals show a related dual action. Whether they hydroxylate the allylic position or add to the C=C bond depends on polar and steric factors [11,54].

EPOXIDATION

As described above, in at least one ω-hydroxylase, alkene epoxidation and ω-hydroxylation appear to be dual functions of the same enzyme [39–41]. This duality is in part duplicated by trifluoroperoxyacetic acid; and this duality has been regarded as a strong point in favor of peracid models for hydroxylases [18].

Epoxidation has not been found for amine oxides and aminium radicals except in photochemical reactions of pyridine oxides [18,24]. Such aromatic amine oxides are not successful in hydroxylations of the type of Eqs. (4) and (5) [10]. Further, the hydroxylases certainly do not act photochemically so that any photochemical reaction cannot be a model for hydroxylase action.

REMOTE HYDROXYLATION

The microbiological conversions of progesterone to 11α-hydroxyprogesterone [33] and camphor to 5-*exo*-hydroxycamphor [34] are shown in Eqs. (9) and (10). They share a common feature, namely, that hydroxylation takes place at the secondary carbon that is most remote from electronegative substituent(s). This is obvious in the case of camphor. In progesterone, it becomes more apparent if the unsaturated ketone is written in the contributing dipolar form, as shown in Eq. (9).

This type of selectivity is termed "remote" selectivity. There are several other examples of this in steroid hydroxylations [33], and it has been exhaustively demonstrated in various camphor and borneol derivatives [34]. Remote selectivity is not restricted to natural metabolites. Cyclododecanone is hydroxylated by a fungi at C-6 and C-7, and cyclopentadecanone is

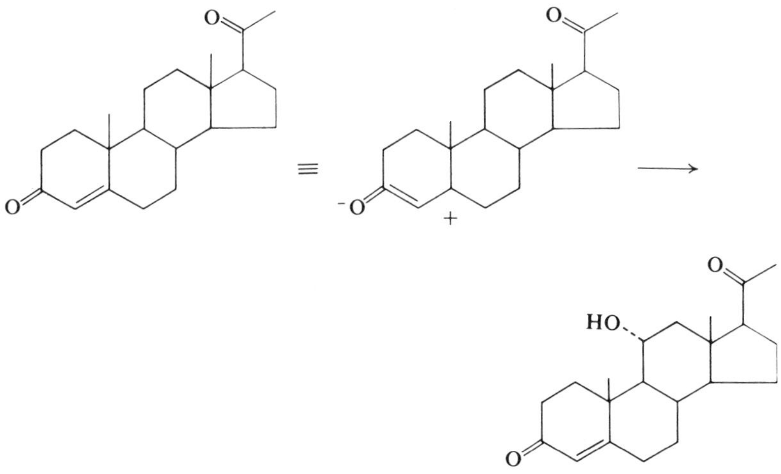

(9)

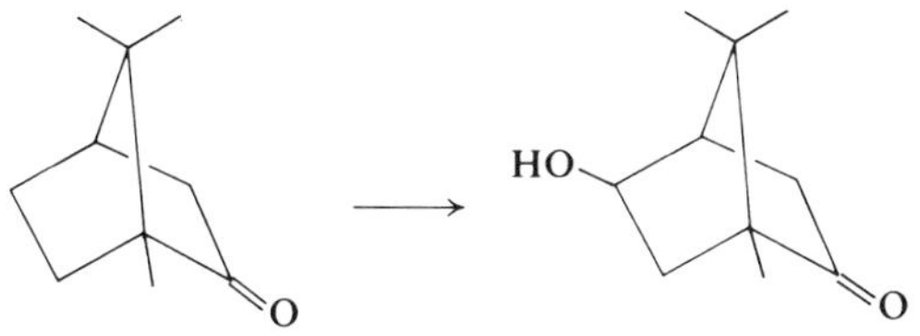

(10)

hydroxylated at C-8 [61]. Both exemplify hydroxylation at the position most remote from the keto group.

This type of selectivity is exactly duplicated by aminium radical reactions. 1-Octanol is chlorinated with 92% selectivity for the 7 position [13] and hydroxylated with 70% selectivity for the same 7 position [9]. These reactions exemplify attack at the most remote secondary position, and a variety of C_5–C_{10} linear species exhibit comparable behavior [11,12]. More pertinent to the enzymatic examples cited above is the exclusive chlorination of cyclohexanol at C-4 [10].

STEREOSPECIFICITY

Several elegant studies have demonstrated that the hydroxylations occur with high stereospecificity. Although alkanes were not used, it is assumed that the results with 2-hexadecanol [62], methyl stearate [63], and decanoic acid [44] are due to the same enzyme that effects terminal hydroxylation of alkanes [1]. It is remarkable that the enzyme can convert the nonpolar —$(CH_2)_n CH_3$ to chiral —$CHOHCH_3$.

Fermentation of racemic 2-hexadecyl acetate gave 2,15-hexadecanediol of specific rotation closely similar to that of (+)-2-hexadecanol, showing that the product was an equimolar mixture of (*S,S*)- and (*R,S*)- 2,15-hexadecanediol [62]. This result shows that hydroxylation was completely stereospecific. What is perhaps of equal importance, it shows that the secondary alcohol group at the other end of the molecule did not influence the stereospecificity since both (*R*)- and (*S*)-2-hydroxyl led to (*S*)-15-hydroxyl. This adds support to the view that only the —$(CH_2)_nCH_3$ enters the enzyme and that the enzyme is specific for this group.

The methyl stearate studies involved monotritium labeling at C-17, and the D, L, and DL forms were all studied [63]. The studies were facilitated by the fact that the action of *Torulopsis gropengiesseri* on methyl stearate gives a 30% yield of methyl 17-hydroxystearate and only a 3% yield of methyl 18-hydroxystearate. The dramatic result was that the L and DL forms showed extensive loss of ^3H on hydroxylation, whereas the D form lost relatively little. This shows not only that hydroxylation occurs with high stereospecific retention of configuration, but that 17-ketooctadecanoates and 16- or 17-octadecenoates are not intermediates. The absence of C=C intermediates was further shown by the retention of 17-^3H in forming methyl 18-hydroxystearate.

The studies on decanoic acid complemented the studies on methyl stearate not only by the use of rat liver enzyme in place of a yeast, but because with decanoic acid there was predominantly ω-hydroxylation (92% 10-hydroxy and 8% 9-hydroxy) whereas in the methyl stearate there was 30% 17-OH and 3% 18-OH. The results were closely parallel in that hydroxylation occurred with retention of configuration and no hydrogens were lost other than the one replaced by hydroxyl [44].

The duplication of such stereospecificities has not been attempted in model systems. Any model system would have to have the capacity for extensive structure in the reagent in order to incorporate the requisite chiral cavities. Again, the aminium radicals would seem to have the most potential for such structure.

The enzymatic stereospecificities have been regarded as supporting oxenoid models in which an oxygen atom is inserted into a CH bond [18–23]. Some very simple systems accomplish this insertion. For example, CF_3CO_3H gave stereoisomeric 1-hydroxy products from *cis*- and *trans*-1,2-dimethylcyclohexane [18,22].

However, retention of configuration by the enzyme does not necessitate an oxygen insertion mechanism. In oxygen insertion, a single chiral active site would suffice. If a free-radical intermediate (or alkyl cation or anion) were involved, two chiral steps would be required. One would be in breaking the CH bond and the other in forming the CO bond. The attraction of the oxygen insertion mechanism is to offer the economy of reducing two chiral steps to one. This attraction is not sufficient to make the oxygen insertion path obligatory.

STERIC SELECTIVITIES

ω-Hydroxylase is able to select terminal methyl over terminal *tert*-butyl for hydroxylation. Fermentation of 2,2-dimethylhexadecane gave several products, all of which were in accord with oxidation at C-15 and C-16 rather than C-1 [64].

Selectivities in chlorinations with aminium radicals vary widely with the structure of the intermediate aminium radical. With more hindered examples such as *N*-chloro-2,2,6,6-tetramethylpiperidine, selectivity for terminal methyl over terminal *tert*-butyl could be achieved [56]. 2,2-Dimethylbutane and 2,2-dimethylpentane gave about 10–20:1 ratios for ω-methyl chlorination over chlorination at C-1 (after statistical correction). However, the secondary hydrogens on 2,2-dimethylpentane were chlorinated, and there can be no doubt that spatial effects in the 2,2,6,6-tetramethylpiperidinium radical are not great enough to avoid predominant substitution on secondary hydrogen in a long-chain example such as 2,2-dimethylhexadecane.

Substitution on tertiary hydrogen is favored by most oxidizing agents [65]. In contrast, ω-hydroxylases and hindered aminium radicals totally avoid the tertiary position. ω-Hydroxylase hydroxylated both terminals of 2-methylhexadecane without oxidizing the tertiary C-2 [64]. It also oxidized 2-methylbutane to 2-methylbutanoic acid [66] and 2-methylhexane to 2- and 5-methylhexanoic acid [67]. In aminium radical chemistry, isopentane could be chlorinated with total elimination of tertiary chlorination by using *N*-chloro-2,2,6,6-tetramethylpiperidine [56].

In contrast to the high ω and ω-1 selectivities shown in most studies, one set of studies using rat liver microsomes reported commonplace selectivities. The tertiary–secondary ratio was 3 for isopentane and 7 for methylcyclohexane [14–16]. The secondary–primary ratio was 24 for methylcyclohexane [14,15], > 50 for pentane [15], and > 100 for butane [16]. Even the ratio of 2-pentanol to 3-pentanol (from pentane) was 2.6. These selectivities are similar to those shown in oxidations with peracids, and this was advanced as support for oxenoid models for hydroxylases [16]. Our view is that there are at least two categories of hydroxylases. One exhibits ω and ω-1 selectivity, and the other shows the more conventional tertiary > secondary > primary pattern.

CENTRAL ATTACK

The conversion of stearic to oleic acid combines the high structural selectivity of removing the 9,10-hydrogens with the stereoselectivity of producing a cis double bond [38,68]. Although the reaction is not hydroxylation, it does involve oxidation of unactivated CH bonds at sites remote from functional groups.

Two interesting models have been reported that give modest selectivity for central substitution. Breslow's photochemical intramolecular cyclization of alkyl 4-benzoylbenzoates was conducted on the C_{14}, C_{16}, C_{18}, and C_{20} 1-alkanols. The initial products were dehydrated and oxidized to produce 1,x-alkanediols. Although the second alcohol group was distributed over several carbons, selectivity for certain regions of the chain was evident. This was attributed to the available conformations for the intramolecular reaction as dictated by size, shapes, and chain lengths [69]. The C_{14} alcohol gave 49% C-12 and 22% C-13 substitution. The C_{16} alcohol gave 85% substitution at C-11–C-14 with ~60% at C-14. The C_{18} alcohol gave 85% substitution at C-11–C-16. The C_{20} alcohol gave >90% at C-11–C-16 [69].

The second model also involves reaction via selected conformations of substrate, but in this case the conformations are dictated by the character of the solvent. When the C_{16} palmitamide is chlorinated by $(2\text{-pr})_2NCl$ via $(2\text{-pr})NH^+$, 80% of the chloro substitution occurs on C-7–C-11 [55]. Since it had been established that there is extreme polar selectivity for attacking positions remote from the electronegative substituent, protonated amide in this case, the simplest interpretation is that the C_{16} chain has folded back on itself so that C-8–C-10 are now most remote from the amide group. This stabilization of such a folded-back conformation is facilitated by the solvent, 84% H_2SO_4, which is actually the fused salt, H_3O^+ HSO_4^- [70]. The tendency of the alkyl chain to minimize contact with this polar fused salt must be very strong. Large energy differences could now exist between conformations and reaction could occur largely via a single conformation or a set of similar folded-back conformations.

With the C_{12} lauramide, remote chlorination is observed in 70% acetic acid plus 30% H_2SO_4, whereas central chlorination (88% on C-6–C-10) is observed in 84% H_2SO_4 [55]. This result emphasizes that the solvent is dictating the conformation that reacts.

ROLE OF Fe

The ω-hydroxylase from the microsomal component of rat liver has been isolated and the activity shown to be associated with cytochrome P-450

[46–49]. Like all cytochromes, P-450 contains combined Fe capable of inter-converting between the II and III states. Combined Fe in the form of rubre-doxin was found in a ω-hydroxylase from bacteria [36,71–73]. Other enzymes that oxidize unactivated CH bonds also contain Fe such as the dehydrogenase that converts stearic acid to oleic acid [38,68], the 11β-hydroxylase that acts on steroids [38], the enzymes responsible for the conversion of cholesterol to pregnenolone [38], and camphor 5-hydroxylase [34]. Most other oxygenases also contain Fe [35].

The roles of cytochromes and rubredoxins have been largely ascribed to electron transfers because some of these are slow and the components in a sequential chain can be spectroscopically characterized. Such studies, how-ever elegant, do not elucidate the nature of the CH bond breaking and CO bond making, and these are the critical steps in hydroxylation from a chemical viewpoint.

The question arises as to whether the ubiquitous presence of Fe is related to its demonstrated ability to cleave N—O and O—O bonds and to generate radicals that can attack CH bonds. The model studies suggest that this is likely.

SUMMARY

The development of aminium radical chemistry provides models that resemble ω-hydroxylases. The models have the potential for complex structure. This has led to steric selectivities resembling those of ω-hydroxylases and offers the potential to duplicate the stereospecificities. The aminium radicals and trifluoroperoxyacetic acid have the ability to hydroxylate alkanes while simultaneously being without action on alcohols, carboxylic acids, and amides. This ability arises from polar effects. It is attractive to speculate that ω-hydroxylases use these same polar effects in conjunction with spatial effects to achieve their unique stereospecific ω and ω-1 hydroxylations.

REFERENCES

1. D. F. Jones and R. Howe, *J. Chem. Soc. C* pp. 2801–2833 (1968).
2. G. J. E. Thijsse and A. C. van der Linden, *Antonie van Leeuwenhoek* **29**, 89 (1963); R. Huybretope and A. C. van der Linden, *ibid.* **30**, 185 (1964).
3. P. E. Kolattukudy and L. Hankin, *J. Gen. Microbiol.* **54**, 145 (1969).
4. M. J. Klug and A. J. Markovetz, *Biotechnol. Bioeng.* **11**, 327 (1969); *J. Bacteriol.* **93**, 1847 (1967); **96**, 1116 (1968); A. J. Markovetz, M. J. Klug, and F. W. Forney, *ibid.* **93**, 1289 (1967).
5. J. E. Stewart W. R. Finnerty, R. E. Kallio, and D. P. Stevenson, *Science* **132**, 1254 (1960).

6. R. Makula and W. R. Finnerty, *J. Bacteriol.* **95**, 2108 (1968).
7. K. R. Dunlap and J. J. Perry, *J. Bacteriol.* **96**, 318 (1968).
8. J. R. L. Smith, R. O. C. Norman, and A. G. Rowley, *J. Chem. Soc., Perkin Trans. 1* p. 566 (1973).
9. N. Deno and D. G. Pohl, *J. Am. Chem. Soc.* **96**, 6680 (1974).
10. N. Deno, E. J. Jedziniak, L. A. Messer, and E. S. Tomezsko, unpublished results.
11. F. Minisci, *Synthesis* **2**, 1 (1972).
12. N. Deno, *Methods Free-Radical Chem.* **3**, 93–153 (1972).
13. N. Deno, W. E. Billups, R. Fishbein, C. Pierson, R. Whalen, and J. C. Wyckoff, *J. Am. Chem. Soc.* **93**, 438 (1971).
14. V. Ullrich and H. Staudinger, *Handb. Exp. Pharmakol.* **28**, 251 (1971).
15. U. Frommer and V. Ullrich, *Z. Naturforsch., Teil B* **26**, 322 (1971).
16. V. Ullrich, *Angew. Chem., Int. Ed. Engl.* **11**, 701 (1972).
17. C. Walling, *Acc. Chem. Res.* **8**, 125 (1975).
18. G. Hamilton, *in* "Molecular Mechanisms of Oxygen Activation" (O. Hayaishi, ed.), Chapter 27. Academic Press, New York, 1974.
19. J. R. Giaun and G. Hamilton, *J. Am. Chem. Soc.* **88**, 1584 (1966).
20. G. Hamilton, B. S. Rebner, and T. M. Hellman, *Adv. Chem. Ser.* **77**, 15 (1968).
21. J. W. Weller and G. Hamilton, *Chem. Commun.* p. 1390 (1970).
22. G. Hamilton, J. R. Geacin, T. M. Hellman, M. E. Snook, and J. W. Weller, *Ann. N.Y. Acad. Sci.* **212**, 4 (1973).
23. M. Snook, Ph.D. Thesis, Pennsylvania State University, University Park (1971).
24. D. M. Jerina, D. R. Boyd, and J. W. Daly, *Tetrahedron Lett.* p. 457 (1970).
25. J. B. Davis, "Petroleum Microbiology." Am. Elsevier, New York, 1967.
26. J. A. Byron and S. Beastall, *in* "Microbiology" (P. Heppel, ed.), p. 73. Inst. Petrol., London, 1971.
27. C. Ratledge, *Chem. Ind. (London)* p. 843 (1970).
28. C. W. Bird and P. M. Molton, *Top. Lipid Chem.* **3**, 125 (1972).
29. A. J. Markovetz, *Crit. Rev. Microbiol.* **1**, 225 (1971).
30. E. J. McKenna and R. E. Kallio, *Annu. Rev. Microbiol.* **19**, 183 (1965).
31. A. C. van der Linden and G. J. E. Thijsse, *Adv. Enzymol. Relat. Areas Mol. Biol.* **27**, 469 (1965).
32. M. J. Klug and A. J. Markovetz, *Adv. Microbiol. Physiol.* **5**, 1 (1971).
33. G. S. Fonken and R. A. Johnson, "Chemical Oxidations with Microorganisms," p. 2. Dekker, New York, 1972.
34. I. C. Gunsalus, J. R. Meeks, J. D. Lipscomb, P. Debrunner, and E. Münck, *in* "Molecular Mechanisms of Oxygen Activation" (O. Hayaishi, ed.), Chapter 14, Academic Press, New York, 1974.
35. O. Hayaishi, ed., "Molecular Mechanisms of Oxygen Activation." Academic Press, New York, 1974.
36. J. A. Peterson and M. J. Coon, *J. Biol. Chem.* **243**, 329 (1968).
37. S. Orrenius and L. Ernster, *in* "Molecular Mechanisms of Oxygen Activation" (O. Hayaishi, ed.), p. 229. Academic Press, New York.
38. M. Hamberg, B. Samuelsson, I. Björkhem, and H. Danielsson, *in* "Molecular Mechanisms of Oxygen Activation" (O. Hayaishi, ed.), pp. 30–76. Academic Press, New York.
39. S. W. May and B. J. Abbott, *J. Biol. Chem.* **248**, 1725 (1973).
40. S. W. May and R. D. Schwartz, *J. Am. Chem. Soc.* **96**, 4031 (1974).
41. S. W. May, R. D. Schwartz, B. J. Abbott, and O. R. Zaborsky, *Biochim. Biophys. Acta* **403**, 245 (1975).

42. C. W. Bird and P. M. Molton, *Top. Lipid Chem.* **3**, 139 (1972).
43. C. W. Bird and P. M. Molton, *Top. Lipid Chem.* **3**, 153 (1972).
44. E. Heinz, A. P. Tulloch, and J. F. T. Spencer, *J. Biol. Chem.* **244**, 882 (1969).
45. R. D. McCarthy, *Biochim. Biophys. Acta* **84**, 74 (1964).
46. A. Y. H. Lu, K. W. Junk, and M. J. Coon, *J. Biol. Chem.* **244**, 3714 (1969).
47. A. Y. H. Lu and M. J. Coon, *J. Biol. Chem.* **243**, 1331 (1968).
48. K. Ichihara, E. Kusunose, and M. Kusunose, *Biochim. Biophys. Acta* **202**, 560 (1970).
49. E. Kusunose, K. Ichihara, and M. Kusunose, *FEBS Lett.* **11**, 23 (1970).
50. S. Bergström, B. Bergström, N. Tryding, and G. Westöö, *Biochem. J.* **58**, 604 (1954).
51. H. Den, *Biochim. Biophys. Acta* **98**, 462 (1965).
52. O. Stoppe, K. Try, and L. Eldjarn, *Biochim. Biophys. Acta* **144**, 271 (1967).
53. M. Hamberg and B. Samuelsson, *J. Am. Chem. Soc.* **91**, 2177 (1969).
54. R. S. Neale, *Synthesis* **1**, 1 (1971).
55. N. Deno and E. J. Jedziniak, *Tetrahedron Lett.* p. 1259 (1976).
56. N. Deno, D. G. Pohl, and H. J. Spinelli, *Bioorg. Chem.* **3**, 66 (1974).
57. J. T. Groves and M. Van der Puy, *J. Am. Chem. Soc.* **96**, 5274 (1974).
58. N. Deno, K. A. Eisenhardt, D. G. Pohl, H. J. Spinelli, and R. C. White, *J. Org. Chem.* **39**, 520 (1974).
59. W. L. Miller and J. L. Gaylor, *J. Biol. Chem.* **245**, 5369 and 5375 (1970).
60. W. L. Miller, D. R. Brody, and J. L. Gaylor, *J. Biol. Chem.* **246**, 5147 (1971).
61. M. J. Ashton, A. S. Bailey, and E. R. H. Jones, *J. Chem. Soc., Perkin Trans. I* p. 1665 (1974).
62. D. F. Jones and R. Howe, *J. Chem. Soc. C* p. 2801 (1968).
63. D. F. Jones, *J. Chem. Soc.* p. 2827 (1968).
64. D. F. Jones, *J. Chem. Soc. C* p. 2809 (1968).
65. L. J. Chinn, "Selection of Oxidants in Synthesis," Chapter 2. Dekker, New York, 1971.
66. J. Takahashi and J. W. Foster, *Bacteriol. Proc.* p. 86 (1966).
67. G. J. E. Thijsse and A. C. van der Linden, *Antonie Van Leeuwenhoek* **27**, 171 (1961).
68. K. Bloch, *Acc. Chem. Res.* **2**, 193 (1969).
69. R. Breslow and M. Winnik, *J. Am. Chem. Soc.* **91**, 3083 (1969).
70. N. Deno and R. W. Taft, Jr., *J. Am. Chem. Soc.* **76**, 244 (1954).
71. M. Kusunose, E. Kusunose, and M. J. Coon, *J. Biol. Chem.* **239**, 1374 and 2135 (1964).
72. J. A. Peterson, D. Basu, and M. J. Coon, *J. Biol. Chem.* **241**, 5162 (1966).
73. J. A. Peterson, M. Kusunose, E. Kusunose, and M. J. Coon, *J. Biol. Chem.* **242**, 4334 (1967).

CHAPTER

5

Intramolecular Nucleophilic Attack on Esters and Amides

Thomas H. Fife

INTRODUCTION

Enzymatic catalysis is generally believed to proceed with formation of an enzyme–substrate (ES) complex which breaks down to products (P) and free enzyme [Eq. (1)]. In the ES complex, the substrate is held in close proximity

$$E + S \; \rightleftharpoons \; ES \; \longrightarrow \; E + P \tag{1}$$

to functional groups in the active site. A postbinding reaction in the active site is therefore closely similar to a chemical intramolecular reaction. If a functional group attacks the substrate in a nucleophilic reaction, then a substituted enzyme intermediate is formed. Abundant evidence points to the fact that in hydrolysis of esters and amides catalyzed by α-chymotrypsin, nucleophilic attack of serine-195, assisted by histidine-57, gives an acyl serine derivative as in (1) or a kinetic equivalent [1–3]. As a consequence, it has been important to study simple chemical intramolecular nucleophilic reactions of functional groups commonly present in enzyme active sites in order to gain qualitative and quantitative understanding of the factors influencing such reactions. In this chapter, recent work from our laboratory dealing with intramolecular nucleophilic attack on esters and amides by hydroxyl, amino, and sulfhydryl nucleophiles in sterically similar systems is

(1)

described. At the time when this work was initiated, a number of intra-molecular nucleophilic reactions had been intensively studied [1–6], but extremely important questions remained to be answered, among which were the following:

1. What are the maximum rate enhancements possible for nucleophilic reactions in sterically favorable situations?

2. What are the relative efficiencies of different nucleophilic groups in sterically similar systems?

3. What types of nucleophiles and mechanisms are effective when the leaving group is poor as in the case of an aliphatic ester? There had been few kinetic studies with such compounds even though substrates for α-chymotrypsin include such esters.

4. How can one account in chemical terms for the magnitude of the rate constants observed with esteratic enzymes such as α-chymotrypsin? This question was especially difficult because of lack of information regarding the rates and mechanism of intramolecular alcoholysis of esters.

It is now possible to give some partial answers to these questions.

CARBAMATE ESTERS

The carbonyl of carbamate esters is highly deactivated by resonance interaction with the adjoining nitrogen [Eq. (2)], which reduces the partial positive charge on the carbonyl carbon. Thus, in bimolecular nucleophilic reactions of these esters considerable bond formation with the nucleophile is required to attain the transition state, resulting in a near maximum loss of translational and rotational entropy of the bimolecular nucleophile. Intra-molecular nucleophilic reactions of carbamate esters are therefore important reactions to study because maximum rate enhancements might be obtained in comparison with corresponding bimolecular reactions.

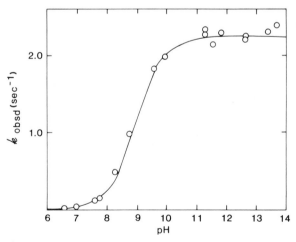

$$R\!-\!\overset{\overset{\displaystyle CH_3}{|}}{\underset{\cdot\cdot}{N}}\!-\!\overset{\overset{\displaystyle O}{\|}}{C}\!-\!OR' \quad\longleftrightarrow\quad R\!-\!\overset{\overset{\displaystyle CH_3}{|}}{\underset{+}{N}}\!\!=\!\!\overset{\overset{\displaystyle O^-}{|}}{C}\!-\!OR' \qquad (2)$$

(2)

$$\qquad\qquad\qquad\qquad\qquad\qquad\qquad\qquad\qquad\qquad\qquad\qquad\qquad C\!=\!O \;+\; \qquad\;-\!O^- \qquad (3)$$

Since the hydroxyl group of serine is a nucleophile in reactions catalyzed by α-chymotrypsin, knowledge of the manner in which oxygen nucleophiles can participate in chemically similar reactions is of crucial importance. In Fig. 1 a plot is shown of k_{obsd} vs. pH for the quantitative cyclization of phenyl N-(2-hydroxyphenyl) N-methylcarbamate **(2)** to N-methyl-2-benzoxazolinone [Eq. (3)] at 25°C [7]. The sigmoidal pH–rate constant profile shows that participation is by the ionized species, and the pK_{app} of 9.0 is reasonable for

Fig. 1 Plot of k_{obsd} for ring closure of phenyl N-(2-hydroxyphenyl) N-methylcarbamate to N-methyl-2-benzoxazolinone and phenoxide ion at 25°C in H_2O with $\mu = 0.5\,M$ (with KCl). [Reprinted with permission from J. E. C. Hutchins and T. H. Fife, *J. Am. Chem. Soc.* **95**, 2282 (1973). Copyright by the American Chemical Society.]

a phenolic hydroxyl group. The efficiency of a nucleophile in an intramolecular reaction is often expressed as the "effective molarity" of the neighboring group. This "effective molarity" is obtained by comparing the rate constant for maximum participation of the intramolecular nucleophile (sec^{-1}) with the second-order rate constant for the analogous bimolecular reaction proceeding by the same mechanism (M^{-1} sec^{-1}). The ratio of rate constants has units of molarity and can be considered to be the concentration of the bimolecular nucleophile required to give a pseudo first-order rate constant of the same magnitude as that obtained in the intramolecular reaction. A second-order rate constant for bimolecular alcoholysis of phenyl N-methyl N-phenyl-carbamate by an alcohol of pK_a 9.0 was obtained by extrapolation of a plot of the logarithms of the rate constants vs. pK_a for reaction of alcohol nucleophiles. The effective molarity of the phenoxide ion of (2) was thereby calculated to be 3×10^8 M. Intramolecular nucleophilic attack by a neighboring carboxylate ion on phenyl esters also gives rate enhancements of 10^7–10^8 M [8]. These are the largest effective molarities that have been observed to date in reactions of esters. The carbamate nitrogen of (2) was methylated to preclude elimination of phenol to give an isocyanate. Such elimination occurs readily with aryl carbamate esters [9]; however, in the case of (3), intramolecular phenoxide ion attack is so favorable that it is the only reaction observed at pH values below 11. Cyclization of (2) is faster than that of (3) by a factor of approximately 10.

(3)

Intramolecular nucleophilic attack by a hydroxymethyl group is a facile process in cyclization of the p-nitrophenyl and ethyl carbamate esters (4) and (5) [10]. Plots of log k_{obsd} vs. pH are linear with slopes of 1.0, thereby

(4), R = —NO$_2$

(5), R = —Et

demonstrating apparent hydroxide ion catalysis. Buffer catalysis could not be detected in cyclization of any of the carbamates (2)–(5). Consequently, nucleophilic attack by the oxide ion must occur as in Eq. (4). The second-order rate constant k_{OH} for cyclization is given in Eq. (5), where K_a is the dissociation constant of the hydroxymethyl group and K_w is the ion product of water.

$$(4)$$

The magnitude of k_{OH} in cyclization of the nitrophenyl ester (4) is 3×10^5 times greater than k_{OH} for hydroxide ion-catalyzed hydrolysis of p-nitrophenyl

$$k_{OH} = k_r K_a / K_w \qquad (5)$$

N-methyl N-phenylcarbamate and 10^5-fold greater than the second-order rate constant for transesterification of that ester by pentaerythritol. Similarly, k_{OH} for cyclization of (5) is 10^6 times larger than k_{OH} for hydroxide ion-catalyzed hydrolysis of ethyl N-methyl N-phenylcarbamate. Thus, an aliphatic alcohol is a potent intramolecular nucleophile toward carbamate esters in reactions proceeding through the oxide ion.

That large effective molarities ($10^5–10^8$ M) are general for negatively charged intramolecular nucleophiles was demonstrated [11] by the cyclization of (6), in which the neighboring group is sulfhydryl. As in the case of (2), the pH–rate constant profile is sigmoidal with $pK_{app} = 8.7$, and buffer catalysis is not observed. Cyclization must occur through attack of the thiol anion as in Eq. (6). Interpolation of a plot of $\log k_B$ vs. pK_a for bimolecular attack of thiols on p-nitrophenyl N-methyl N-phenylcarbamate gave the second-order rate constant for attack of a thiol of $pK_a = 8.7$. The effective molarity of the neighboring thiol of (6) was in that manner calculated to be 1.4×10^5 M. The rate constant for maximum participation of the neighboring group of (6) at 25°C is considerably larger than that for (2) (24 sec^{-1} compared with 2.2 sec^{-1}). Consequently, even though the leaving group of (6) is the p-nitrophenolate ion, which should give a rate constant 10–100 times greater

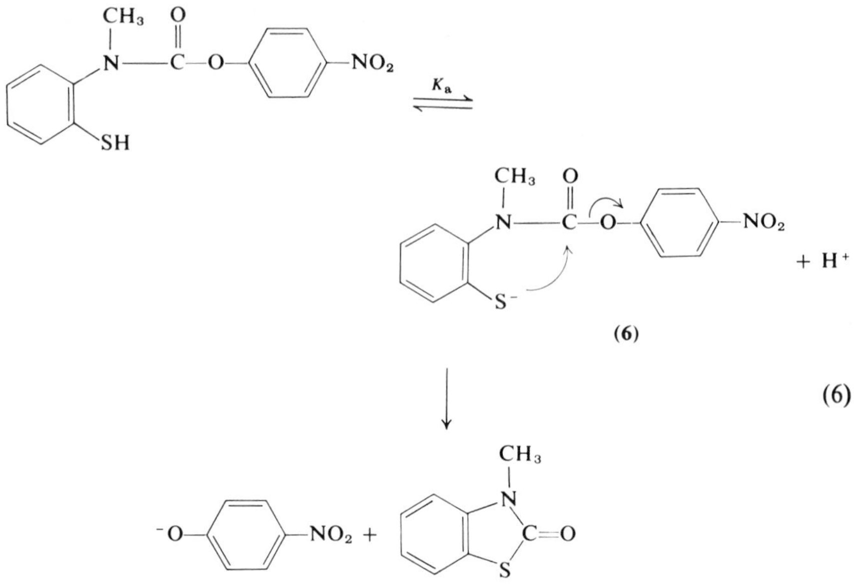

(6)

(6)

than phenolate (the ratio of second-order rate constants for hydroxide ion-catalyzed hydrolysis of phenyl and p-nitrophenyl N-methyl N-phenylcarbamate is 100), it is apparent that the smaller effective molarity of the thiol of **(6)** is due predominantly to relatively more favorable bimolecular reactions of thiols than alcoholate ions with the unsubstituted reference ester. This is reflected in the small dependence of the rate constants for bimolecular thiolysis on the pK_a of the nucleophile [11,12].

The neutral amino group of **(7)** is also a highly efficient intramolecular nucleophile [Eq. (7)] [11]. A pK_{app} of 2.7, identical with the measured thermodynamic pK_a, was obtained from the sigmoidal pH–rate constant profile of Fig. 2. Bimolecular attack of amines on phenyl N-(4-aminophenyl) N-methylcarbamate was too slow to be conclusively demonstrated, but a lower limit of

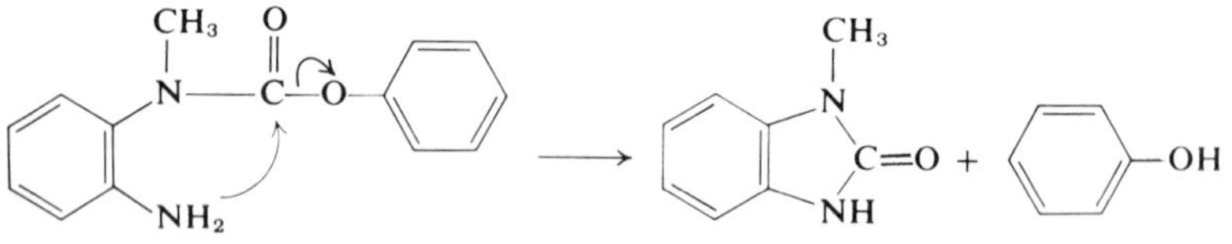

(7)

the effective molarity of the amine group of **(7)** was calculated to be 3×10^8 M. Similarly, an effective molarity of 10^8 M was determined from study of bimolecular aminolysis of p-nitrophenyl N-methyl N-phenylcarba-

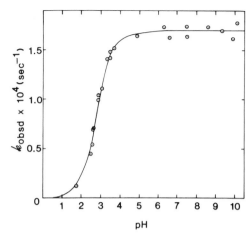

Fig. 2 Plot of k_{obsd} for cyclization of phenyl N-(2-aminophenyl) N-methylcarbamate (**7**) vs. pH at 50°C in H_2O with $\mu = 0.5$ M (KCl). [Reprinted with permission from T. H. Fife, J. E. Hutchins, and M. S. Wang, *J. Am. Chem. Soc.* **97**, 5878 (1975). Copyright by the American Chemical Society.]

mate and making a reasonable allowance for the different leaving group. Highly efficient intramolecular reactions can therefore be obtained with neutral as well as anionic nucleophiles. The largest effective molarity that had previously been observed for a neutral amine nucleophile was 5×10^3 M in the case of the dimethylamino group of p-nitrophenyl γ-dimethylamino-butyrate [13]. The imidazole group of p-nitrophenyl γ-(4-imidazoyl) butyrate has an effective concentration of only 9.4 M [14].

Reduced ground-state solvation of an intramolecular nucleophile as compared with the analogous bimolecular nucleophile might be an important factor in explaining the large effective molarities found for anionic nucleophiles. Before a nucleophile can attack it must be desolvated in a process that is energetically unfavorable with anions [15]. An intramolecular nucleophile· might not, however, be as highly solvated in the ground state as a bimolecular nucleophile if the steric situation is such that water molecules cannot fit between the nucleophile and the reaction center. Relative ease of desolvation could certainly be significant in reactions of many nucleophiles in determining relative efficiency, but it cannot be of general importance in view of the large rate enhancement obtained with the neutral amine nucleophile of (**7**).

Intramolecular reactions owe much of their efficiency to relatively favorable ΔS^* values because translational entropy of the nucleophile is not lost as it is in the bimolecular comparison [1–3]. Page and Jencks [16] have calculated that rate enhancements of 10^8 M can be ascribed to this factor alone. Steric compression of the nucleophile and the reaction center apparently produces rate enhancements that can be attributed to changes in ΔH^*. For example, tetramethylsuccinanilic acid cyclizes to an anhydride 1200-fold faster than succinanilic acid, and this rate difference is explained by a more favorable ΔH^* for the tetramethyl derivative [17]. Neighboring phenoxide ion attack in cyclization of the carbamate ester (2) owes much of its great efficiency to a relatively favorable ΔS^* (-3.0 eu). In comparison, ΔS^* for hydroxide ion-catalyzed hydrolysis of phenyl N-methyl N-phenylcarbamate is -28 eu [18]. Intramolecular attack by the neutral amine group of (7) has a ΔS^* of -23.8 eu. If the rate enhancement in that intramolecular reaction (10^8 M) were to be completely attributed to a favorable ΔS^*, then bimolecular aminolysis of the reference ester would necessarily have a ΔS^* of at least -60 eu, which seems unlikely. The ΔS^* for aminolysis of p-nitrophenyl N-methyl N-phenyl-carbamate by cyclohexylamine is -33 eu. Thus, at least part of the efficiency of the intramolecular reaction of (7) is probably due to a relatively favorable ΔH^* in a system where steric compressional effects are apparently absent. Clearly, more abundant data will be required before intramolecular reactions can be confidently considered to be understood. In particular, activation parameters must be measured for an extensive series of intramolecular reactions and their bimolecular counterparts in which large effective molarities (10^5–10^8 M) have been determined.

The rate enhancements of 10^5–10^8 M observed in the cyclization of (2)–(7) are very likely close to the maximum enhancements possible in nucleophilic reactions of hydroxyl, sulfhydryl, and amine nucleophiles. This is because of the deactivated carbonyl of carbamate esters and the severe restriction of rotational degrees of freedom of the nucleophile and carbonyl in these aromatic derivatives. Aliphatic carbamate esters [esters with an aliphatic R group in Eq. (2)] would have a still further deactivated carbonyl, so somewhat larger rate enhancements might be anticipated, but it is unlikely that steric fit can be greatly improved over compounds (2)–(7) without introducing steric compressional effects. It will be noted that with (4) and (5) a six-membered ring transition state is formed rather than the kinetically more favored five-membered ring of (2), (6), and (7). Reactions of aliphatic carbamate esters are currently being studied in this laboratory.

ALIPHATIC CARBOXYLIC ESTERS AND AMIDES

Hydroxyl Group Participation

Belke et al. [19] investigated the cyclization of 2-hydroxymethylbenzamide to phthalide. Both general base and general acid catalysis were reported. The slope of a Brönsted plot of log k_B for general base catalysis vs. the pK_a of the catalyst changes from 1.0 to 0.2 with increasing pK_a, implying a change in rate-determining step. Strict general base catalysis was found in the cyclization of ethyl 2-hydroxymethylbenzoate (8) and ethyl 2-hydroxymethyl-4-nitro-benzoate (9) to phthalide and 5-nitrophthalide [20] shown in Eq. (8). The D_2O solvent isotope effect for imidazole-catalyzed cyclization ($k_{Im}^{H_2O}/k_{Im}^{D_2O}$)

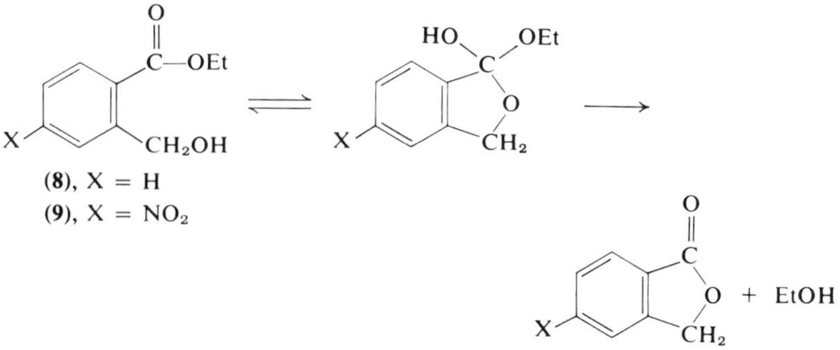

(8), X = H
(9), X = NO₂

is 3.46 for (8) and 3.15 in the case of the 4-nitro derivative (9). The fact that the second-order rate constants in D_2O are smaller than those in H_2O indicates that proton transfer is occurring in the critical transition state. Since imidazole catalysis is observed in the intramolecular alcoholysis of these compounds, it can be inferred that aliphatic esters and amides of 2-hydroxymethylbenzoic acid represent reasonable models for acylation of α-chymotrypsin by ester and amide substrates. A mechanism involving concerted proton transfer and bond making in formation of a tetrahedral intermediate is shown in (10). This mechanism is analogous to that given in

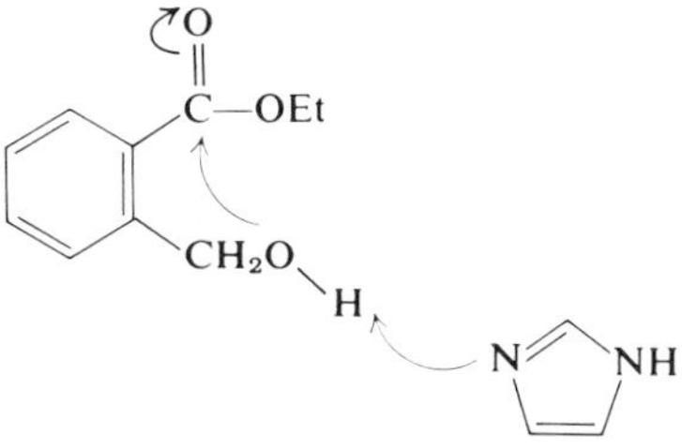

(10)

(1) for α-chymotrypsin, but, kinetically equivalent possibilities exist.

The Brönsted plot for general base-catalyzed cyclization of (9) is presented in Fig. 3. The value of β is 0.97 ($r = 0.995$), within error of unity. A Brönsted coefficient of unity shows that proton transfer is diffusion controlled in the thermodynamically unfavorable direction and rate limiting in a process not

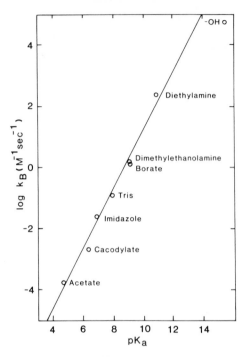

Fig. 3 Brönsted plot of log k_B vs. the pK_a of the catalyzing base in the cyclization of ethyl 2-hydroxymethyl-4-nitrobenzoate to 5-nitrophthalide at 30°C in H_2O ($\mu = 0.5\ M$).

concerted with bond making or breaking [21]. Proton abstraction from the hydroxymethyl group of (9) by the general base [Eq. (9)] cannot be rate

$$\text{(9)}$$

determining. Transfer of a proton to imidazole would have a second-order rate constant given by Eq. (10), where K_a is the dissociation constant of the

$$k_1 = k_{-1}K_a/K_{Im} \tag{10}$$

hydroxymethyl group and K_{Im} is the dissociation constant of imidazolium ion. Reasonably assuming that K_a has a value of 10^{-14}–10^{-16} M and that k_{-1} would be the rate constant of a diffusion-controlled reaction (10^{10} M^{-1} sec^{-1}), k_1 is calculated to be 10–10^3 M^{-1} sec^{-1}, which is several orders of magnitude greater than the experimentally determined second-order rate constant (10^{-2} M^{-1} sec^{-1}). Similarly, the rate constant calculated for imidazolium ion-catalyzed cyclization of the anionic species to form a tetrahedral intermediate is considerably greater than the experimental rate constant. Consequently, general base catalysis of the formation of a tetrahedral intermediate can be ruled out as the rate-determining step of the reaction. Proton transfer from a zwitterionic tetrahedral intermediate (11) resulting from attack of the neutral hydroxymethyl group [Eq. (11)] would be diffusion controlled with all general bases including water and would give a Brönsted coefficient of zero rather than 1.0. General base catalysis very likely

(11)

involves proton abstraction by base from a neutral tetrahedral intermediate (12) or the kinetically equivalent general acid catalysis in (13).

The Brönsted coefficient of 0.87 for general base-catalyzed cyclization of (8) is sufficiently less than unity that a concerted reaction should be regarded as a possibility for that compound, although proton transfer must be appreciable in the transition state. Decomposition of a tetrahedral intermediate must be rate limiting in cyclization of (9), but this is not necessarily the case with (8). Nitro group substitution would have a complex effect on the cyclization reaction in that nucleophilic attack at the carbonyl would be facilitated and acidity of the hydroxymethyl group would be increased, but expulsion of ethoxide ion from the tetrahedral intermediate would be more difficult than with (8).

The pH–rate constant profiles for (8) and (9), obtained by extrapolation to zero buffer concentration, are shown in Fig. 4. Hydronium ion and hydroxide ion catalysis are occurring. Second-order rate constants, k_{OH}, for the apparent hydroxide ion-catalyzed reactions differ by only a factor of 5, probably because of the partially compensating effects of nitro substitution noted above. The k_{OH} value for cyclization of (8) (10^4 M^{-1} sec^{-1} at 30°C) is $\sim 10^5$ times greater than k_{OH} for hydroxide ion-catalyzed hydrolysis of ethyl benzoate. Once again it can be seen that an alcohol function is a powerful intramolecular nucleophile toward the ester carbonyl.

Lactonization of phenyl 4-hydroxybutyrate is catalyzed by acetate and phosphate buffers in what was considered to possibly be general base-catalyzed reactions [22]. Imidazole catalysis of phenoxide ion release proceeds with nucleophilic attack by imidazole at the carbonyl of the ester. Consequently, in this nonrigid system where the leaving group is a phenol, general base-catalyzed intramolecular alcoholysis is not sufficiently advantageous to compete with nucleophilic attack by imidazole. Rearrangement of 2-hydroxymethyl-4-nitrophenyl trimethylacetate to its benzyl ester counterpart has,

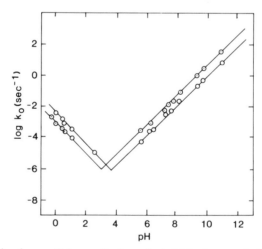

Fig. 4 Plot of log k_0 vs. pH for cyclization of ethyl 2-hydroxymethylbenzoate (○) and ethyl 2-hydroxymethyl-4-nitrobenzoate (◯) to phthalide and 5-nitrophthalide at 30°C in H_2O with $\mu = 0.5$ M. The points were obtained by extrapolation to zero buffer concentration.

however, recently been reported [23] to be catalyzed by imidazole with a D_2O solvent isotope effect of 2.4, suggesting a general base mechanism. Thus, it is chemically reasonable that acylation of α-chymotrypsin by both alkyl and aryl esters utilizes general base catalysis by histidine-57.

Intramolecular alcoholysis is of comparable efficiency with esters and amides of 2-hydroxymethylbenzoic acid in comparison with hydroxide ion-catalyzed hydrolysis of ethyl benzoate or benzamide (10^5 rate enhancement). However, amides are poorer substrates for α-chymotrypsin than esters, leading to the conclusion that either steric factors are not optimal in the active site for attack of serine-195 on amide substrates or that acylation of the enzyme by esters involves relatively advantageous mechanistic features. Apparently, the rates of acylation of the enzyme by amide substrates can be completely explained [20,24,25] in terms of an intracomplex alcoholysis with intramolecular catalysis by histidine-57 without invoking unknown chemistry (charge relay, etc.). Neighboring imidazole need only possess an effective molarity of $\sim 20\ M$ in such a reaction [24].

The rate constant for acylation of α-chymotrypsin by the specific ester substrate N-acetyl-L-tyrosine ethyl ester is 4000 sec^{-1} [3]. In order to obtain a comparable pseudo-first-order rate constant in the imidazole-catalyzed cyclization of (8) ($k_{Im} = 8.75 \times 10^{-3}\ M^{-1}\ sec^{-1}$) imidazole would have to be present at a concentration of $\sim 400,000\ M$. An effective molarity of $400,000\ M$ is not unreasonable for a functional group in an intramolecular reaction, but in view of the lack of chemical information on the effective molarity of an intramolecular general base in intramolecular alcoholysis of esters it should be considered that other factors may be important in the enzymatic reaction with ester substrates.

The pH–log k_{obsd} profile for cyclization of 3-amino-2-hydroxymethyl-benzamide (14) to 4-aminophthalide [Eq. (12)] is shown in Fig. 5 [25]. An inflection in the profile at pH 3.5 corresponds closely to the measured thermodynamic pK_a of 3.80 for the 3-amino group. The pH-independent reaction from pH 3.5 to 10 is at least 10^3 times faster than the corresponding reaction for 2-hydroxymethylbenzamide [19] or the 4-, 5-, or 6-amino-substituted compounds. Also, there is no buffer catalysis in contrast with the

$$(12)$$

(14)

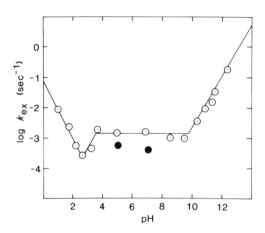

Fig. 5 Plot of log k_{ex} for 3-amino-2-hydroxymethylbenzamide (rate constant obtained by extrapolation to zero buffer concentration) vs. pH at 30°C and $\mu = 0.5\ M$ in 50% dioxane–H_2O (v/v) (○) and 50% dioxane–D_2O (●). The rate constants are for appearance of product (4-aminophthalide).

marked general base catalysis observed in the case of 2-hydroxymethyl-benzamide or the other amino-substituted 2-hydroxymethylbenzamides ($\beta = 0.4$). As a consequence, it is probable that the 3-amino group is participating in cyclization of **(14)**. The D_2O solvent isotope effect (k_{H_2O}/k_{D_2O}) for the pH-independent reaction is 2.82, signifying that proton transfer is occurring in the critical transition state. A Stuart–Briegleb model of **(14)** shows that the steric situation is unfavorable for the concerted reaction **(15)** in which the hydroxymethyl group must approach the carbonyl. If the reaction is concerted, which might reasonably be surmised from Brönsted coefficients less than unity for general base catalysis with the other amino-substituted compounds, then general base catalysis must involve one or more molecules of solvent **(16)**. The negative ΔS^* of -23 eu suggests that the transition state is highly ordered in accord with **(16)**.

The 3-amino group of **(14)** has an effective molarity of 15 M in comparison

(15)   **(16)**

with bimolecular general base catalysis by an amine of pK_a 3.5 in cyclization of 4-amino-2-hydroxymethylbenzamide. If the steric fit of the neighboring group was better a much larger value would be anticipated. Thus, there is little doubt that the rate constant for acylation of α-chymotrypsin by amides is explainable by known chemistry.

Sulfhydryl Group Participation

An excellent opportunity is provided by the 2-substituted benzoate system for comparing the efficiency of various nucleophiles toward the ester carbonyl in a system where steric effects are quite similar and where the groups are held in proximity without steric compressional effects. A plot of log k_{obsd} vs. pH for cyclization of *tert*-butyl 2-mercaptomethylbenzoate (**17**) [Eq. (13)] to

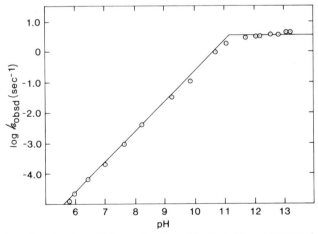

benzothiolactone (Fig. 6) has a slope of 1.0, indicating hydroxide ion catalysis at pH values below 11.1 [26]. The break in the plot at pH 11.1 most likely corresponds to the pK_a of the thiol group, although this value is somewhat high for an aliphatic thiol [27]. Unfortunately, the rapid pH-independent reaction ($k_0 = 4.4$ sec^{-1} at 30°C) precluded attempts to measure the

Fig. 6 A plot of log k_{obsd} vs. pH for cyclization of *tert*-butyl 2-mercaptomethylbenzoate to benzobutyrothiolactone at 30°C and $\mu = 0.5$ M in H$_2$O.

thermodynamic pK_a of (17). While the hydroxide ion-catalyzed reactions of (8) and (17) occur at nearly the same rate ($k_{OH} = 10^4$ M^{-1} sec^{-1}), the reactions differ in that pronounced general base catalysis is not observed in cyclization of (17). In half-neutralized ethanolamine buffers, k_{obsd} increased only 6% as buffer concentration was increased from 0 to 0.5 M. In imidazole buffers a 19% increase occurs in the same concentration range. Larger apparent catalysis was observed in piperidine buffers but proved to be predominantly a specific salt effect. Buffer effects are small when ionic strength is maintained constant with tetramethylammonium chloride rather than KCl. Cyclization to thiolactone is probably taking place with preequilibrium ionization of the thiol group, as in Eq. (14). The difference in susceptibility to general base catalysis in the cyclization of (8) and (17) could result from

$$(14)$$

different mechanisms or simply from the point for hydroxide ion lying on or close to a line of unit slope in a Brönsted plot for cyclization of (17).

Amino Group Participation

A plot of log k_0 vs. pH for cyclization of methyl 2-aminomethylbenzoate (18) to phthalimidine is given in Fig. 7 [28]. At high pH the slope is 1.0, but at

$$(15)$$

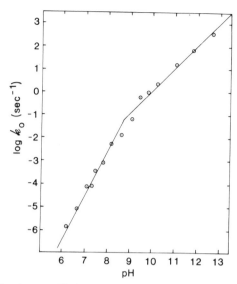

Fig. 7 Plot of $\log k_0$ vs. pH for cyclization of methyl 2-aminomethylbenzoate to phthalimidine at 30°C and $\mu = 0.5$ M with KCl in H_2O. The points were obtained by extrapolation to zero buffer concentration. [Reprinted with permission from T. H. Fife and B. R. De Mark, *J. Am. Chem. Soc.* **98**, 6978 (1976). Copyright by the American Chemical Society.]

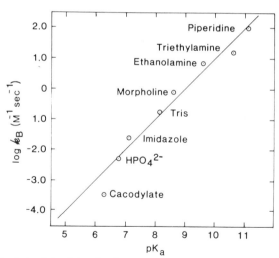

Fig. 8 A Brönsted plot of $\log k_B$ for general base-catalyzed cyclization of methyl 2-aminomethylbenzoate to phthalimidine at 30°C vs. the pK_a of the conjugate acid of the general base catalyst. [Reprinted with permission from T. H. Fife and B. R. De Mark, *J. Am. Chem. Soc.* **98**, 6978 (1976). Copyright by the American Chemical Society.]

pH 8.6, which is probably the pK_a of the amino function, there is an abrupt change to a slope of 2.0. The pH–log rate constant profile shows that hydroxide ion only catalyzes the reaction of the neutral species as depicted in Eq. (15). General base catalysis is also observed. Hence, the equation for k_{obsd} is Eq. (16), where K_a is the dissociation constant of the conjugate acid

$$k_{obsd} = [k_{OH}(OH^-) + k_B(B)]\left(\frac{K_a}{K_a + a_H}\right) \tag{16}$$

of the amine group. A Brönsted plot of log k_B vs. pK_a (Fig. 8) has a slope of 1.02 ($r = 0.99$), within error of unity. Thus, in the general base-catalyzed reaction, proton transfer is not concerted with bond making or breaking. The reaction could therefore entail rate-determining ionization of the amino group [Eq. (17)] or rate-determining proton transfer to or from a tetrahedral

(17)

intermediate [(19) or (20)].

(19)   (20)

Bimolecular aminolysis of esters has been extensively investigated [29–31]. Satterthwait and Jencks [31] proposed that the rate-determining step at high pH in aminolysis of aliphatic esters is an internal proton transfer mediated by water within a zwitterionic intermediate (21) formed by attack of neutral amine on the ester [Eq. (18)]. General base catalysis results from proton

$$
\underset{\text{(21)}}{R-\overset{\overset{\displaystyle O}{\|}}{C}-OR' + R''NH_2} \rightleftharpoons R-\overset{\overset{\displaystyle O^-}{|}}{\underset{\underset{\displaystyle ^+NH_2R''}{|}}{C}}-OR'
$$

$$
\Bigg\downarrow \tag{18}
$$

$$
\text{products} \longleftarrow R-\overset{\overset{\displaystyle OH}{|}}{\underset{\underset{\displaystyle NHR''}{|}}{C}}-OR'
$$

abstraction from (21) by base to give an anionic tetrahedral intermediate (22) [Eq. (19)].

$$
R-\overset{\overset{\displaystyle ^-O}{|}}{\underset{\underset{\displaystyle \overset{+}{H_2NR''}}{|}}{C}}-OR' + B \longrightarrow R-\overset{\overset{\displaystyle ^-O}{|}}{\underset{\underset{\displaystyle NHR''}{|}}{C}}-OR' + BH^+ \tag{19}
$$

$$
\text{(22)}
$$

It is clear that the proton switch in Eq. (18) cannot be rate limiting in the intramolecular aminolysis of (18) since that would result in a pH-independent region in the pH–rate constant profile, which is not observed.

Intramolecular aminolysis of (18) is mechanistically different than the bimolecular aminolysis reactions of aliphatic esters that have been studied to date. Mechanism is, of course, a function of transition state structure, and there appears to be no *a priori* reason why they should be the same in corresponding intramolecular and bimolecular reactions. In view of the analogy between chemical intramolecular reactions and the intracomplex reactions of enzymes, speculations on enzymatic mechanisms must be based on the chemistry of intramolecular reactions. Many speculations in the literature are, in fact, based on an assumed similarity to mechanisms of bimolecular reactions.

The rate constants k_{OH} for apparent hydroxide ion catalysis are almost identical for cyclization of (8), (9), (17), and (18) ($\sim 10^4$ M^{-1} sec^{-1} at 30°C). Also, the second-order rate constants for general base catalysis are quite similar for (8), (9), and (18). For example, the second-order rate constants at 30°C for imidazole-catalyzed cyclization are 8.75×10^{-3}, 2.29×10^{-2}, and 2.49×10^{-2} M^{-1} sec^{-1}, respectively, even though the neighboring alcohol and amine groups must differ greatly in basicity. Thus, the rate constants for cyclization are nearly independent of the identity of the nucleophilic group.

This could be due to compensation of effects, but it could also point to a common rate-limiting step in these cyclizations other than nucleophilic attack. If there is a common rate-limiting step in the general base-catalyzed reactions, it must be proton transfer to or from a tetrahedral intermediate. The apparent hydroxide ion-catalyzed reaction could have rate-limiting C—O bond breaking in decomposition of the tetrahedral intermediate since proton abstraction from a neutral intermediate by ($^-$OH) would be diffusion controlled and thermodynamically favorable. If a common rate-determining step in these reactions of esters with aliphatic alcohol leaving groups is leading to similar rates of cyclization, then the value of the equilibrium constant K_{eq} for preequilibrium cyclization [Eq. (20)] must be independent of the nature of the nucleophilic group. An alternative possibility, which cannot at present be ruled out, is that the reactions occur with different mechanisms that coincidentally in these examples lead to similar rate constants.

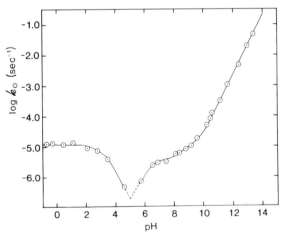

$$\text{(20)}$$

Another 2-aminomethylbenzoic derivative of interest in which the leaving group is poor is the amide (23) [26]. Figure 9 shows a plot of $\log k_0$ vs. pH for cyclization of (23) to phthalimidine [Eq. (21)]. At high pH, hydroxide ion catalysis occurs (the slope of the plot is 1.0). General base catalysis is observed

Fig. 9 Plot of $\log k_0$ vs. pH for cyclization of 2-aminomethylbenzamide to phthalimidine at 30°C and $\mu = 0.5\ M$ with KCl in H_2O. The points were obtained by extrapolation to zero buffer concentration.

$$\tag{21}$$

in the reaction at pH values above the pK_a of (23) by an extensive series of general bases with a Brönsted coefficient, β, of 0.4. Consequently, proton transfer may be concerted with bond making or breaking. A reasonable reaction scheme is depicted in Eq. (22), assuming that breakdown of a tetrahedral intermediate is rate determining at high pH as in the cyclization of 2-hydroxymethylbenzamide [32]. Concerted general base catalysis could

$$\tag{22}$$

result from partial proton transfer from the neutral intermediate (24) or the more likely, kinetically equivalent proton donation to the leaving group. These possibilities are shown in (25) and (26). General base-catalyzed formation of a tetrahedral intermediate could entail partial proton transfer from the amine function as it attacks the carbonyl. Thus, the amide is probably cyclizing at high pH through the same reaction pathways as the corresponding methyl ester, with the difference that proton transfer is rate limiting with the methyl ester whereas C—N bond breaking is concerted with proton transfer in the case of the amide. Brönsted coefficients of 0.4 for general base-catalyzed cyclization of 4- and 6-amino-2-hydroxymethylbenzamides [25] again suggest a concerted reaction in cases where a hydroxymethyl group acts as a nucleophile toward an amide. The k_{OH} values for these compounds, for 2-hydroxymethylbenzamide, and for (23) are nearly identical ($k_{OH} \sim 1$ $M^{-1} sec^{-1}$ at 30°C). Thus, as with the aliphatic esters, the rate constants are nearly independent of the nature of the nucleophilic group but are dependent on the leaving group, that is, whether the compound is an ester or an amide.

The pH–rate constant profile shows clearly that both the neutral and protonated species are undergoing cyclization. The pH-independent portion of the pH–rate constant profile below the pK_a of the amine (9.2) shows the presence of one proton in the transition state. Since there is also a proton in the ground state, this will result in a pH-independent reaction. The downward bend in the profile at approximately pH 7 can be considered evidence for a change in rate-determining step. Curvature was detected in plots of k_{obsd} vs. imidazole concentration, showing that the rate-limiting step also changes with increasing buffer concentration. While the identity of the rate-determining steps has not as yet been rigorously established, they can be tentatively ascribed to formation of a tetrahedral intermediate at low pH and decomposition of a tetrahedral intermediate to products at high pH. It would reasonably be expected that formation of a tetrahedral intermediate would be retarded at low pH where the amino group is predominantly in the protonated form. At the same time, C—N bond breaking to give products should be favored at low pH because decomposition of a protonated tetrahedral intermediate should be facile in comparison with breakdown of the anionic intermediate required at high pH.

CONCLUSIONS

In summary, the following conclusions are possible in regard to intramolecular nucleophilic attack on esters and amides:

1. Maximum rate enhancements in intramolecular nucleophilic reactions of esters are of the order of 10^8–10^9 M in comparison with corresponding bimolecular reactions. Enhancements larger than this are possible, but it appears probable that effects such as steric compression will then be required.

2. Anionic nucleophiles appear in general to be superior to neutral nucleophiles in intramolecular reactions of aryl esters, although the great

efficiency of the intramolecular reaction of (7) shows that this will not necessarily be observed in all cases. Phenoxide and hydroxymethyl nucleophiles have a greater effective molarity than neighboring sulfhydryl, but this is brought about by relatively more favorable bimolecular reactions of thiol nucleophiles.

3. When the leaving group is an aliphatic alcohol the rates of intramolecular displacement are independent of the nucleophilic group in sterically similar systems. In these reactions, proton transfer steps are rate determining when hydroxymethyl and aminomethyl nucleophiles are involved. To the extent that (8), (9), and (18) represent reasonable models for reaction of α-chymotrypsin with aliphatic esters, it can be concluded that proton transfer steps are also important in the enzymatic reaction.

4. When the leaving group is an amide the rates of intramolecular displacement are also independent of the nucleophilic group, but in contrast with aliphatic esters proton transfer and bond making or breaking are probably concerted.

5. The magnitude of the rate constants for acylation of α-chymotrypsin by amide substrates is explainable in terms of intracomplex alcoholysis along with intramolecular catalysis by histidine-57. Additional factors may, however, be required in a quantitative account of acylation with ester substrates.

REFERENCES

1. T. C. Bruice and S. J. Benkovic, "Bioorganic Mechanisms." Benjamin, New York, 1966.
2. W. P. Jencks, "Catalysis in Chemistry and Enzymology." McGraw-Hill, New York, 1969.
3. M. L. Bender, "Mechanisms of Homogeneous Catalysis from Protons to Proteins." Wiley (Interscience), New York, 1971.
4. T. C. Bruice, in "The Enzymes" (P. D. Boyer, ed.), 3rd ed., Vol. 2, Chapter 4. Academic Press, New York, 1970.
5. A. J. Kirby and A. Fersht, *Prog. Bioorg. Chem.* 1, 1 (1971).
6. T. H. Fife, *Adv. Phys. Org. Chem.* 11, 1 (1975).
7. J. E. C. Hutchins and T. H. Fife, *J. Am. Chem. Soc.* 95, 2282 (1973).
8. T. C. Bruice and A. Turner, *J. Am. Chem. Soc.* 92, 3422 (1970).
9. M. L. Bender and R. B. Homer, *J. Org. Chem.* 30, 3975 (1965); A. Williams, *J. Chem. Soc., Perkin Trans. 2* p. 808 (1972).
10. J. E. C. Hutchins and T. H. Fife, *J. Am. Chem. Soc.* 95, 3786 (1973).
11. T. H. Fife, J. E. C. Hutchins, and M. S. Wang, *J. Am. Chem. Soc.* 97, 5878 (1975).
12. J. W. Ogilvie, J. T. Tildon, and B. S. Strauch, *Biochemistry* 3, 754 (1964).
13. T. C. Bruice and S. J. Benkovic, *J. Am. Chem. Soc.* 85, 1 (1963).
14. T. C. Bruice and J. M. Sturtevant, *J. Am. Chem. Soc.* 81, 2860 (1959).
15. K. D. Gibson and H. Scheraga, *Proc. Natl. Acad. Sci. U.S.A.* 58, 420 (1967).
16. M. I. Page and W. P. Jencks, *Proc. Natl. Acad. Sci. U.S.A.* 68, 1678 (1971).

17. T. Higuchi, L. Eberson, and A. K. Herd, *J. Am. Chem. Soc.* **88**, 3805 (1966).
18. I. Christenson, *Acta Chem. Scand.* **18**, 904 (1964).
19. C. J. Belke, S. C. K. Su, and J. A. Shafer, *J. Am. Chem. Soc.* **93**, 4552 (1971).
20. T. H. Fife and B. M. Benjamin, *J. Am. Chem. Soc.* **95**, 2059 (1973); *Bioorg. Chem.* **5**, 37 (1976).
21. M. Eigen, *Angew. Chem., Int. Ed. Engl.* **3**, 1 (1964); for a clear explanation of why a proton transfer reaction may be considered diffusion-controlled in both directions see W. P. Jencks, "Catalysis in Chemistry and Enzymology," pp. 207–208. McGraw-Hill, New York, 1969.
22. B. Capon, S. T. McDowell, and W. V. Raftery, *J. Chem. Soc., Perkin Trans. 2* p. 1118 (1973).
23. D. W. Griffiths and M. L. Bender, *Bioorg. Chem.* **4**, 84 (1975).
24. K. N. G. Chiong, S. D. Lewis, and J. A. Shafer, *J. Am. Chem. Soc.* **97**, 418 (1975).
25. T. H. Fife and B. M. Benjamin, *J. Chem. Soc., Chem. Commun.* p. 525 (1974).
26. T. H. Fife and B. R. De Mark, unpublished work.
27. M. M. Kreevoy, E. T. Harper, R. E. Duvall, H. S. Wilgus, and L. T. Ditsch, *J. Am. Chem. Soc.* **82**, 4899 (1960).
28. T. H. Fife and B. R. De Mark, *J. Am. Chem. Soc.* **98**, 6978 (1976).
29. T. C. Bruice and M. F. Mayahi, *J. Am. Chem. Soc.* **82**, 3067 (1960); W. P. Jencks and J. Carriuolo, *ibid.* p. 675; W. P. Jencks and M. Gilchrist, *ibid.* **88**, 104 (1966); T. C. Bruice, A. Donzel, R. W. Huffman, and A. R. Butler, *ibid.* **89**, 2106 (1967); W. P. Jencks and M. Gilchrist, *ibid.* **90**, 2622 (1968).
30. G. M. Blackburn and W. P. Jencks, *J. Am. Chem. Soc.* **90**, 2638 (1968).
31. A. C. Satterthwait and W. P. Jencks, *J. Am. Chem. Soc.* **96**, 7018 (1974).
32. T. Okuyama and G. L. Schmir, *J. Am. Chem. Soc.* **94**, 8805 (1972).

CHAPTER

6

Chemical Aspects of
Lipoxygenase Reactions

Morton J. Gibian and Ronald A. Galaway

INTRODUCTION

Lipoxygenase (EC 1.13.11.12) is present in a wide variety of plants, the richest sources being legumes. The reaction catalyzed [Eq. (1)] involves the oxygenation of a 1,4-*cis,cis*-pentadiene on a long-chain fatty acid to a

$$(1)$$

1-hydroperoxy-2,4-*trans,cis*-pentadiene. Previous reviews by Axelrod [1] and by Tappel [2] have thoroughly covered the diversity of sources, the earlier history, and the metabolic role (on which there is no general agreement) of the enzyme. Certain animal tissues apparently also have this activity in association with (or as part of) the prostaglandin synthetase complex [3,4], which shows a rather different specificity and may or may not be similar in terms of structure and pathway. Since the soybean enzyme has been studied the most, this review is largely restricted to that lipoxygenase. In addition, the emphasis is on chemical aspects and the pathway of action.

Lipoxygenase action on substrate [Eq. (1)] can be coupled to oxidation of a variety of other easily oxidized materials (discussed in some detail later). There is also a secondary catalysis of decomposition of the primary product hydroperoxide, which results in chain scission, carbonyl or epoxide formation, and/or dimerization. These reactions occur primarily under anaerobic conditions and have been discussed by Galliard [5].

There are four isozymes of lipoxygenase in soybean [6–8]. The major one (lip-1, optimum activity at pH 9) was the main constituent of Theorell's crystalline lipoxygenase [2]. The molecular weight is close to 102,000 [7–10], and the amino acid composition [8,9] shows four buried cysteines with four [9] to six [1] cystines.

All four isozymes contain one iron atom per enzyme molecule (10^5 daltons) [11–14a], this iron having remained undiscovered for many years. It is a nonheme, non-acid-labile sulfide iron; its potential role is discussed later. There have been reports of subunits of lipoxygenase [10] and also fragments of varying size upon denaturation [15], but these remain unconfirmed [1]. If the iron plays a catalytic role in the reaction (which is almost certain since there are no other cofactors and the iron changes valence during catalysis) there should be one active site per molecule of enzyme.

The two best assay methods for lipoxygenase are polarographic O_2 uptake measurement and spectrophotometric observation of diene conjugation using linoleate (9,12-cis,cis-octadecadienoate) as substrate. Both methods are direct, continuous, and precise [1,16]. Measurement of hydroperoxide formation by the ferrous ion–thiocyanate method has not been reproducible and is laborious. The coupled oxidation of carotenes or other cooxidants, while a continuous and experimentally straightforward method, suffers from the fact that it is not the primary reaction, has potentially unknown variables, and is linear with enzyme only over a narrow range of enzyme concentrations [16–18].

Reaction Specificity

Substrates for lipoxygenase may be straight-chain fatty acids, esters, alcohols, hydroxamates [1], sulfates [19], or halides [20] containing a cis,cis-1,4-pentadiene unit, the acids being best. Holman *et al.* studied a series of variously positioned diunsaturated C_{18} acids [21]. The best rates result when the unsaturation begins at the sixth carbon atom from the methyl (ω) end of the chain. Additional unsaturation on either side of the ω-6,9 system does not have a great effect on the efficacy of reaction, nor does the total chain length for medium- and longer-chain fatty acids. (There are specificities within these groups, but all are within the same order of magnitude.) The requirement for the ω-6,9 is not highly stringent, in that a wide variety of methylene-interrupted octadecadienoic acids showed some activity. The unsaturation must be olefinic, since octadeca-9-en-12-ynoic acid gave a very low rate of oxygenation. Interestingly, among C_{18} dienoic acids, the rate of the 13,16-dienoate was second only to that of linoleate (the 9,12-dienoate). These studies did not separate binding from catalysis, and there are few data on binding or competitive inhibition by nonsubstrate fatty acids so that it is impossible at this point to know whether these are cases of poor binding, nonproductive binding, or a combination.

The positional specificity of oxygenation has been investigated for linoleic and a variety of other acids. It is now clear that lipoxygenase-1 produces, essentially, solely 13-L_S-hydroperoxy-9,11-cis,trans-octadecadienoic acid [22–26] from linoleate. Some 9-hydroperoxy product is usually found, but this stems from either a small amount of solution autoxidation [27], some isozymic impurity [25,26], or a rearrangement of the product hydroperoxide during workup and analysis [28]. Hamberg and Samuelsson [22] investigated a wide range of good substrate unsaturated dienoic acids and found that all formed hydroperoxide at the ω-6 position. Most recently, however, Egmond et al. [29] examined the product from 13,16-octadecadienoic acid and found it to be 85% the 17-hydroperoxide (ω-3), optically active. The specificity thus seems to be considerably broader than had previously been believed. Lipoxygenases from corn germ [30] and potato [31] produce the 9-isomer almost exclusively, while others give varying specificity with linoleic acid [32,33].

The stereochemical aspects of the reaction have been investigated using both linoleic and 8,11,14-eicosatrienoic acids. In the soybean lipoxygenase reaction only the pro-S hydrogen is removed from the substrate molecule at C-11 or C-13, respectively, for these two substrates [22,34]. Corn germ lipoxygenase removes the pro-R hydrogen from C-11 of linoleate [34]. The abstraction is accompanied by a significant discrimination for the lighter isotope, which was observed upon examination of remaining substrate in mixtures of tritiated samples of both of the above substrates [22,34]. Subsequent kinetic analysis using deuterated linoleate showed that the pro-S hydrogen at C-11 manifests a k_H/k_D of about 9 on the catalytic rate constant (K_m was unaffected) measured in turnover [35].

With the soybean enzyme, O_2 enters both linoleate and 8,11,14-eicosatrienoate stereospecifically at ω-6 to give the L_S-hydroperoxide, as confirmed in two independent studies [22,34]. The double bond that moves always ends up trans. Apparently the substrate must be rather rigidly held at the enzyme active site with highly directed hydrogen removal and oxygen addition on the carbon chain with no free chain rotation. The bisallylic hydrogen is removed in, or before, the rate-limiting step—certainly early in the reaction.

Equation (2) depicts the net changes that occur in the catalytic reaction; it does not imply any reaction mechanism, but summarizes the stereochemical

$$\text{(structure diagram)} \qquad (2)$$

results just discussed. If the geometry of the substrate is fixed at the active site during the entire conversion, the O_2 must enter from the *opposite* side of the molecule from which the pro-*S* hydrogen is removed in order to produce the *trans*-L_S-hydroperoxide. It is not at all clear that the conformation is fixed during the total conversion, but neither is it unlikely.

CHEMICAL ANALOGS OF THE REACTION

Using standard bond dissociation energies [36] one can calculate that Eq. (1) is strongly exothermic ($\Delta H^0 \simeq -21$ kcal/mole). While there is certainly

$$D[(C=C)_2CH\!-\!\!-\!H] - D(C=C\!-\!C=C\!-\!\!-\!OO\cdot) - D(ROO\!-\!\!-\!H) = \Delta H^0$$
$$(+84) \qquad - \qquad (+15) \qquad - \qquad (+90) = -21 \text{ kcal/mole}$$

an entropy decrease since two molecules go to one [and, for $H_2 + O_2 \rightarrow H_2O_2(g)$, $\Delta S^0 \simeq -25$ eu [37]], it could not reasonably be sufficiently unfavorable to give a ΔG^0 of greater than approximately -10 kcal/mole. It is not surprising that several formally identical and a variety of closely analogous nonenzymatic transformations are well known.

Upon radical initiation, 1,4-pentadienes in the presence of O_2 produce conjugated 1,3-diene hydroperoxides by autoxidation [Eqs. (3)–(5)] [38]. The conjugated diene results because of its greater stability than the methylene-interrupted diene, obviously manifest in reaction (4). The isomerization of

$$\text{RCH}=\text{CH}-\text{CH}_2-\text{CH}=\text{CHR}'$$
$$\text{SH}$$

$$\Big\downarrow +\text{In}\cdot \qquad (3)$$

$$[\text{RCH}=\text{CH}-\overset{\cdot}{\text{C}}\text{H}-\text{CH}=\text{CHR}' \longleftrightarrow \text{R}\overset{\cdot}{\text{C}}\text{H}-\text{CH}=\text{CH}-\text{CH}=\text{CHR}' \longleftrightarrow$$
$$\text{3-S}\cdot \qquad\qquad\qquad\qquad \text{1-S}\cdot$$
$$\text{RCH}=\text{CH}-\text{CH}=\text{CH}-\overset{\cdot}{\text{C}}\text{HR}']$$
$$\text{5-S}\cdot$$

$$\diagup +\text{O}_2 \qquad (4)$$

$$[\text{3-SOO}\cdot + \text{1-SOO}\cdot + \text{5-SOO}\cdot] \xrightarrow[\ \ ]{} \text{3-SOOH} + \text{1-SOOH} + \text{5-SOOH} \qquad (5)$$
$$\text{(Minor)} \qquad \text{(Major)} \qquad \text{SH} \quad \text{S}\cdot$$

the double bond leads largely to the more stable *trans*-olefin, so that the product is structurally the same as in the lipoxygenase reaction, although racemic. This process is very rapid for linoleate, methyl linoleate, linolenate, and 1,4-pentadiene itself [38–40]. Initiation can be effected by a variety of radical-generating methods. This autoxidation mechanism had been seriously considered for lipoxygenase [2], but at this juncture it can be ruled out on a variety of very firm grounds (including, *inter alia*, stereochemistry, kinetics, and effect of potential inhibitors). This does not imply, however, that one or more of the steps above may not be *related* to the enzyme process.

While ground-state triplet oxygen is not reactive toward olefins (such as substrates for lipoxygenase), there is a relatively low lying, electronically excited singlet state that is [41]. Indeed, one of the best known reactions of $^1O_2$ is the conversion of an olefin bearing allylic hydrogens to the rearranged allylic hydroperoxide [e.g., Eq. (6)] [42]. The reaction is stereospecific, with

$$
\underset{H_3C}{\overset{H_3C}{>}}C=C\underset{CH_3}{\overset{CH_3}{<}} + {}^1O_2 \longrightarrow \underset{H_2C}{\overset{H_3C}{>}}C-\underset{\underset{OOH}{|}}{\overset{\overset{CH_3}{|}}{C}}-CH_3 \tag{6}
$$

the hydrogen being removed from the same face of the molecule to which the oxygen adds [41,42]. This is opposite to that which formally obtains for the lipoxygenase reaction. Since the lowest lying $^1O_2$ form is ~ 22 kcal/mole above the ground state, this amount of energy must either be provided in a chemical reaction that generates $^1O_2$, or be transferred directly to the oxygen molecule if ground-state oxygen is to produce $^1O_2$. While this is certainly achievable in photochemical reactions at normal temperatures, it is very difficult to envision the attainment of such a state directly from O_2 in a normal enzymatic process.

Organometallic and Anion Reactions

At low temperature under very dilute conditions, Grignard reagents directly oxygenate to hydroperoxide salts [Eq. (7)], which may then be hydrolyzed to hydroperoxides [43]. This reaction is fairly general for many

$$
RMgX + O_2 \longrightarrow ROO^-MgX^+ \tag{7}
$$

organometallics [44], although products can be further destroyed by complex chemistry. Formally, this could be a direct O_2 addition, but recent evidence [45, 46] is that the path is actually via Eqs. (8) and (9),

$$
R\cdot + O_2 \longrightarrow RO_2\cdot \tag{8}
$$

$$
RO_2\cdot + RMgX \longrightarrow RO_2MgX + R\cdot \tag{9}
$$

probably initiated by Eq. (10).

$$RMgX + O_2 \longrightarrow R\cdot + MgX^+O_2^{\overline{\cdot}} \tag{10}$$

G. Russell and his group have studied a wide range of anion reactions using strong base in aprotic dipolar solvent to generate the carbanions [47]. Only the most stable carbanions (i.e., from the most acidic conjugate acids, e.g., β-dicarbonyl compounds) are inert to molecular oxygen. There are two groups of reactants. (a) Triphenyl- and diphenylmethane are among those compounds that oxygenate to ROO^- as rapidly as carbanion is formed [$(C_6H_5)_3C\cdot$ reacts with O_2 more slowly than does $(C_6H_5)_3C^-$]. For this group of molecules Russell favors a route involving electron transfer followed by coupling [Eq. (11)], which yields no *free* radicals but allows for spin inter-

$$R^- + O_2 \longrightarrow [R\cdot O_2^{\overline{\cdot}}] \longrightarrow RO_2^- \tag{11}$$

conversion of oxygen [46]. (b) On the other hand, some molecules, such as fluorene (the hydrocarbon best studied of the second group) undergo a free-radical chain oxygenation which can be catalyzed by electron acceptor molecules (usually nitroaromatics). These latter are not required but produce higher rates. The path is as follows (π is the catalyst):

$$Ar_2CH^- + O_2 \longrightarrow Ar_2CH\cdot + O_2^{\overline{\cdot}}$$
$$Ar_2CH^- + \pi \longrightarrow Ar_2CH\cdot + \pi^{\overline{\cdot}}$$
$$Ar_2CH\cdot + O_2 \longrightarrow Ar_2CHOO\cdot$$
$$Ar_2CHOO\cdot + Ar_2CH^- \longrightarrow Ar_2CHOO^- + Ar_2CH\cdot$$
$$Ar_2CHOO\cdot + O_2^{\overline{\cdot}} \longrightarrow Ar_2CHOO^- + O_2$$
$$\pi^{\overline{\cdot}} + O_2 \longrightarrow \pi + O_2^{\overline{\cdot}}$$

Under appropriate conditions all these reactions can yield hydroperoxides.

Metal-Catalyzed Processes

We will not attempt to cover the array of inorganic chemistry that could be considered as potentially fundamental to the lipoxygenase reaction. The most widely examined processes involve H_2O_2 oxidations of organic compounds [e.g., 48–50] which lead primarily to monooxygenase-type products. Hamilton [49] has discussed the model and enzymatic processes at some length and feels that they are often ionic (or polar) reactions of either coordinated O or O_2 on substrate. A variety of catalytic processes utilizing O_2 lead to oxidations but usually not to direct dioxygenations. For these, a typical key intermediate would be

$$[Fe(II)\cdots O_2 \longleftrightarrow Fe(III)\cdots O_2^{\overline{\cdot}}]$$

which can lead to direct nucleophilic reactions, give $O_2^{\overline{\cdot}}$, or in some cases be redox active. Additionally, the metal can simply serve as a "shuttle" for electrons, an interesting proposal [51] that has not been shown to operate in any chemical transformations with O_2 but that has excellent analogies in other areas of inorganic chemistry. With the recent advances in the chemistry of dioxygen adducts of transition-metal complexes [52] the proposed pathways should begin to be directly testable.

Neither free $HO_2\cdot$ nor $O_2^{\overline{\cdot}}$ is an attractive intermediate for the lipoxy-genase reaction. While $HO_2\cdot$ is reactive toward olefins, in order to produce the observed products (no H_2O_2 is formed) one would conclude that the autoxidation scheme [Eqs. (3)–(5)] is much more likely if $HO_2\cdot$ is ever formed since it initiates this process. Superoxide ion does not react with olefins, but metal-bound $O_2^{\overline{\cdot}}$ would be much more like $HO_2\cdot$ and should be reactive.

Finally, there are certain metal ions [Co(III), Mn(III)] that directly oxidize olefins and aromatics via one-electron transfers [Eqs. (12) and (13)] that then lead to further reactions [53]. This is brought up in a direct proposal at the end of this review.

$$Co(III) + -CH{=}CH- \longrightarrow -\overset{\cdot}{C}H-\overset{+}{C}H- + Co(II) \qquad (12)$$

$$Co(III) + ArH \longrightarrow ArH^{+\cdot} + Co(II)$$

$$(13)$$

$$Ar\cdot + H^+$$

IRON CONTENT AND STATE

That lipoxygenase contains 1 gm-atom of nonheme iron per mole of protein (by atomic absorption analysis) was reported first by Chan in 1972 [11] and was confirmed in three independent studies [12–14] during the next year. Inhibition by dithizone and aromatic o-dithiols [14a], synergistic inhibition by o-phenanthroline and mercaptans with concomitant appearance of an Fe complex [13–14a], and formation of iron complexes upon denatura-tion [12,14a] also indicated the presence of an iron atom. There is no acid-labile sulfur and no heme [14,14a]. Previous detection of the metal had been thwarted by the lack of inhibition by EDTA, CN^-, and other complexing agents coupled with very low ultraviolet (uv), visible, and electron spin resonance (esr) spectra for the native enzyme. Certainly, the availability of purified enzyme was a prime factor in allowing the identification of lipoxy-genase as an iron enzyme.

The iron atom is very tightly bound (as evidenced by the difficulty of removal), and to date there has been no report of successful reconstitution of active enzyme by readdition of iron after its removal. Apparently oxidative denaturation precedes release of iron in the cases known. Denaturation by urea, sodium dodecyl sulfate, and other simple agents does not facilitate iron removal [12].

Absorption, fluorescence, and spin resonance spectroscopy have recently been employed to examine the state and functional significance of the iron atom. Axelrod [14,14a], while finding the native enzyme to be esr silent, reported that addition of substrate and O_2 produced a signal at $g = 6$. The laboratories at Utrecht and Rome have examined the esr at 15K [54] and the absorption and emission spectra of lipoxygenase under various conditions above 5°C [55–58]. Native lipoxygenase exhibits only very weak tail absorption in the near uv, is transparent in the visible, and has only tryptophan fluorescence (328 nm) upon 280 nm excitation [55]. The anaerobic addition of linoleate gave no changes in any of these properties (there is a very weak native esr spectrum that was attributed to impurities), but the addition of the 13-L-hydroperoxide gave rise to substantial esr resonances at $g = 7.5$, 6.2, 5.9, 4.3, and 2.0 [54], a new near-uv band at 330 nm ($\varepsilon \approx 10^3 \ M^{-1} \ cm^{-1}$), and a quenching of the fluorescence [56]. The same effects occur upon addition of linoleate plus O_2, which produces product. When excess linoleate is added anaerobically to this esr-active enzyme form the signals and spectra return to those of native lipoxygenase (presumably because of complete consumption of product hydroperoxide), and addition of O_2 (after thawing for the esr case) results in the reappearance of the esr lines, the 330 nm band, and the fluorescence quenching. Unfortunately, the actual stability of this esr-active form was not reported.

The most recent report [58] indicates that there are at least two enzyme states in addition to native lipoxygenase; these are the 330 nm-absorbing state just described and another that is formed upon addition of a higher amount of product hydroperoxide than that used above. Addition of further 13-hydroperoxide to the 330 nm material which had been isolated by Sephadex chromatography produced a purple solution ($\lambda_{max} = 570$ nm, $\varepsilon \approx 10^3$ $M^{-1} \ cm^{-1}$) which had a greatly enhanced esr signal at $g = 4.3$ and reduced signals at other g values. This purple material returned to the yellow (330 nm) form fairly rapidly upon standing. It was reported that the yellow form shows the same kinetic lag (see next section) as does the native enzyme [59], but no catalytic experiments were reported using the 570 nm solution.

In the formation of the 330 nm form, it was found [56] that high concentrations of H_2O_2 had the same effect as did the 13-hydroperoxide but that the 9-hydroperoxide was totally ineffective on lipoxygenase-1. Hydrogen peroxide however, destroys the enzyme activity. Potassium iodide also quenches the fluorescence, especially in the absence of O_2, possibly because I^- competes for the O_2 site [57].

These experiments indicate that the native enzyme probably contains iron in the low-spin Fe(II) (d^6) state which can be converted to the Fe(III) or Fe(IV) level with accompanying spectral changes. Since these conversions are caused by product hydroperoxide and are reversed by substrate (and product is known to activate the enzyme; see next section) it is likely that these redox changes are catalytically significant.

The iron in lipoxygenase seems to be rather unique in biochemical systems, and at this point there is no clear chemical precedent for the kind of ligands on protein side chains that could produce this unusual esr-silent, almost colorless species.

DIRECT STEADY-STATE KINETICS

Since the turnover kinetics are complicated by a variety of experimental difficulties and, beyond this, are obviously complex, the situation is not yet very clear.

Lipoxygenase-1 has maximum activity at pH 8.5–9, while the other isozymes show various maxima between 5.5 and 7 [1]. It also has apparent higher specific activity toward linoleate than do the others (two to five times), but less toward unsaturated triglycerides, and there is a great difference in the efficacy of β-carotene-coupled oxidation between the forms. The work discussed below is essentially all on soybean lipoxygenase-1.

In order to help solubilize substrate fatty acids, organic solvent (often ethanol) and detergent (typically Tween 20) have frequently been used, but both are inhibitory and change the pH dependence [16,60]. Their use has now been discontinued or rigidly controlled by workers studying the kinetics. Lipoxygenase is apparently insensitive to substrate micelle formation: while the critical micelle concentration of linoleate is not far from its K_m and thus in the middle of the concentration range usually employed [60,61], double reciprocal plots show no break when the medium is controlled. This is an interesting matter for speculation, since the free substrate concentration must not be varying directly with total concentration over this range.

At low concentrations lipoxygenase adheres to and/or is inactivated by glass surfaces, leading to progressive diminution in rate and much lower total conversions than at higher E_0 when stock enzyme is diluted into a reaction mixture to $\sim 10^{-9}$ M (0.1 μg/ml) [60,62].

Kinetic Lag

An appreciable lag (induction) phase appears in direct lipoxygenase kinetics which is abolished by low levels ($\leq 5\,\mu M$) of product 13-L-hydroperoxide [63,63a] but not by 9-hydroperoxide [54] nor H_2O_2 and only very poorly by 1-octyl hydroperoxide [63,63a]. Smith and Lands found increased lags when glutathione (GSH) and GSH-peroxidase were added to lipoxygenase reactions, neither having an effect separately [63,63a]. Since GSH-peroxidase catalyzes the reaction of hydroperoxides with GSH to give the corresponding alcohols and GSSG, the apparent effect is to reduce product *in situ* to its alcohol, thereby destroying the product activation. Using the 13-L-alcohol of the linoleate product directly (produced by $NaBH_4$ reduction of isolated enzyme product), we have shown that not only does the alcohol not diminish the lag, but it inhibits added peroxide from doing so [62].

It had earlier been proposed [63a] that product is required as a reactant in each turnover event, but the evidence is only that product is needed in very low concentration in the initial stages of turnover. In light of the recent spectroscopic results (previous section), it is easiest to interpret this as an initial product activation of enzyme, probably via change of the iron oxidation level by specific reaction with the hydroperoxide group.

Turnover Kinetics

Over and above the wall effect, reaction with substrate does not proceed to completion but rather ceases before all O_2 or S is consumed. The rate slows down as product builds up, and finally ceases. In recent work, we have found that this is primarily a *reversible* inactivation due to the presence of product [62], rather than an irreversible self-destruction of enzyme during catalysis [63,63a]. It may also be an enzyme redox change. Along with the product activation, this deactivation has also hampered turnover studies.

Egmond *et al.* [59] have recently reported on kinetic aspects of the turnover of linoleate by lipoxygenase. The rate of diene production is the same, including the lag, for native and yellow (esr-active) enzyme forms (see previous section). As O_2 was lowered with respect to linoleate (each varied in turn as the other was held constant) rate curves became increasingly sigmoidal. The K_m (app) for linoleate was 20 μM, in good agreement with the results of others [60]. Strong substrate inhibition was observed, manifest as an apparent decrease in O_2 binding. The authors postulated a binding site for each substrate (O_2 and linoleate) and a regulatory site for product or substrate. The situation is clearly complex and requires more work.

Finally, as was described earlier, lipoxygenase exhibits a large kinetic isotope effect in turnover on k_{cat} for the hydrogen removed in catalysis [35].

Inhibition

Metal ion reagents are discussed above. A number of conventional anti-oxidants have long been known to inhibit lipoxygenase [2], and this was often taken as evidence for a radical process. However, Vanderhoek and Lands have shown that the inhibitory effectiveness does not follow "typical" anti-oxidant tendencies [64], nor does it follow the redox potentials [16,65]. In fact, the I_{50} values are not at all out of line with expected competitive inhibition based on the structural variations among these compounds. Aromatic hydrocarbons are inhibitors, and examination of the Vanderhoek list shows that increasing bulk causes poorer inhibition, which may indicate a selectivity of binding to the site.

Studies (in which often limited concentration ranges of inhibitor and substrate were used so that assessment of the nature of inhibition and associated potency is not clear) showed that many fatty acids acted as inhibitors. *trans,trans*-Linoleate, 10,12-linoleate, oleate, and octanoate were decreasingly inhibitory in that order [66]. 9-*cis*,12-*trans*-Linoleate [2] and monohydric alcohols of various lengths [67] were also inhibitory to various degrees. Blain and Shearer [68] screened a range of acetylenic and olefinic compounds of different chain length (alcohols, acids, and others) and found that many long-chain unsaturated compounds had at least some inhibitory effect and then found that several acetylenic compounds were more or less competitive, very strong inhibitors (the best being the acetylenes corresponding to lino-leate, linolenate, and arachidonate, the best substrates for lipoxygenase). It was subsequently reported that both the 9,12-diynoic C_{18} and 5,8,11,14-tetraynoic C_{20} acids are irreversible inhibitors of the enzyme [69].

COUPLED REACTIONS AND PRODUCT DECOMPOSITION

A wide range of substances are cooxidized when present during lipoxygenase oxygenation of one of its natural substrates. These processes can be truly coupled in the sense that no enzymatic oxidation occurs in the absence of primary substrate or O_2 nor in the presence of hydroperoxide product alone. Among cooxidants that have been studied are polyolefins (carotenes, bixin, etc.), some readily oxidized aromatic molecules (luminol, 1,3-diphenylisobenzofuran, certain hydroquinones), and cholesterol [1]. This is a representative list, certainly not a comprehensive one. It is clear that during catalysis some highly reactive species is being generated that either directly oxidizes or catalyzes the oxidation of the cooxidant. On the basis of these cooxidations and related physical experiments, various types of species (including oxygen, substrate, and product-derived moieties) have been proposed as the key intermediate in lipoxygenase catalysis. In this section we discuss a few of the recent studies that have helped to narrow down the range of possibilities for the nature of the cooxidation. We also briefly discuss the lipoxygenase-catalyzed decomposition of its hydroperoxide product.

The most widely studied cooxidants are members of the carotene group [17]. β-Carotene is inefficiently cooxidized by lipoxygenase-1 (about 1 for every 35 linoleates, [16]) but much more efficiently by some of the other lipoxygenases [70]. In the case of lip-1, the β-carotene rate followed the linoleate dependence on concentration and pH and was significantly inhibited by radical retarders (butylated hydroxytoluene and Santoquin) at levels at which there was no effect on the linoleate oxygenation [16]. There was no solvent D_2O kinetic isotope effect [71] on the primary or secondary reaction (in contradiction to another report [72]). It seems that this reaction is an autoxidation of β-carotene initiated by some radical-type species generated during linoleate turnover.

Using a rather elegant probe, Smith has examined a variety of biological systems that were known to cause oxidation of organic compounds [73]. Chemical work shows that cholesterol when reacted with singlet oxygen gives solely cholest-6-ene-5α-hydroperoxide initially, while radical attack produces the epimeric 7-cholesterol hydroperoxides, superoxide results in no oxidation itself, and H_2O_2 (and organic peroxides) give cholesterol oxides [74]. Lipoxygenase gives the radical product of cholesterol and also the radical product of cholest-4-en-3β-ol [73,74].

While tetracyclone and 1,3-diphenylisobenzofuran are cooxidized by lipoxygenase, the product of the latter, at least, is different from that with singlet oxygen [75].

Superoxide dismutase, which catalyzes the disproportionation of super-oxide to H_2O_2 and O_2, had virtually no effect at very high concentration (3 μM) on either the direct reaction of linoleate or the coupled oxidation of β-carotene in our hands and those of others [76] (although there is again some disagreement on this point [72]). Cytochrome c (oxidized) reacts rapidly with superoxide [77] yet was not subject to coupled reaction in the linoleate–lipoxygenase system (although it was slowly reduced by the product [16]).

Product Decomposition

A variety of products are formed when linoleyl hydroperoxide is incubated with lipoxygenase anaerobically [5,78,79]. It is clear that the decomposition is faster when the solution is anaerobic, but it occurs in air as well [79,80]. The products are different when *both* substrate and product are present, and, again, there is evidence that the process is then more rapid. In any case, the process is clearly catalyzed by lipoxygenase [81]. (There is also a nonenzymatic, much slower decomposition, and there seem to be enzymes that are specific fatty acid hydroperoxide-decomposing-systems [78,82].) Several years ago it was found that when 13-L-hydroperoxide, linoleate, and lipoxygenase (but no two alone, and not the 9-hydroperoxide) are mixed anaerobically, free radicals are formed, as evidenced by their being trapped by a nitroso compound to form a nitroxyl radical with linoleate attached [83]. The site of attachment was the 13 (or possibly 9) position of the added substrate (not the product) as shown by the esr spectrum of suitably deuterated materials. Apparently this experiment was under exactly the conditions that produce maximum hydroperoxide decomposition and dimer formation, so that it bears on that aspect of the lipoxygenase system and may or may not be relevant to the primary catalysis of substrate oxygenation.

In fact, some of the coupled oxidations may be of a similar nature. It is known that product reacts with the enzyme, even in the presence of oxygen, and that it also inhibits the substrate reaction [62]. This could involve a redox reaction with the catalytic iron, a process that deactivates the enzyme and perhaps generates radicals, giving rise to radical attack on added materials or substrate, thus generating secondary products from the product. The classic Haber–Weiss mechanism [Eq. (14)] [48] for H_2O_2 decomposition operates

$$Fe(II) + ROOH \longrightarrow Fe(III) + RO\cdot + OH^-$$

$$RO\cdot + A\text{---}H \longrightarrow A\cdot + ROH \tag{14}$$

$$Fe(III) + A\cdot \longrightarrow Fe(II) + A^+ \longrightarrow products$$

well with organic hydroperoxides, generating oxy radicals that react primarily with other organic compounds. In this case, one would also expect some accompanying irreversible destruction of the enzyme, as apparently occurs [62–63a]. In the presence of O_2, however, the iron is kept primarily in the oxidized form, and such a process should not be the major one, but could accompany it.

PROPOSED MECHANISMS

The following major characteristics of the lipoxygenase reaction are salient in any consideration of pathway:

1. Specificity for position of cis,cis-1,4-diene along the chain and site of O_2 attachment

2. Stereospecificity of H removal and O_2 attack; formation of solely trans-olefin from the rearranged position

3. Large primary kinetic isotope effect for removal of the bisallylic hydrogen

4. Chemical involvement of product (13-hydroperoxide for lip-1) in at least initial activation step

5. Catalyzed decomposition of product (especially in absence of O_2)

6. Native enzyme inactive, esr silent; product activation leading to esr-active enzyme, reversed by substrate (these also accompanied by fluorescence and absorbance changes)

7. Coupled oxidations of various materials, likely intercepting a reaction intermediate, showing characteristics of free-radical reactions

With the discovery that lipoxygenase contains a likely functional iron atom, this enzyme is no longer a unique "cofactorless" oxygenase. A variety of ingenious mechanisms that were proposed before 1973 are only briefly covered here, although the direct involvement of iron in the catalysis is by no means an absolute certainty.

From the list above, it is clear that all steps in this reaction occur on the enzyme surface and that the bisallylic hydrogen is removed in the rate-limiting step (points 1–3). Along with results mentioned earlier, this constitutes compelling evidence against *solution* free-radical autoxidation mechanisms and any real $^1O_2$ process.

Several mechanistic proposals have incorporated an apparent requirement for product by postulating a hydroperoxide involvement in each turnover. These involve a site for product and one for substrate, with a "shuttle" of electrons and reactants while all intermediates remain bound to the enzyme surface. An example is shown in Scheme 1 [84]. One problem here is the

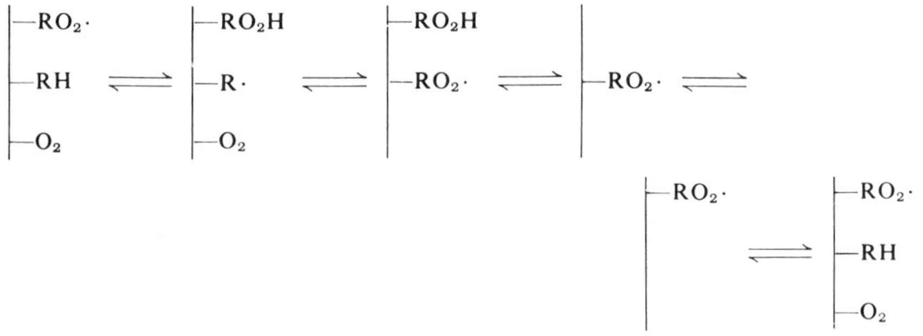

Scheme 1

initial step; once a peroxy radical or substrate radical is generated, all other steps are chemically favorable.

Hamilton [49] has suggested formation of a ternary complex which can then proceed to product via Scheme 2. The first step is quite endothermic, but the overall scheme is attractive *if* iron is not involved and product is required in each catalytic turnover.

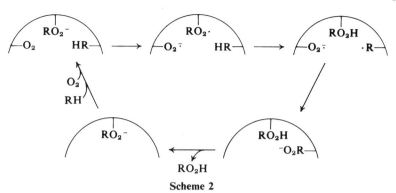

Scheme 2

The spectroscopic results (electronic and esr) are reasonably indicative that iron is involved in the catalytic reaction and, together with the kinetic results, led de Groot *et al.* [54] to propose the mechanism that is depicted, in essence, in Scheme 3. The first step [(1) to (2)] was formulated as requiring O_2, but the

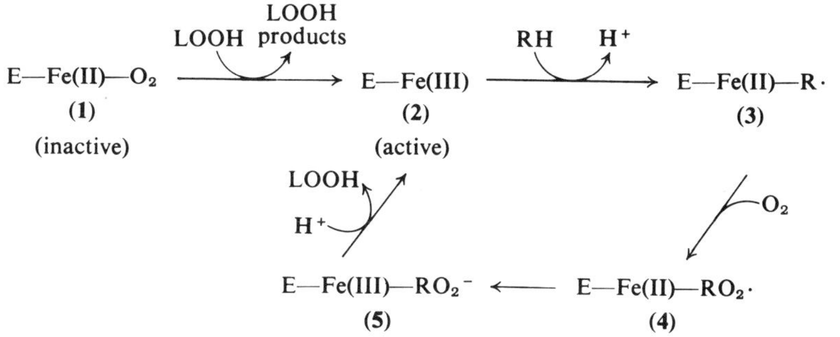

$$E\text{—}Fe(II)\text{—}O_2 \xrightarrow[\text{LOOH products}]{\text{LOOH}} E\text{—}Fe(III) \xrightarrow[\text{RH} \quad H^+]{} E\text{—}Fe(II)\text{—}R\cdot$$

(1)   (2)   (3)

(inactive)   (active)

$$E\text{—}Fe(III)\text{—}RO_2^- \xleftarrow[H^+]{\text{LOOH}} E\text{—}Fe(II)\text{—}RO_2\cdot \quad \xleftarrow{O_2}$$

(5)   (4)

Scheme 3

evidence for the O_2 involvement is weak, and otherwise it is a normal Fe(II) plus hydroperoxide reaction to give Fe(III) and decomposition products [see Haber–Weiss first step, Eq. (14)]. At low ROOH levels this would be stoichiometric and generate little RO·, which by and large could generate products in very small amount out in solution. From intermediate (3) on to product (LOOH) and back to (2) is well-precedented chemistry. A leakage from (3) to E—Fe(II) and free R· was used to explain trace autoxidation, coupled oxidations, and substrate reduction of enzyme which in the presence of O_2 leads back to (1). In the absence of O_2, product LOOH could react with E—Fe(II) via one-electron transfer, thus setting up catalytic peroxide breakdown.

The slow step (because of the kinetic isotope effect) of Scheme 3 must be from (2) to (3), the first substrate-involving reaction. This is the crucial and difficult step in the mechanism. The reaction as written may or may not have mechanistic significance since this is not a known route for ferric complexes and hydrocarbons. We would like to propose an attractive possibility for this process, one that has a chemical basis and utilizes the special structural characteristics required of lipoxygenase substrates.

In the section on metal-catalyzed processes, the electron transfer reactions between certain metals of high oxidation potential [notably Co(III)] and olefins are briefly presented [Eq. (12)]. If the unusual environment of the lipoxygenase iron serves to increase the oxidation potential, after conversion to the ferric form it might function as in Eq. (15), perhaps in a reversible fashion. This process could be very rapid if it were exothermic or nearly so.

$$E-Fe(III) + RCH=CHR' \longrightarrow E-Fe(II) + R\overset{.}{C}H-\overset{+}{C}HR' \qquad (15)$$

We do know, however, that removal of the bisallylic hydrogen occurs in a rate-limiting step, so that Eq. (15) could be followed by a rate-limiting E_1 elimination [Eq. (16)]. Binding of O_2 to the E—Fe(II) enzyme generated in

$$(16)$$

Eq. (15) [or binding of O_2 prior to Eq. (15)] would produce [E—Fe(II)—$O_2 \leftrightarrow$ E—Fe(III)—O_2^-], known to dissociate to either O_2 or O_2^-, corresponding to either formal resonance form. Finally, the process shown in Eq. (17) leads to product and regenerates active enzyme in the Fe(III) state. In the absence of O_2, the substrate radical formed in Eq. (15) or (16) (most likely the latter) can dissociate and cause coupled reactions.

The proposed mechanism [Eqs. (15), (16), binding of O_2, then Eq. (17)] is in accord with each point of the summary at the beginning of this section. The initial product activation is essentially the same as in Scheme 3 and, if substrate is present, Eqs. (15) and (16) follow. In the absence of O_2, this either is nonproductive (i.e., reversible), or else leads to destruction of substrate (observed anaerobically only if product is present). At high product concentration in the presence of O_2, there is a competitive inhibition for the substrate site, product being largely inactive toward ferric iron (but not entirely so [85]).

This mechanism is *similar* in form to that of Scheme 3 but is chemically different. It uses known chemistry and the structural features of the substrate

(17)

(namely, the diolefinic functionality). Additional chemical information, structural elucidation of the iron environment on the enzyme, and probably rapid kinetic studies will be necessary before the pathway can be adequately tested, but at this time it represents a more rational working hypothesis for the enzymatic process than was possible a few years ago.

ACKNOWLEDGMENTS

We are grateful to the National Science Foundation (GP-34389, BMS-03481) and the Public Health Service (GM-21903) for research grants that supported work reported here. One of us (M.J.G.) also acknowledges the support of his sabbatical by NIH and the hospitality of the Department of Chemistry, Harvard University, 1975–1976.

REFERENCES

1. B. Axelrod, *Adv. Chem. Ser.* **136**, 324–348 (1974).
2. A. L. Tappel, *in* "The Enzymes" (P. D. Boyer, H. Lardy, and K. Myrbäck, eds.), 2nd rev. ed., Vol. 8, Part B, pp. 275–283. Academic Press, New York, 1963.
3. M. Hamberg and B. Samuelsson, *J. Biol. Chem.* **242**, 5344 and 5536 (1967); for leading references, see B. Samuelsson and R. Paoletti, eds., "Advances in Prostaglandin and Thromboxane Research," Vols. 1 and 2. Raven, New York, 1976.
4. D. Nugteren, *Biochim. Biophys. Acta* **380**, 299 (1975).

5. T. Galliard, *in* "Recent Advances in the Chemistry and Biochemistry of Plant Lipids" (T. Galliard and E. Mercer, eds.), Chapter 11. Academic Press, New York, 1975.
6. J. Christopher, E. Pistorius, and B. Axelrod, *Biochim. Biophys. Acta* **284**, 54 (1972).
7. W. Verhue and A. Francke, *Biochim. Biophys. Acta* **284**, 43 (1972).
8. A. Yamamoto, K. Yasumoto, and H. Mitsuda, *Agric. Biol. Chem.* **34**, 1169 (1970).
9. J. Christopher, E. Pistorius, and B. Axelrod, *Biochim. Biophys. Acta* **198**, 12 (1970).
10. F. Stevens, D. Brown, and E. Smith, *Arch. Biochem. Biophys.* **136**, 413 (1970).
11. H. W.-S. Chan, *11th World Congr. Int. Soc. Fat Res., 1972.*
12. M. Roza and A. Francke, *Biochim. Biophys. Acta* **327**, 24 (1973).
13. H. W.-S. Chan, *Biochim. Biophys. Acta* **327**, 32 (1973).
14. E. K. Pistorius and B. Axelrod, *Fed. Proc., Fed. Am. Soc. Exp. Biol.* **32**, 544 (1973).
14a. E. K. Pistorius and B. Axelrod, *J. Biol. Chem.* **249**, 3183 (1974).
15. W. Grosch, B. Höxer, H. Stan, and J. Schormüller, *Fette, Seifen, Anstrichm.* **74**, 16 (1972).
16. R. A. Galaway, Ph.D. Thesis, University of California, Riverside (1975).
17. S. Grossman, A. Ben-Aziz, P. Budowski, I. Ascarelli, A. Gertler, Y. Birk, and A. Bondi, *Phytochemistry* **8**, 2287 (1969).
18. R. Holman, *Methods Biochem. Anal.* **2**, 113 (1955).
19. J. Allen, *Chem. Commun.* **16**, 609 (1969).
20. J. Blain and G. Shearer, *J. Sci. Food Agric.* **16**, 373 (1965).
21. R. Holman, P. Egwim, and W. Christie, *J. Biol. Chem.* **244**, 1149 (1969).
22 M. Hamberg and B. Samuelsson, *J. Biol. Chem.* **242**, 5329 (1967).
23. A. Dolev, W. Rohwedder, and H. Dutton, *Lipids* **2**, 28 (1967); C. Eriksson and K. Leu, *ibid.* **6**, 144 (1971).
24. G. Veldink, J. Vliegenthart, and J. Boldingh, *Biochem. J.* **120**, 55 (1970); *Biochim. Biophys. Acta* **202**, 198 (1970).
25. J. Christopher and B. Axelrod, *Biochem. Biophys. Res. Commun.* **44**, 731 (1971).
26. J. Christopher, E. Pistorius, F. Regnier, and B. Axelrod, *Biochim. Biophys. Acta* **289**, 82 (1972).
27. D. Zimmerman and B. Vick, *Lipids* **5**, 392 (1970).
28. H. W.-S. Chan, C. T. Costaras, F. Prescott, and P. Swoboda, *Biochim. Biophys. Acta* **398**, 347 (1975).
29. M. Egmond, G. Veldink, J. Vliegenthart, and J. Boldingh, *Biochim. Biophys. Acta* **409**, 399 (1975).
30. H. Gardner and D. Weisleder, *Lipids* **8**, 678 (1970).
31. T. Galliard and D. Phillips, *Biochem. J.* **124**, 431 (1971).
32. C. Chang, W. Esselman, and C. Clagett, *Lipids* **6**, 100 (1971).
33. H. Gardner, D. Christianson, and R. Kleiman, *Lipids* **8**, 271 (1973).
34. M. Egmond, J. Vliegenthart, and J. Boldingh, *Biochem. Biophys. Res. Commun.* **48**, 1055 (1972).
35. M. Egmond, G. Veldink, J. Vliegenthart, and J. Boldingh, *Biochem. Biophys. Res. Commun.* **54**, 1178 (1973).
36. K. U. Ingold, *Acc. Chem. Res.* **2**, 1 (1969).
37. "JANAF Interim Thermochemical Tables." Dow Chemical Company, Thermal Laboratory, Midland, Michigan, 1960.
38. For a review of autoxidation, see W. G. Lloyd, *Methods Free-Radical Chem.* **4**, 1 (1973).
39. S. Bergstrom, *Ark. Kemi* **21A**, 1 (1946).

40. J. A. Howard and K. U. Ingold, *Can. J. Chem.* **45**, 793 (1967).
41. D. R. Kearns, *Chem. Rev.* **71**, 395 (1971).
42. C. S. Foote, *Acc. Chem. Res.* **1**, 104 (1968).
43. C. Walling and S. A. Buckler, *J. Am. Chem. Soc.* **77**, 6032 (1955).
44. A. G. Davies and B. P. Roberts, *in* "Free Radicals" (J. Kochi, ed.), Vol. 1, Chapter 10. Wiley, New York, 1973.
45. C. Walling and A. Cioffari, *J. Am. Chem. Soc.* **92**, 6609 (1970).
46. J. F. Garst, *in* "Free Radicals" (J. Kochi, ed.), Vol. 1, Chapter 9 (p. 529ff, discusses these reactions in detail). Wiley, New York, 1973.
47. G. A. Russell, A. G. Bemis, E. J. Geels, E. G. Janzen, and A. J. Moye, *Adv. Chem. Ser.* **75**, 174 (1968).
48. C. Walling, *Acc. Chem. Res.* **8**, 125 (1975).
49. G. A. Hamilton, *in* "Molecular Mechanisms of Oxygen Activation" (O. Hayaishi, ed.), Chapter 10, and reference therein. Academic Press, New York, 1974.
50. G. A. Hamilton, *Adv. Enzymol.* **32**, 55 (1969).
51. A very recent example of one elegant system is J. T. Groves and G. A. McCluskey, *J. Am. Chem. Soc.* **98**, 859 (1976).
52. F. Basolo, B. M. Hoffman, and J. A. Ibers, *Acc. Chem. Res.* **8**, 384 (1975), and references therein.
53. J. Kochi, *in* "Free Radicals" (J. Kochi, ed.), Vol. 1, Chapter 11, p. 642ff. Wiley, New York, 1973.
54. J. J. M. C. de Groot, G. A. Veldink, J. F. G. Vliegenthart, J. Boldingh, R. Wever, and B. F. van Gelder, *Biochim Biophys. Acta* **377**, 71 (1975).
55. A. Finazzi-Agro, L. Avigliano, G. A. Veldink, J. F. G. Vliegenthart, and J. Boldingh, *Biochim. Biophys Acta* **326**, 462 (1973).
56. M. R. Egmond, A. Finazzi-Agro, P. M. Fasella, G. A. Veldink, and J. F. G. Vliegenthart, *Biochim. Biophys. Acta* **397**, 43 (1975).
57. A. Finazzi-Agro, L. Avigliano, M. R. Egmond, G. A. Veldink, and J. F. G. Vliegenthart, *FEBS Lett.* **52**, 73 (1975).
58. J. J. M. C. de Groot, G. J. Garssen, G. A. Veldink, J. F. G. Vliegenthart, and J. Boldingh, *FEBS Lett.* **56**, 50 (1975).
59. M. R. Egmond, M. Brunori, and P. M. Fasella, *Eur. J. Biochem.* **61**, 93 (1976).
60. J. C. Allen, *Eur. J. Biochem.* **4**, 201 (1968).
61. F. Orthoefer and L. Dugan, *J. Sci. Food Agric.* **24**, 357 (1973).
62. M. J. Gibian and R. A. Galaway, *Biochemistry* **15**, 4209 (1976).
63. W. Smith and W. Lands, *Biochem. Biophys. Res. Commun.* **41**, 846 (1970).
63a. W. Smith and W. Lands, *J. Biol. Chem.* **247**, 1038 (1972).
64. J. Vanderhoek and W. Lands, *Biochim. Biophys. Acta* **296**, 382 (1973).
65. N. Uri, *Autoxid. Antioxid.* **1**, 157 (1961); L. Reich and S. Stivala, "Autoxidations of Hydrocarbons and Polyolefins," pp. 206–207. Dekker, New York, 1969.
66. R. Holman and O. Elmer, *J. Am. Oil Chem. Soc.* **24**, 127 (1947).
67. H. Mitsuda, K. Yasumoto, and A. Yamamoto, *Arch. Biochem. Biophys.* **118**, 664 (1967).
68. J. Blain and G. Shearer, *J. Sci. Food Agric.* **16**, 373 (1965).
69. D. Downing, J. Barve, F. Gunstone, F. Jacobsberg, and M. KenJie, *Biochim. Biophys. Acta* **280**, 343 (1972).
70. F. Weber, G. Laskawy, and W. Grosch, *Z. Lebensm.-Unters. -Forsch.* **155**, 142 (1974).
71. R. Nilsson and D. Kearns, *J. Am. Chem. Soc.* **94**, 1030 (1972).

72. C. Richter, A. Wendel, U. Weser, and A. Azzi, *FEBS Lett.* **51**, 300 (1975).
73. J. Teng and L. Smith, *J. Am. Chem. Soc.* **95**, 4060 (1973); L. Smith and J. Teng, *ibid.* **96**, 2640 (1974).
74. L. Smith, M. Kulig, and J. Teng, *Abstr. 1st Chem. Congr. North Am. Cont.*, *1975* Biol. Abstract No. 75 (1976).
75. J. Baldwin, J. Swallow, and H. Chan, *Chem. Commun.* p. 1407 (1971); *Proc. Robert A. Welch Found. Conf. Chem. Res.* **15**, 220–223 (1973).
76. A. Finazzi-Agro, C. Giovagnoli, P. DeSole, L. Calabrese, G. Rotilio, and B. Mondovi, *FEBS Lett.* **21**, 183 (1972); A. Finazzi-Agro, P. DeSole, G. Rotilio, and B. Mondovi, *Ital. J. Biochem.* **22**, 217 (1974).
77. J. McCord and I. Fridovich, *J. Biol. Chem.* **243**, 5753 (1968).
78. H. Gardner, *J. Agric. Food Chem.* **23**, 129 (1975).
79. G. Streckert and H. Stan, *Lipids* **10**, 847 (1975).
80. T. Sanders, H. Pattee, and J. Singleton, *Lipids* **10**, 568 (1975).
81. G. Garssen, J. Vliegenthart, and J. Boldingh, *Biochem. J.* **130**, 435 (1972).
82. D. Christianson and H. Gardner, *Lipids* **10**, 448 (1975).
83. J. De Groot, G. Garssen, J. Vliegenthart, and J. Boldingh, *Biochim. Biophys. Acta* **326**, 279 (1973).
84. M. Hayano, *in* "Oxygenases" (O. Hayaishi, ed.), p. 181. Academic Press, New York, 1962.
85. D. Swern, ed., "Organic Peroxides." Wiley (Interscience), New York, 1972.

CHAPTER

7

The Multiplicity of the Catalytic Groups in the Active Sites of Some Hydrolytic Enzymes

E. T. Kaiser and Y. Nakagawa

INTRODUCTION

In the context of this volume it seems appropriate to reflect on the remarkable progress seen in the last decade in the description of the molecular events involved in enzymatic catalysis. There is no doubt that a major factor contributing to the progress of this important area of bioorganic chemistry is the explosive growth of our knowledge of the three-dimensional structures of enzymes. With the less complex enzymes, at least, it is becoming commonplace to analyze the course of the reactions catalyzed in terms of the involvement of specific residues in the catalytic events.

From the standpoint of the mechanistically oriented bioorganic chemist, if one considers the more complex systems such as enzymes containing subunits capable of cooperative interactions or membrane-bound enzymes, a nearly virgin territory beckons. While enormous effort has been expended, for instance, on the quantitative description of cooperative effects in enzymatic catalysis, in general, explanations have not yet been advanced to account for the effects in terms of molecular mechanisms like those that have been written for the reactions of relatively simple enzymes such as chymotrypsin, papain, carboxypeptidase, and lysozyme. It takes little foresight to realize that the definition of the molecular events involved in the catalytic action of complex enzymes in the language of the organic chemist will be a field of bioorganic chemistry in which much progress is likely to be made during the next decade.

In view of these considerations a great deal of the research effort in our laboratory is now being concentrated on complex enzymes. Still, it must be conceded that numerous problems remain even with the simpler systems. For example, it has been by no means clear how much one can tamper with the nature of the catalytic groups at the active site and still retain a species capable of acting as an efficient enzymatic catalyst. Furthermore, we have not arrived at the point where we can say that the organic chemical models for even the most well understood enzymes like α-chymotrypsin really simulate both the binding and catalytic processes adequately.

In this chapter the focus is on our continuing attempts to gain a better understanding of pepsin, one of the relatively simpler enzymes we are studying. Besides the intrinsic interest in developing a mechanistic hypothesis for the action of this enzyme, through our research on pepsin we have been able to explore a fundamental question in enzymology. Can more than one catalytic apparatus function in the active site of a given hydrolytic enzyme? In other words, is it possible that multiple combinations of active-site residues are responsible in some cases for the catalytic behavior of a hydrolytic enzyme?

Because pepsin, the principal enzymatic constituent of gastric juice, was the first individual enzyme to be recognized as such and the second to be crystallized, it occupies a prominent position in the development of the field of enzymology [1-3]. Despite this long history, much less is known about the mechanism of pepsin action than is the case for other hydrolytic enzymes like the serine [4,5] or cysteine proteinases [6]. Thus, although several of the mechanistic pathways that have been proposed for pepsin-catalyzed hydrolytic reactions include the postulation of the intermediacy of covalent enzyme–substrate complexes (acyl enzymes and/or amino enzymes) and the arguments supporting the existence of covalent complexes seem reasonable, the direct detection of these species has posed a challenging problem until the present time. Furthermore, only recently has the three-dimensional structure of pepsin been examined at a high degree of resolution [7].

Since most peptidases have their pH optima near pH 7, the unusually low pH optima seen for the action of pepsin on many peptide substrates make studies on the mechanism of action of this enzyme especially interesting. The pH optimum for the peptidase activity of pepsin is often observed at about pH 2, although a variation in the value of the optimum from approximately pH 1.8 to 4.4 has been found in some instances. An exceptionally large number of free carboxyl groups is present in the pepsin molecule, and a combination of chemical modification data together with kinetic results has provided strong evidence that carboxyl groups play important roles in the catalytic action of this enzyme.

In the initial investigations that we performed on pepsin our efforts were concentrated on developing an understanding of the action of the enzyme in the hydrolysis of synthetic dipeptide substrates [8–10]. Our later studies have been concerned principally with the pepsin-catalyzed hydrolysis of sulfite esters [11–13], compounds found only a few years ago to be very reactive pepsin substrates [14]. Although the structures of the sulfite ester substrates are clearly quite different from those of the more usual peptide substrates of pepsin, the sulfites have certain features that make them particularly useful in mechanistic investigations. In the pH region where pepsin is active they are uncharged (neutral), and in many instances they are rapidly hydrolyzed by the enzyme. Also, the hydrolysis of sulfite esters derived from phenols can often be easily observed spectrophotometrically. Thus, it has been very attractive to pursue the elucidation of the catalytic process involved in the pepsin-catalyzed hydrolysis of sulfite esters with the ultimate goal of clarifying the relationship between the peptidase and the sulfite esterase activities of pepsin.

Much of the current research on pepsin in other laboratories is concentrated in two areas. In one type of endeavor, the very important secondary interactions between pepsin and its peptide and its oligopeptide substrates, which affect reaction rates enormously, are being examined [1–3]. In the other, the characteristics of the transpeptidase reactions catalyzed by pepsin are being scrutinized in the hope that evidence can be obtained that will bear on the possible mechanistic roles of acyl enzyme or amino enzyme intermediates [15–21]. In the pepsin-catalyzed hydrolysis of sulfite esters secondary interactions of the substrates with the enzyme do not assume the importance that they have with peptide substrates, and the catalytic effects of the active-site residues of the enzyme can be examined in a very direct fashion. Of course, to get the most complete picture of the way in which pepsin acts one has to have a good understanding of both the reactivity of active-site functional groups and the secondary interactions of the enzyme–substrate complex.

RESULTS AND DISCUSSION

Our first investigation of the pH dependency of the hydrolysis of sulfite esters catalyzed by pepsin was concerned with the symmetrical ester bis-p-nitrophenyl sulfite [11,12]. The enzyme-catalyzed hydrolysis of this compound proceeded several orders of magnitude more rapidly than that of other known sulfite ester substrates over the pH range in which pepsin is active. The pH dependency of k_{cat}/K_m for the reaction of bis-p-nitrophenyl sulfite with pepsin showed a bell-shaped profile, and the pK values calculated for the

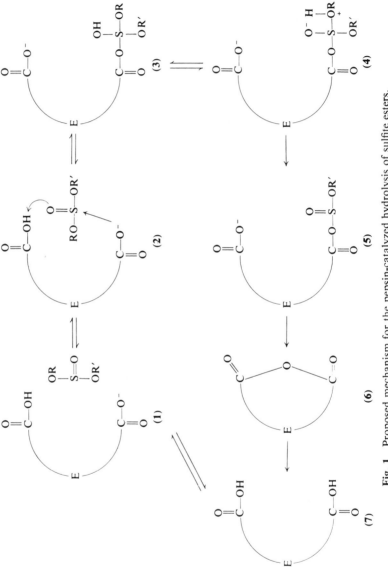

Fig. 1 Proposed mechanism for the pepsin-catalyzed hydrolysis of sulfite esters.

ionizing groups on the enzyme responsible for this behavior were $pK_{E_1} = 0.8$ and $pK_{E_2} = 5.2$. The similarity between these results and those reported for neutral peptide substrates [22–25] suggested that there are certain mechanistic features common to both the peptidase and the sulfite esterase activities of pepsin. From the kinetic data available and from a consideration of the usual interpretation of pH–rate profile measurements like those we have made for bis-p-nitrophenyl sulfite [12], it seemed reasonable to postulate [13] that two carboxyl groups on the free enzyme, one in its acidic form and the other in its basic form, are involved in each process. The mechanism for the sulfite esterase action of pepsin which we proposed on this basis is shown in Fig. 1. A prominent feature of this mechanism is the postulation of the formation of a mixed anhydride species (5) as a reactive intermediate.

(8)   (9)

(10)

In the hope that the examination of the reactions of pepsin with unsymmetrical sulfite esters, containing alcoholate moieties with widely differing leaving tendencies, might lead to the discovery of systems in which the enzyme-bound intermediate anhydrides (5) could be directly observed, we studied the interaction of these compounds with pepsin. However, the asymmetry at the sulfur in the unsymmetrical sulfites poses an interesting complication. When pepsin discriminates between the enantiomers of the racemic unsymmetrical sulfites, kinetic measurements on the hydrolysis of these compounds can involve the observation of the simultaneously occurring reactions of the two enantiomers present, which may proceed at different rates. In some instances, such as the pepsin-catalyzed hydrolysis of phenyl tetrahydrofurfuryl sulfite, one enantiomer reacts much faster with the enzyme than does the other, and it has been possible to isolate the unreactive enantiomer [26]. We have found, however, that in the pepsin-catalyzed hydrolysis of phenyl p-nitrophenyl sulfite (9) and methyl p-nitrophenyl sulfite (10) the discrimination between enantiomers is far smaller.

Specifically, our measurements have shown that over the pH range 3.0–4.2 the value of the rate parameter k_{cat}/K_m for the more rapidly reacting enantiomer of (9) is roughly eight times larger than that for the slowly reacting enantiomer. Our data show that the value for pK_{E_2} for both the rapidly and slowly hydrolyzed enantiomers of (9) are close to the pK_{E_2} value of 5.2 seen for (8). We have made similar observations for the pH dependence for the pepsin-catalyzed hydrolysis of one enantiomer of (10). Although our results need additional refinement, it appears, surprisingly, that the value of k_{cat}/K_m for the hydrolysis of the other enantiomer of (10) is nearly pH independent over the pH range 2.6–5.1. Because of this, at pH values near 5 there is little enzymatic discrimination between the two enantiomers of (10), while at pH 2.6 one enantiomer reacts three times more rapidly than the other. Despite the complications due to the enantiomeric specificity of pepsin, we are currently examining the possibility that in mixed organic–aqueous solvents where we have found that the enzyme can function very effectively we may be able at subzero temperatures to use the unsymmetrical sulfites to detect spectrophotometrically the formation and decomposition of anhydride species like structure (5) of Fig. 1.

In parallel with our enzymatic studies we have examined the catalytic effects of model carboxylate buffers on the hydrolysis of sulfite esters [27]. For example, in the monocarboxylate ion-catalyzed hydrolysis of diphenyl sulfite a substantial Brönsted β value (0.8) was seen. Also, for the acetate-catalyzed reaction a small kinetic solvent isotope effect, $(k_{OAc})^{H_2O}/(k_{OAc})^{D_2O} = 1.3$, was found. These observations led to the suggestion that the carboxylate ion-catalyzed reactions proceed by way of anhydride intermediates (11) [Eq. (1)] related to the proposed enzyme-bound intermediate (5) of Fig. 1. Support for this proposal that the nucleophilic attack of model carboxylate ions on sulfite esters leads to the formation of mixed anhydride intermediates has been obtained by the detection of the acetohydroxamic acid produced when diphenyl sulfite, bis-p-nitrophenyl sulfite (8), and phenyl p-nitrophenyl sulfite (9) are solvolyzed in acetate buffer in the presence of hydroxylamine [Eq. (2)]. In the case of diphenyl sulfite, measurement of the rate of aceto-hydroxamic acid production has indicated that under the conditions employed the rate-controlling step is mixed anhydride formation.

Extension of our kinetic measurements to the catalytic action on diphenyl sulfite of other oxygen nucleophiles including hydroxide ion, carbonate ion, and water revealed that over a pK range of about 17 units the rate constants fall on essentially the same Brönsted plot as that obtained for the mono-carboxylate species, suggesting that a very wide range of nucleophiles catalyzes sulfite ester hydrolysis by a nucleophilic pathway [28]. Due to the operation of α effects, the points obtained for hydroxylamine and for anti-α-morphilinoacetophenone oxime showed rate enhancements of 86- and 133-fold, respectively. In the reactions of the sulfite esters diphenyl sulfite and phenyl p-nitrophenyl sulfite with hydroxylamine the direct detection of

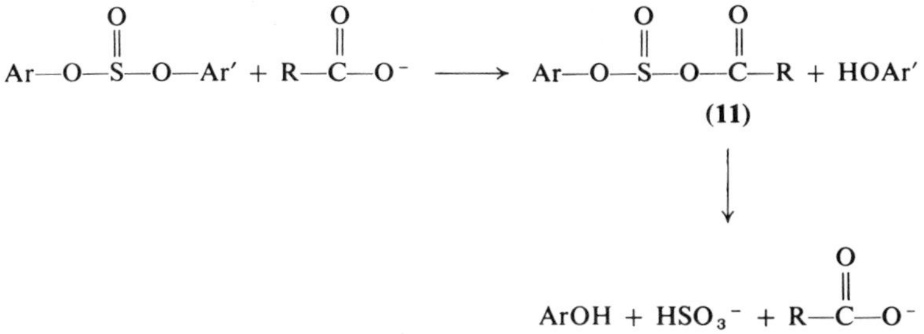

$$\underset{\substack{\| \\ O}}{Ar\!-\!O\!-\!S\!-\!O\!-\!Ar'} + \underset{\substack{\| \\ O}}{R\!-\!C\!-\!O^-} \longrightarrow \underset{\substack{\| \quad\; \| \\ O \quad O}}{Ar\!-\!O\!-\!S\!-\!O\!-\!C\!-\!R} + HOAr'$$

(11)

(1)

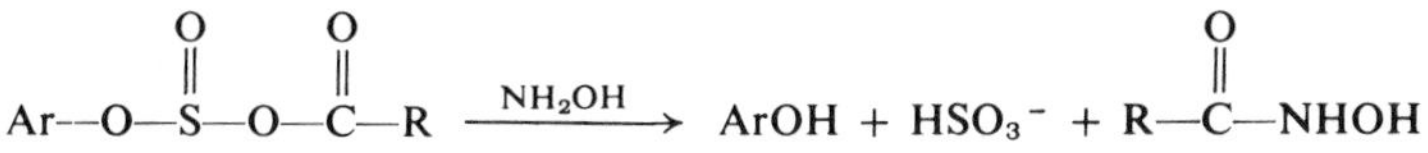

$$ArOH + HSO_3^- + \underset{\substack{\| \\ O}}{R\!-\!C\!-\!O^-}$$

$$\underset{\substack{\| \quad\; \| \\ O \quad O}}{Ar\!-\!O\!-\!S\!-\!O\!-\!C\!-\!R} \xrightarrow{\;NH_2OH\;} ArOH + HSO_3^- + \underset{\substack{\| \\ O}}{R\!-\!C\!-\!NHOH} \quad (2)$$

intermediates [Eq. (3)] has been accomplished [27]. The rate constants for intermediate formation and decomposition were obtained by a computer-assisted kinetic analysis in the case of diphenyl sulfite and by the observation of the reaction course at different wavelengths in the case of phenyl *p*-nitrophenyl sulfite (9). The rate constants measured for the hydroxylamine-catalyzed decomposition of the intermediates formed from the two sulfite esters were the same within experimental error, which indicates, not surprisingly, in combination with the observation that *p*-nitrophenol is released in the first step of the reaction of (9) with hydroxylamine that the intermediates have identical structures (12a) or (12b). While at first glance it might seem that if the intermediate had the structure shown in formula (12b), containing a sulfur–nitrogen bond, it would not be very labile, the probability appears high that it could rapidly react via hydroxylamine catalysis to give phenol by an elimination process involving the ionization of the proton on its nitrogen atom. In the case of the solvolysis of phenyl *p*-nitrophenyl sulfite in *N,O*-dimethylhydroxylamine buffer, by the use of kinetic and synthetic experiments the reaction has been shown to proceed through the formation of phenyl methoxymethylamidosulfite (13). The demonstration that reactive intermediates could be detected in model solvolytic reactions of sulfite esters encouraged us to search for conditions under which intermediates formed by nucleophilic attack at the active-site enzymatic carboxylate groups could be detected directly. As is discussed shortly, we have had success in this research recently.

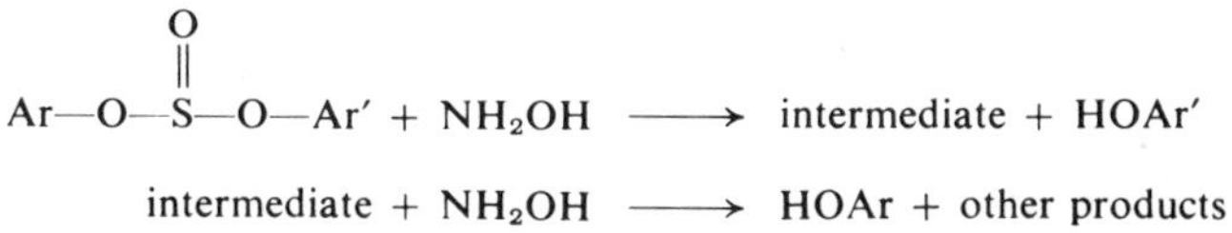

$$\underset{\substack{\| \\ O}}{Ar\!-\!O\!-\!S\!-\!O\!-\!Ar'} + NH_2OH \longrightarrow intermediate + HOAr'$$

intermediate + NH_2OH \longrightarrow HOAr + other products

(3)

Turning now from the model system back to the enzymatic reactions, one of the questions we have felt to be crucial concerns the relationship between

(12a)   (12b)

(13)

the active sites of pepsin acting as a sulfite esterase and as a peptidase, a topic only briefly alluded to above. Three principal lines of evidence have been adduced in support of the hypothesis that the active-site requirements for the sulfite esterase action of the enzyme are the same as those for its peptidase action. (a) Slowly hydrolyzed peptides binding to the active site of pepsin have been shown to act as competitive inhibitors toward the sulfite esterase activity of the enzyme, and the inhibition constants obtained with these peptides correspond closely with the Michaelis constants calculated from their pepsin-catalyzed hydrolysis [10,14]. (b) As mentioned before, the pH dependency of the rate parameter k_{cat}/K_m for the pepsin-catalyzed hydrolysis of the reactive sulfite ester substrate bis-p-nitrophenyl sulfite agrees fairly well with the pH dependency of this parameter for the hydrolysis of neutral dipeptides like N-acetyl-L-phenylalanylamide [24]. (c) The diazocarbonyl reagent N-diazoacetyl-DL-norleucine methyl ester, a reagent that causes the inactivation of pepsin as a peptidase also is known to inactivate the enzyme as a catalyst for the hydrolysis of diphenyl sulfite and methyl phenyl sulfite at pH 2 [14].

On further examination of the third line of evidence cited above, however, we discovered that pepsin esterified at active-site carboxylate groups by treatment with either the diazoketone α-diazo-p-bromoacetophenone (14) [pepsin M–(14)] in the presence of cupric ion [29] or the epoxide 1,2-epoxy-3-(p-nitrophenoxy)propane (15) [pepsin M–(15)] [30,31] at pH 5 and 25°C, retaining less than 1% activity toward the peptide substrate hemoglobin [32], remains very active over a range of pH values as a catalyst for the hydrolysis of sulfite ester substrates including bis-p-nitrophenyl sulfite, phenyl p-nitrophenyl sulfite, and methyl p-nitrophenyl sulfite.

Under conditions of enzyme in excess, pseudo first-order kinetics were observed and no evidence for enantiomeric specificity was seen when the hydrolysis of the unsymmetrical sulfite ester methyl p-nitrophenyl sulfite (10) catalyzed by pepsin M–(14) or pepsin M–(15) was monitored. The k_{cat}/K_m values measured for the action of the modified pepsin species ascended from the more acidic to the less acidic reaction conditions. In the case of pepsin

M–(14) an ionization with $pK_a = 3.9$ was measured from the kinetic studies on methyl p-nitrophenyl sulfite and $(k_{cat}/K_m)_{lim} = (9.1 \pm 2.4) \times 10^2$ M^{-1} sec^{-1} was found. For pepsin M–(15) with this substrate, $pK_a = 4.3$ and $(k_{cat}/K_m)_{lim} = (1.7 \pm 0.2) \times 10^3$ M^{-1} sec^{-1}. As described above, in some contrast to the behavior of the modified enzymes, native pepsin showed significant enantiomeric specificity in its reaction with (10) at low pH values, although at pH 5 this specificity was not seen. Also, measurements on the native pepsin-catalyzed hydrolysis of the fast-reacting enantiomer of methyl p-nitrophenyl sulfite indicated that the k_{cat}/K_m values in this case decreased at the higher pH values.

(14)   (15)

Although the pH dependencies of the kinetic behavior toward sulfite esters of pepsin modified by other diazocarbonyl reagents have not been examined by us as extensively as that of pepsin M–(14), we have found at least qualitative similarities using other modified pepsins. The amino acid residue in pepsin modified by the diazocarbonyl reagents is known to be Asp-215 [29,33–35] and that modified by 1,2-epoxy-3-(p-nitrophenoxy)propane is Asp-32 [30,31,34]. From our results with the modified pepsin species it is clear that the β-carboxyl groups of neither of these two aspartates, which are believed to be important active-site residues in the peptidase action of pepsin, are essential to the catalytic action of the enzyme on sulfite ester substrates. A possible interpretation of the sigmoidal (k_{cat}/K_m)–pH profiles obtained with pepsin M–(14) and pepsin M–(15) is that only a single active-site carboxylate group is necessary for the hydrolysis of sulfite esters. Going back to the mechanism of Fig. 1, therefore, whether acidic catalysis by a carboxyl group, as implied by structure (2), is important to the sulfite esterase action of pepsin is questionable. The ionization of this carboxyl group might be reflected in the kinetics of reaction primarily because of its inhibitory effect on catalysis. In any event, although the (k_{cat}/K_m)–pH dependencies for the action of the modified pepsins and that of pepsin itself on a substrate like methyl p-nitrophenyl sulfite are quite different, the $(k_{cat}/K_m)_{lim}$ values found in our work with the former species lie between the values obtained for the hydrolysis of the fast- and slow-reacting enantiomers of this compound catalyzed by the native enzyme.

At this point, several possible hypotheses consistent with our results might be proposed. One is that the carboxylate group(s) required for the sulfite esterase activity of pepsin differs from that (those) necessary for peptidase activity in spite of the evidence summarized above that the active sites for the two types of activities overlap. A second possibility is that carboxyl groups in pepsin in addition to those of Asp-32 and Asp-215 can function as effective catalytic species in sulfite ester hydrolysis. Perhaps when Asp-32 and Asp-215 are modified there is somewhat of a change in the geometry of the active site which brings another carboxyl group into play, at least as far as sulfite esters are concerned. Yet another hypothesis might be that despite the inactivation of pepsin as a peptidase due to the esterification of Asp-32 and Asp-215 the β-carboxylate functions of these residues in their unmodified forms are not catalytic participants in the peptidase action of the enzyme. If this postulate were correct, then the differences between the (k_{cat}/K_m)–pH profiles for the action of pepsin M–(14) and M–(15) on methyl p-nitrophenyl sulfite (10) and the action of the native enzyme on a variety of substrates [1] would suggest that the esterification of a carboxyl group in the vicinity of the active site causes a substantial alteration in the pK_a for the ionization of the catalytically active carboxyl function.

Faced with a wealth of intriguing explanations for the catalytic activity of the modified pepsins on sulfite esters, we decided to examine as directly as we could the question of whether covalent intermediates might be detected in the native enzyme catalysis of sulfite ester hydrolysis and what residues might be involved in the production of these intermediates. As mentioned before, our success with the use of hydroxylamine as a trapping agent for the detection of anhydride intermediates in the catalytic action of model carboxylate species prompted us to employ this approach in the enzymatic reactions as well.

The nucleophile trapping approach, however, is not without its difficulties. In analogy to Eq. (2), we hoped that hydroxylamine might attack the carbonyl function of the enzyme–substrate mixed anhydride [structure (5) of Fig. 1] as illustrated in Eq. (4), resulting in the introduction of a hydroxamic acid group at a catalytically important carboxylate residue of the enzyme. The possibility has to be considered, however, that there might be a preference for hydroxylamine to attack the sulfite sulfur in the enzyme-bound mixed anhydride, in which case hydroxylamine might not be introduced into the enzyme. Also, as seen in Eq. (3), hydroxylamine can react directly with sulfite esters, and if this reaction competes too effectively with the process of Eq. (4), then again no labeling of the enzyme might be observed when hydroxylamine is incubated with pepsin and sulfite esters. Last, of course, despite the analogy with the catalysis of sulfite ester hydrolysis by model carboxylate ions, it must be considered that before doing the trapping experiments with the enzyme there was no truly compelling argument that the enzymatic reactions proceeded through mixed anhydride species. Despite these problems, we have been able to find conditions under which nucleophile trapping by hydroxylamine in the pepsin-catalyzed hydrolysis of sulfites was achieved with facility.

$$
\underset{\textbf{(5)}}{E}\left\{\begin{array}{l} \overset{O}{\overset{\|}{C}}-O^- \\[6pt] \\ \text{Hydroxylamine} \\[6pt] \\ \overset{}{C}-O-\overset{}{S}-OR' \\ \overset{\|}{O}\quad\overset{\|}{O} \end{array}\right. \longrightarrow \quad E\left\{\begin{array}{l} \overset{O}{\overset{\|}{C}}-O^- \\[6pt] \\ \\[6pt] \\ \overset{}{C}-NHOH \\ \overset{\|}{O} \end{array}\right. \quad + R'O-\overset{\overset{\displaystyle O}{\|}}{\underset{\underset{\displaystyle O}{\|}}{S}}-O^- \qquad \textbf{(4)}
$$

In the experiments performed, various ratios (2.5–11.7) of phenyl tetra-hydrofurfuryl sulfite (16) to pepsin were employed, and these mixtures were incubated with 0.01 M hydroxylamine in 2-(N-morpholino)ethanesulfonic acid buffer, pH 5.3. The duration of single incubation experiments was 30 min, and for multiple incubation experiments additional substrate was added at 30-min intervals. In a single incubation experiment at a molar ratio of (16) to pepsin = 2.5 about 20% of the peptic activity in the hydrolysis of hemoglobin was lost and concomitantly 1 mole of hydroxamate per mole of pepsin was found. This result is consistent with the reaction illustrated in Eq. (4), in which the potent nucleophile hydroxylamine is postulated to attack the mixed anhydride intermediate (5) of Fig. 1. With increases in the number of incubations performed with phenyl tetrahydrofurfuryl sulfite (16) or increases in the concentration of (16), the peptic activity toward hemo-globin was observed to decrease with a concomitant increase in the number of enzyme-bound hydroxamate groups produced, up to a maximum value of 3 to 4 moles per mole of pepsin. At pH 5 the more highly modified pepsin species retained about 30–40% of the activity of the native enzyme toward diphenyl sulfite.

$$
\text{⬡}-O-\overset{\overset{\displaystyle O}{\|}}{S}-O-CH_2-\text{⬠}
$$

(16)

In the next stage of our work we sought to determine the relationship of the carboxyl groups involved in hydroxamate formation to those identified in earlier chemical modification experiments as important residues. Following the procedure of Gross and Morell [36], the hydroxamate-containing pepsin species obtained in the hydroxylamine trapping experiments performed were subjected to Lossen rearrangement, followed by acid hydrolysis. With all samples studied, the amino acid analyses performed on the hydrolyzates revealed that only 2,3-diaminopropionic acid was formed. No 2,4-diamino-butyric acid was detected. These degradation results show that the inter-mediates formed in the reaction of phenyl tetrahydrofurfuryl sulfite with pepsin and trapped with hydroxylamine must have been formed at the β-carboxyl groups of aspartate residues.

Having shown that aspartate residues are involved in anhydride formation in the pepsin-catalyzed hydrolysis of phenyl tetrahydrofurfuryl sulfite, we focused our studies on the identification of the particular residues taking part in catalysis. In a typical experiment, a solution of the hydroxamate-containing pepsin (3.3 moles of hydroxamate per mole of pepsin) obtained from two incubations of phenyl tetrahydrofurfuryl sulfite (16) with pepsin [mole ratio of (16) to pepsin in each incubation = 2.2] was digested by native pepsin at pH 3.5 and subjected to gel filtration through Sephadex G-25. Subsequent high-voltage electrophoresis and descending paper chromatography led to the detection of three hydroxamate-containing peptides. One peptide con-tained Asp, Thr, Ser, Gly, Val, Ile, Phe. This amino acid analysis is consistent with the composition of a fragment of pepsin (Val-29 → Ser 35) containing the active-site residue Asp-32, the group known to be esterified by 1,2-epoxy-3-(p-nitrophenoxy)propane (15) [30,31,34]. The composition of the second peptide was Asp_1, Thr_1, Ser_1, Gly_3, Ala_1, Glu_1, Val_1, Ile_1, Leu_1. This seems to correspond to a fragment of pepsin (Gly-208 → Leu-220) containing the active-site residue Asp-215, the group modified by diazocarbonyl reagents [29,33,35,37]. (While this sequence contains a cysteine residue, and a cysteine peak was observed in the amino acid analysis of the peptide fragment, the peak height was too small for computation. The color factor for cysteine is one-half that for other amino acids except proline. Furthermore, the analysis values obtained for serine and threonine were uncorrected ones, and therefore this peptide fragment might be a somewhat longer chain than that indicated here.)

We have drawn three main conclusions from our experiments. (a) Our observation of hydroxamate formation in the pepsin-catalyzed hydrolysis of phenyl tetrahydrofurfuryl sulfite constitutes the first direct demonstration that anhydrides are intermediates in the hydrolysis of sulfite esters. Thus, support is provided for this aspect of the mechanism of Fig. 1, and this finding is in good agreement with circumstantial evidence that acyl enzymes [15–18] (anhydrides) may be intermediates in at least some pepsin-catalyzed transpeptidation reactions. (b) The correspondence of two of the aspartate residues involved in hydroxamate formation to those aspartates (Asp-32 and Asp-215), the esterification of which is known to inactivate pepsin as a peptidase, not only shows that these residues are indeed direct participants in the hydrolysis of sulfite esters but also strongly supports the hypothesis that the active site of pepsin as a sulfite esterase overlaps with that for its action as a peptidase. (c) Most importantly, our observation that as many as 3–4 moles of hydroxamate can be incorporated per mole of pepsin shows that anhydride formation can occur at several carboxyl groups in pepsin-catalyzed sulfite ester hydrolysis. This finding is consistent with the discovery discussed above that pepsin esterified either at Asp-32 [pepsin M–(15)] or at Asp-215 [pepsin M–(14)] can function as a catalytically active species toward sulfite esters.

An intriguing problem that has emerged from the nucleophile trapping experiments is to determine why pepsin in which hydroxamate groups have been introduced at the active site carboxyls can still be quite effective as a catalyst for hemoglobin hydrolysis, as mentioned briefly above. One explanation of this observation might be that, just as we have shown for the hydrolysis of sulfite esters, a carboxyl group other than the β-carboxyls of Asp-32 or Asp-215 can participate in the peptidase action of pepsin, at least when the latter carboxyl groups have been altered by hydroxamate formation. The reasons why the modified species pepsin M–(14) and M–(15) are not active as peptidases even though they can function as sulfite esterases may have to do with unfavorable steric effects of the large esterifying groups, which are not present in the case of the much smaller hydroxamate functions. Alternatively, one must consider the possibility at this point that the hydroxamate groups introduced at the active-site aspartate residues in the nucleophilic trapping experiments can function themselves as catalysts in the hydrolysis of hemo-globin. This hypothesis, now being tested in our laboratory, is closely related to the suggestion made by Gruhn and Bender in connection with their studies on ester hydrolysis catalyzed by N-methylacetohydroxamic acid that the introduction of an N-alkylhydroxamic acid function at the active site of pepsin might lead to an enzyme with new functional group specificity [38].

CONCLUSIONS

The question posed earlier in this chapter has been answered affirmatively. More than one catalytic apparatus can function in the active site of a given hydrolytic enzyme. Our results on the enzymatic hydrolysis of phenyl tetra-hydrofurfuryl sulfite in the presence of hydroxylamine clearly demonstrate that several carboxylate groups in the active site of pepsin can function as reactive nucleophiles in catalyzing the hydrolysis of sulfite esters. Because our observations on the catalytic activity toward hemoglobin of pepsin in which hydroxamate groups have been incorporated at Asp-32 and Asp-215 remain to be explained, we do not know at present whether this conclusion can be extended to the peptidase action of pepsin. However, the related finding by

Suh and Kaiser [39] that carboxypeptidase A can act on different enantiomers of a thiolester substrate through different mechanisms strongly suggests that there may be other cases in which multiple catalytic apparatuses exist in the active site of a hydrolytic enzyme. From the standpoint of attempting to tamper with the nature of the catalytic groups at the active site of an enzyme with retention of efficient enzymatic catalysis, we feel that our observations offer considerable encouragement. Indeed, recently [40] we have found that by the expedient of introducing a covalently bound coenzyme analog either on the periphery or right at the active site of a hydrolytic enzyme we can obtain a modified species capable of carrying out oxidation–reduction reactions.

ACKNOWLEDGMENTS

The support of this research by the National Science Foundation is gratefully acknowledged. We wish to thank also the former graduate students and postdoctoral associates whose research contributions are recorded here.

REFERENCES

1. J. S. Fruton, in "The Enzymes" (P. D. Boyer, ed.), 3rd ed., Vol. 3, pp. 120–164. Academic Press, New York, 1971.
2. J. S. Fruton, in "Proteases and Biological Control" (E. Reich, D. B. Rifkin, and E. Shaw, eds.), pp. 35–50. Cold Spring Harbor Lab., Cold Spring Harbor, New York, 1975.
3. J. S. Fruton, Acc. Chem. Res. 7, 214 (1974).
4. G. P. Hess, in "The Enzymes" (P. D. Boyer, ed.), 3rd ed., Vol. 3, pp. 213–248. Academic Press, New York, 1971.
5. M. L. Bender and F. J. Kézdy, Annu. Rev. Biochem. 34, 49 (1965).
6. A. N. Glazer and E. L. Smith, in "The Enzymes" (P. D. Boyer, ed.), 3rd ed., Vol. 3, pp. 502–546. Academic Press, New York, 1971.
7. N. S. Andreeva, A. A. Fedorov, A. E. Guschchina, N. E. Shutskever, R. R. Riskulov, and T. V. Volnova, Dokl. Akad. Nauk SSSR 228, 480 (1976).
8. E. Zeffren and E. T. Kaiser, J. Am. Chem. Soc. 88, 3129 (1966).
9. E. Zeffren and E. T. Kaiser, J. Am. Chem. Soc. 89, 4204 (1967).
10. E. Zeffren and E. T. Kaiser, Arch. Biochem. Biophys. 126, 965 (1968).
11. S. W. May and E. T. Kaiser, J. Am. Chem. Soc. 91, 6491 (1969).
12. S. W. May and E. T. Kaiser, J. Am. Chem. Soc. 93, 5567 (1971).
13. S. W. May and E. T. Kaiser, Biochemistry 11, 592 (1972).
14. D. Fahrney and T. Reid, J. Am. Chem. Soc. 89, 3941 (1967).
15. T.-T. Wang and T. Hofmann, Biochem. J. 153, 691 (1976).
16. T.-T. Wang and T. Hofmann, Biochem. J. 153, 701 (1976).
17. M. Takahashi and T. Hofmann, Biochem. J. 147, 549 (1975).
18. M. Takahashi, T.-T. Wang, and T. Hofmann, Biochem. Biophys. Res. Commun. 57, 39 (1974).

19. V. K. Antonov, L. D. Rumsh, and A. G. Tikhodeev, *FEBS Lett.* **46**, 29 (1974).
20. A. K. Newmark and J. R. Knowles, *J. Am. Chem. Soc.* **97**, 3557 (1975).
21. M. S. Silver and M. Stoddard, *Biochemistry* **14**, 614 (1975).
22. G. E. Clement, S. L. Snyder, H. Price, and R. Cartmell, *J. Am. Chem. Soc.* **90**, 5603 (1968).
23. G. E. Clement, J. Rooney, D. Zakheim, and J. Eastman, *J. Am. Chem. Soc.* **92**, 186 (1970).
24. A. J. Cornish-Bowden and J. R. Knowles, *Biochem. J.* **113**, 353 (1969).
25. J. L. Denburg, R. Nelson, and M. S. Silver, *J. Am. Chem. Soc.* **90**, 479 (1968).
26. T. W. Reid, T. P. Stein, and D. Fahrney, *J. Am. Chem. Soc.* **89**, 7125 (1967).
27. L.-H. King and E. T. Kaiser, *J. Am. Chem. Soc.* **96**, 1410 (1974).
28. M. C. Rykowski, K. T. Douglas, and E. T. Kaiser, *J. Org. Chem.* **41**, 141 (1976).
29. B. F. Erlanger, S. M. Vratsanos, N. Wasserman, and A. G. Cooper, *Biochem. Biophys. Res. Commun.* **28**, 203 (1967).
30. J. Tang, *J. Biol. Chem.* **246**, 4510 (1971).
31. K. C. S. Chen and J. Tang, *J. Biol. Chem.* **247**, 2566 (1972).
32. M. L. Anson, *J. Gen. Physiol.* **22**, 79 (1938).
33. T. G. Rajagopalan, W. S. Stein, and S. Moore, *J. Biol. Chem.* **241**, 4295 (1966).
34. P. Sepulveda, J. Marciniszyn, Jr., D. Liu, and J. Tang, *J. Biol. Chem.* **250**, 5082 (1975).
35. A. F. Paterson and J. R. Knowles, *Eur. J. Biochem.* **31**, 510 (1972), and references therein.
36. E. Gross and J. L. Morell, *J. Biol. Chem.* **241**, 3638 (1966).
37. Y. Nakagawa, L.-H. King Sun, and E. T. Kaiser, *J. Am. Chem. Soc.* **98**, 1616 (1976).
38. W. B. Gruhn and M. L. Bender, *J. Am. Chem. Soc.* **91**, 5883 (1969).
39. J. Suh and E. T. Kaiser, *Biochem. Biophys. Res. Commun.* **64**, 863 (1975).
40. H. Levine, Y. Nakagawa, and E. T. Kaiser, *Biochem. Biophys. Res. Commun.* (in press).

CHAPTER

8

Multifunctionality and Microenvironments in the Catalytic Hydrolysis of Phenyl Esters

Toyoki Kunitake

INTRODUCTION

This chapter is an account of our recent research efforts, which aim at preparing highly efficient hydrolytic catalysts.

Enzymatic catalysis is in many respects superior to the catalytic action of simpler organic compounds. Among the numerous enzymes known, serine proteases belong to a class of enzymes that have been investigated most extensively. Probably the most important feature of the catalytic mechanism of serine protease action is the presence of the charge-relay system as proposed by Blow *et al.* [1] in 1969. This system is composed of a hydrogen-bonding series among the seryl hydroxyl, histidyl imidazole, and aspartyl carboxylate groups, as shown in Fig. 1.

In the hydrolysis of specific amide substrates, the seryl hydroxyl group attacks the amide carbonyl group with concomitant protonation of the amide nitrogen. The acyl-enzyme intermediate thus formed is hydrolyzed through general base action of the histidyl imidazole. The remarkable efficiency of these processes is currently explained in terms of multifunctionality and the microenvironment of the catalytic site. Therefore, our efforts were directed toward the combined use of these features in synthetic catalysts.

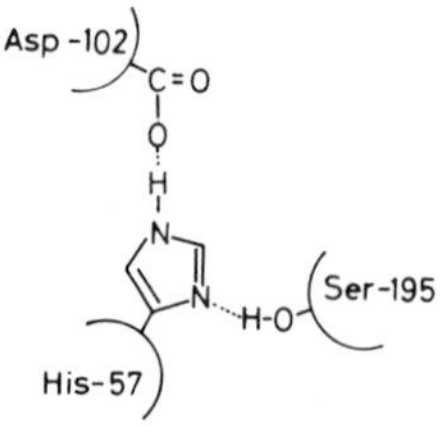

TOYOKI KUNITAKE

Fig. 1   Charge-relay system of serine proteases.

COMPLEMENTARY BIFUNCTIONAL CATALYSIS

Nucleophilic catalysis of the hydrolysis of esters and amides proceeds via the formation and decomposition of the acyl intermediate, and it is efficient only when both the acylation and deacylation processes are fast. This is difficult with a single functional group because a good nucleophile is usually not a good leaving group. Then an efficient catalytic system may be realized by a combination of complementary functional groups: a good nucleophile and a second functional group which assists the decomposition of the acyl intermediate.

An earlier example of complementary bifunctional catalysis is that of Schonbaum and Bender [2]. The catalytic hydrolysis of *p*-nitrophenyl acetate by the *o*-mercaptobenzoic acid (**1**) dianion involves acetyl transfer to the thiolate anion and the subsequent deacylation due to general base catalysis of the neighboring carboxylate group [3]. Several bifunctional catalysts containing thiol and imidazole groups [(**2**)–(**4**)] were similarly employed as papain models [4–7]; however, noteworthy rate enhancements were not reported.

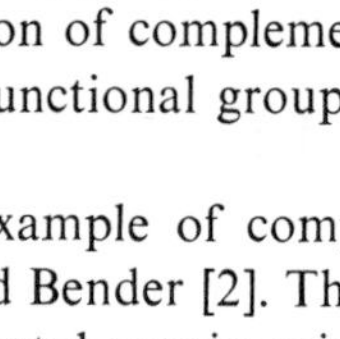

(1) [2]

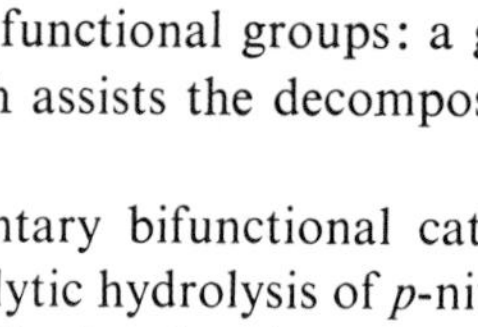

(2) [4,5] 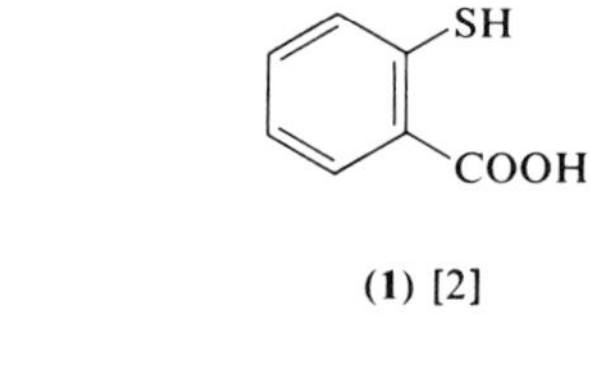
($n = 1, 2$)

(3) [6]
($n = 1, 2$)

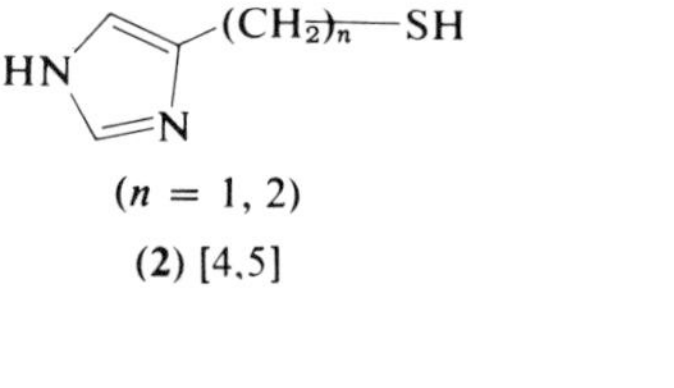

(4) [7]

The hydroxamate anion is known to be a powerful oxygen nucleophile [8], and its combination with second complementary functions would give efficient catalysts. Gruhn and Bender [9] prepared *N*-(2-dimethylaminoethyl)-acetohydroxamic acid (**5**) and examined its catalytic activity in the hydrolysis

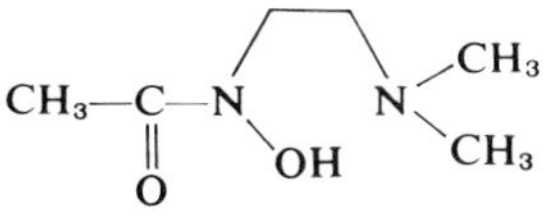

(**5**)

of *p*-nitrophenyl acetate. The initial step of the catalysis is acyl transfer to the hydroxamate anion, and the acetyl hydroxamate intermediate is decomposed by the intramolecular catalysis of the dimethylamino group.

The combination of the oxygen nucleophile and the imidazole function would be an interesting catalytic system in connection with the charge-relay system. Thus, we synthesized several bifunctional catalysts of this type and used them in the hydrolysis of phenyl esters [10–14]. Some of our bifunctional catalysts [(**6**)–(**8**), PHA-MIm-AAm (**10**), PHA-VIm-AAm (**11**)] are shown below, together with a bifunctional cyclodextrin (**9**) of Kitaura and Bender [15].

Small-molecule catalysts

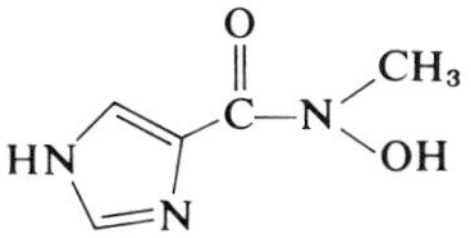

(**6**) [10] 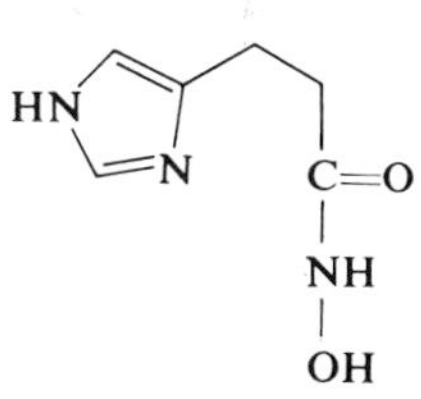   (**7**) [11]

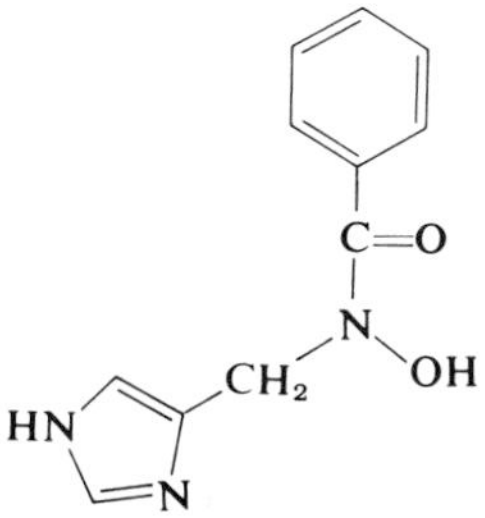

(**8**) [12] 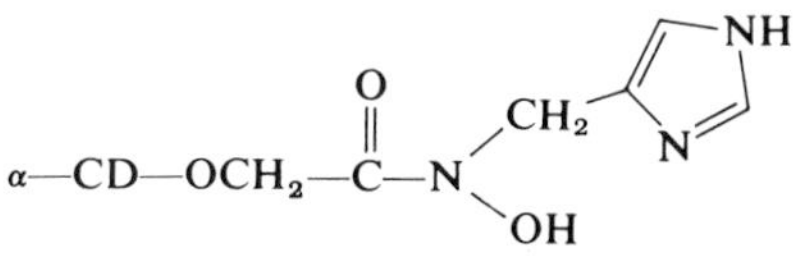   (**9**) [15]

Polymer catalysts

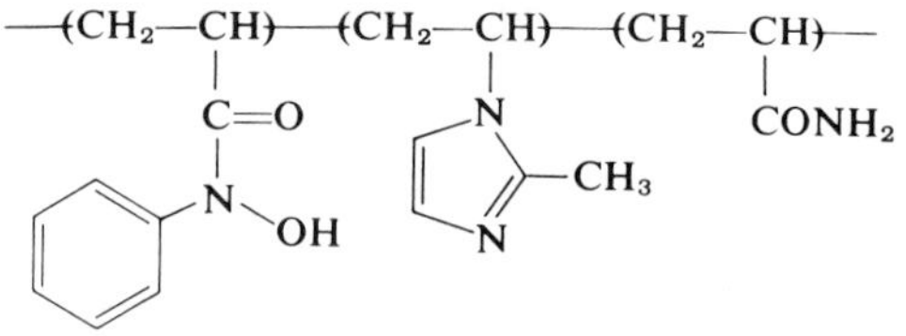

PHA-MIm-AAm [13]
(10)

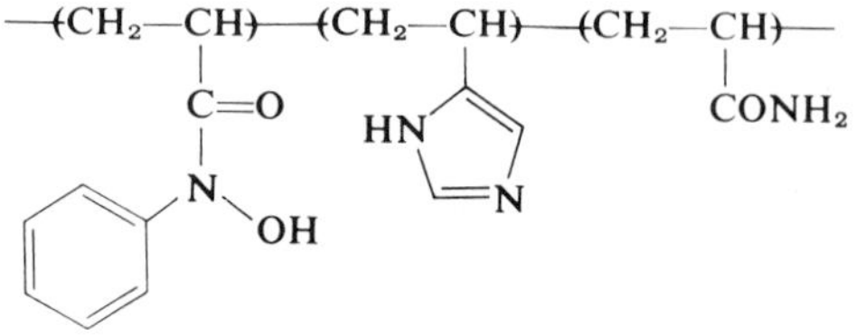

PHA-VIm-AAm [14]
(11)

The general kinetic pattern of the bifunctional catalytic process is now explained using the hydrolysis of p-nitrophenyl acetate (PNPA) by polymer catalyst PHA-VIm-AAm (11) as a typical example [14]. When PNPA is allowed to react with a water-soluble polymer containing only the hydroxamate functional group, the stoichiometric amount of p-nitrophenol is released [Eq. (1)]. The hydrolysis of the acetyl hydroxamate formed is much slower than acylation under conventional conditions (30°C, pH 8–9) [16]. This is true for all the hydroxamate catalysts, unless an acyl-activated substrate is employed [9]. On the other hand, when the polymer contains 5% of the PHA

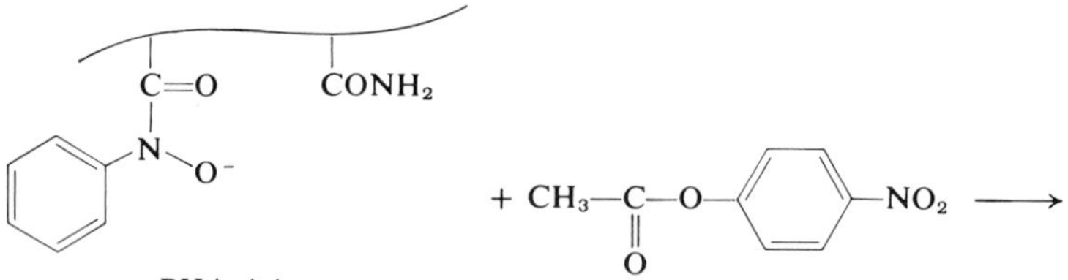

PHA-AAm

(12)

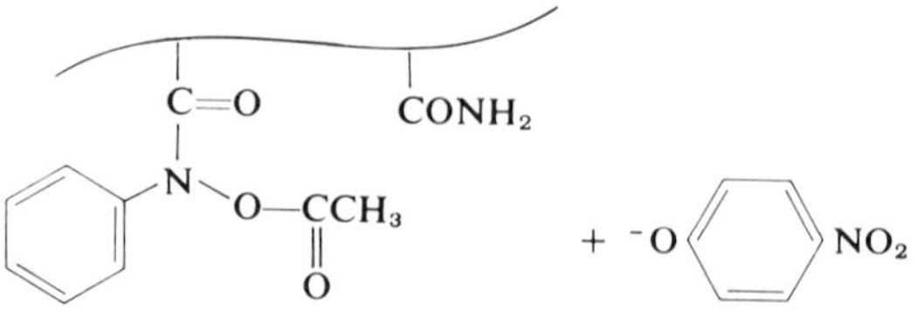

$$+ \ CH_3 - \underset{\underset{O}{\|}}{C} - O - \!\!\!\bigcirc\!\!\!- NO_2 \longrightarrow$$

$$+ \ ^-O - \!\!\!\bigcirc\!\!\!- NO_2 \quad (1)$$

unit and 35% of the VIm unit, pseudo first-order kinetics are observed for the catalytic hydrolysis of PNPA (1.16×10^{-4} M) in the presence of 5.8×10^{-5} M of the PHA unit. Catalysis by the imidazole group alone is much slower. Therefore, the hydroxamate group must be repeatedly used for the catalysis. A detailed kinetic analysis can be performed using burst kinetics.

The catalytic release of p-nitrophenol in the presence of large excesses of PNPA involves the initial rapid liberation and the slower, steady portion, as shown in Fig. 2. Analysis of the burst kinetics of simple nucleophilic catalysis has been performed by Bender and Marshall [17]. In the present system, however, there are two nucleophilic groups which contribute to the p-nitrophenol release:

$$C_{HA} + S \xrightarrow[P_1]{k_a} \text{acetyl-}C_{HA} \longrightarrow C_{HA} + P_2 \qquad (2)$$

$$C_{Im} + S \xrightarrow[P_1]{k_{Im.slow}} \text{acetyl-}C_{Im} + P_2 \qquad (3)$$

Here, C_{HA} and C_{Im} denote the hydroxamate and imidazole groups, respectively; S is the substrate; and P_1 and P_2 are p-nitrophenol and acetic acid, respectively.

The time course of the p-nitrophenol release is given by Eqs. (4)–(7) [18]:

$$[P_1] = A't + B(1 - e^{-bt}) \qquad (4)$$

where

$$A' = \frac{k_a k_d [S]_0 [HA]_0}{k_a [S]_0 + k_d} + k_{Im}[S]_0[Im]_0 \qquad (5)$$

$$B = \frac{k_a^2 [S]_0 [HA]_0}{(k_a [S]_0 + k_d)^2} \qquad (6)$$

$$b = k_a [S]_0 + k_d \qquad (7)$$

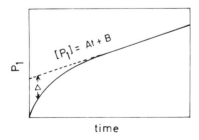

time

Fig. 2   Burst-type liberation of p-nitrophenol (P_1).

The rate constants are therefore given by Eqs. (8)–(10):

$$k_a = \frac{b\sqrt{B}}{[S]_0([HA]_0)^{1/2}}$$ (8)

$$k_d = b - k_a[S]_0$$ (9)

$$k_{Im} = \frac{1}{[Im]_0[S]_0}\left(A' - \frac{k_a k_d [HA]_0 [S]_0}{k_a[S]_0 + k_d}\right)$$ (10)

Delta [Eq. (11)] is defined as the difference in $[P_1]$ between the dashed line (extrapolation of the steady portion) and the burst curve of Fig. 2, and the B and b values are obtained from the intercept and slope, respectively, of the logarithmic plot.

$$\Delta = Be^{-bt}$$ (11)

Scheme 1 depicts the bifunctional catalytic process.

The substrate may be hydrolyzed via acylation (k_a) and deacylation (k_d) of the PHA unit or directly by the action of the VIm unit (k_{Im}). The rate constants for these processes are summarized in Table 1. It is clear that the PHA unit is much more reactive than the VIm unit under the experimental conditions (pH 7–9, 30°C). The deacylation rate of the acetyl-PHA intermediate is enhanced more than 200 times by the introduction of the VIm group. The difference is larger ($\sim 10^3$) at lower pH values. Thus, it is concluded that the predominant course of catalysis with this bifunctional polymer is simple acyl transfer from substrate to the PHA anion followed by hydrolysis of the acetyl hydroxamate intermediate, which is catalyzed by the intramolecular imidazole group.

Similar studies were conducted for a wide variety of bifunctional (hydroxamate and imidazole) catalysts. The catalytic patterns were fundamentally the same as that shown in Scheme 1. The pH–rate profiles of acylation and deacylation are given in Figs. 3 and 4.

TABLE 1

Rate Constants of Monofunctional and Bifunctional Catalysis[a]

Catalyst	pH	$k_{a,obs}$ (M^{-1} sec^{-1})	$10^2 \times k_{d,obs}$ (sec^{-1})	k_{Im} (M^{-1} sec^{-1})
PHA-AAm (12) (monofunctional)	9.10	3.66	0.01	—
PHA-VIm-AAm (11) (bifunctional)	8.04	0.79	0.96	0.03
	8.56	1.69	1.69	0.05
	9.18	3.47	2.76	0.09

[a] Conditions 28.9 v/v% EtOH–H$_2$O, 30°C, $\mu = 0.1$.

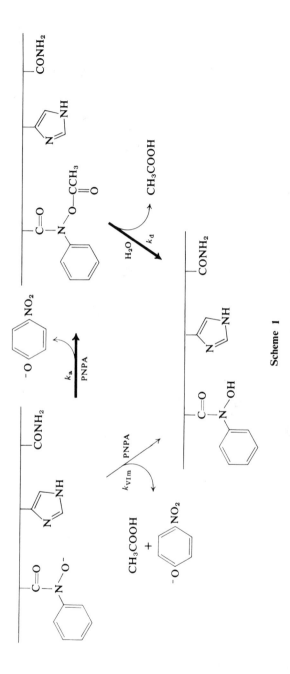

Scheme 1

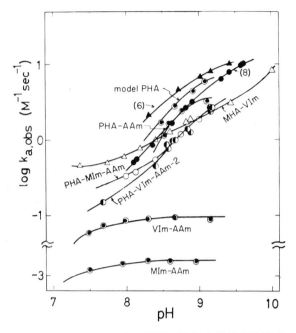

Fig. 3   pH–Rate profile of acylation (30°C, 28.9 v/v% EtOH–H$_2$O, $\mu = 0.1$).

In Fig. 3, the k_a values increase with pH, due to increasing dissociation of the hydroxamic acid group. Furthermore, k_a is not much affected by the presence of the neighboring imidazole group, and similar values were found for polymeric and small-molecule catalysts.

As for the deacylation process (Fig. 4), the k_d value of the bifunctional catalysts is always much larger than those of the monofunctional (hydroxamate) catalyst. It is also noted that the k_d process is more efficient for some polymer catalysts than for the small-molecule counterpart.

MECHANISMS OF ACYLATION AND DEACYLATION

Acylation

In the charge-relay system of serine proteases, the undissociated hydroxyl group is activated by hydrogen bonding with the neighboring imidazole group. An interaction of this sort would be facilitated by the favorable arrangement and/or high local concentration of the functional groups. The bifunctional polymer catalyst PHA-VIm-AAm (**11**) undergoes acylation at the hydroxamate and imidazole sites more or less independently. The lack of concerted action in this system can be attributed to the presence of a third component

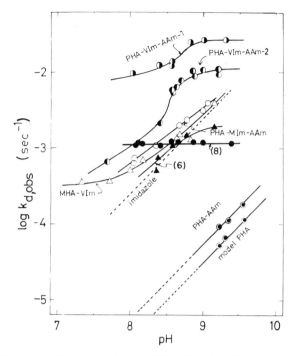

Fig. 4 pH–Rate profile of deacylation (30°C, 28.9 v/v% EtOH–H$_2$O, μ = 0.1).

(acrylamide unit), which was incorporated in order to increase solubility. Therefore, concerted action would be more probable for a water-soluble bifunctional polymer, MHA-VIm (**13**) [18].

$$-(CH_2-CH)_{25}-(CH_2-CH)_{75}- \qquad -(CH_2-CH)_{21}-(CH_2-CH)_{79}-$$

MHA-VIm
(**13**)

MHA-AAm
(**14**)

Figure 5 shows plots of the apparent rate constant of acylation $k_{a,obs}$ of the MHA unit against its degree of dissociation α_{HA}. The plots do not pass the origin in the case of the bifunctional polymer (**13**), in contrast to the plots for the related monofunctional polymer MHA-AAm (**14**). The presence of the positive intercept indicates that the undissociated hydroxamic acid group also possess nucleophilic reactivity. The deuterium solvent isotope effect observed indicates that the hydroxamic acid is activated through general base catalysis of the imidazole group. Thus, the acylation process of the

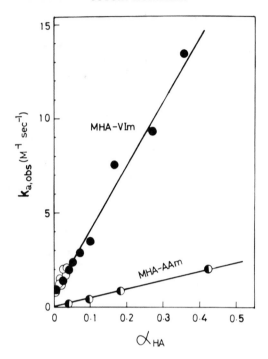

Fig. 5 Dependence of k_a on the dissociation of the hydroxamic acid group (MHA) (30°C, 28.9 v/v% EtOH–H$_2$O, $\mu = 0.1$).

MHA-VIm (**13**) polymer with PNPA substrate involves the following three types (in the absence of the intrachain VIm unit, the undissociated hydroxamic acid is not activated):

Similar results were obtained with other substrates (PNPH, p-nitrophenyl hexanoate; and NABA, 3-nitro-4-acetoxybenzoic acid), as summarized in Table 2. The efficiency of the concerted action (k_{HA}) was 1.5–10% of the simple nucleophilic attack (k_{A^-}). However, k_{HA} is always larger (~ 5–30 times) than k_{Im}. Therefore, direct nucleophilic attack of the imidazole group is less efficient than its general base assistance of the nucleophilic attack of the hydroxamic acid group.

Activation of the undissociated hydroxamic acid is possible also by general base action of the neighboring carboxylate group, as shown below (30°C, 28.9 v/v% EtOH–H_2O) [19]. In this case, the concerted action becomes possible through the favorable arrangement of the functional groups rather than the high local concentration of the basic functional group as in the MHA-VIm (13) polymer.

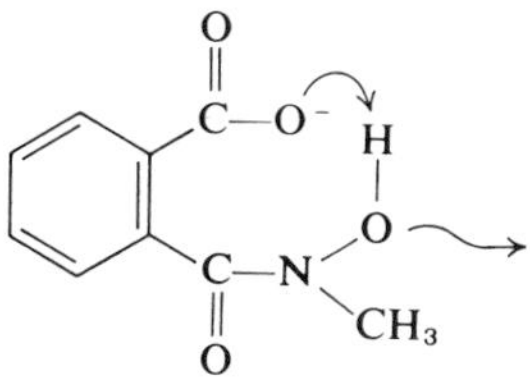

$k_{HA} = 0.22$ M^{-1} sec^{-1}

($k_H/k_D = 2.5$)

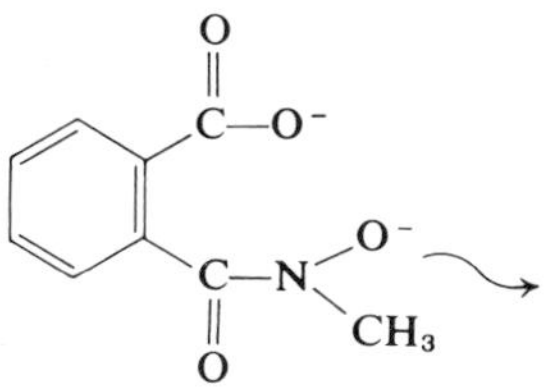

$k_{A^-} = 11.0$ M^{-1} sec^{-1}

($k_H/k_D = 1.0$)

Activation of alcoholic hydroxyl groups has been possible only in an intramolecular reaction. Belke et al. [20] showed that lactonization of 2-hydroxymethylbenzamide is accelerated by general base catalysis of imidazole

TABLE 2

Rate Constants of Acylation[a]

Substrate	Catalyst	k_{A^-} (M^{-1} sec^{-1})	k_{HA} (M^{-1} sec^{-1})	k_{Im}
PNPA	MHA-VIm (13)	36	0.50	0.10
	MHA-AAm (14)	4.8	0	—
PNPH	MHA-VIm (13)	13	1.2	0.04
	MHA-AAm (14)	1.6	0	—
NABA	MHA-VIm (13)	15	1.3	0.10
	MHA-AAm (14)	5.9	0	—

[a] Conditions: 30°C, 28.9 v/v% EtOH–H_2O, $\mu = 0.1$.

[Eq. (12)]. Kirby and Lloyd [21] reported that deacylation of salicyl γ-hydroxybutyrate is 500 times faster than that of phenyl γ-hydroxybutyrate

$$\text{(12)}$$

which does not possess the orthocarboxylate group. The mechanism shown in Eq. (13) was presented on the basis of the deuterium solvent isotope effect ($k_H/k_D = 2.28$).

$$\text{(13)}$$

Deacylation

The acceleration of the deacylation process by intramolecular imidazole can be realized through the general base or nucleophilic mechanism. In the latter, the acetylimidazole group formed by acyl transfer must be further hydrolyzed.

General base   Nucleophilic

The pH–rate profile of deacylation is not straightforward in the case of PHA-VIm-AAm (**11**) polymer (see Fig. 4). There is a sharp rise in k_d at pH 8.5–9, which is close to pK_a of the hydroxamic acid unit. Therefore, the hydroxamate anion may be involved directly or indirectly in the deacylation process. On the other hand, a simpler profile is observed for the MHA-VIm (**13**) polymer, and a kinetic deuterium solvent isotope effect ($k_H/k_D = 1.5$ to 3) was found at pH 8.5–9.0. Therefore, the general base mechanism appears to prevail.

It appears that the general base mechanism is favored in the intramolecular catalysis by imidazole, as shown by the examples shown below [12,22,23]. This is supposedly attributable to the fact that the acetylimidazole intermediate, if formed, would readily revert to the starting compound because of the higher nucleophilicity of the oxyanion group. This problem has been very lucidly discussed in the case of the intramolecular carboxylate catalysis [24].

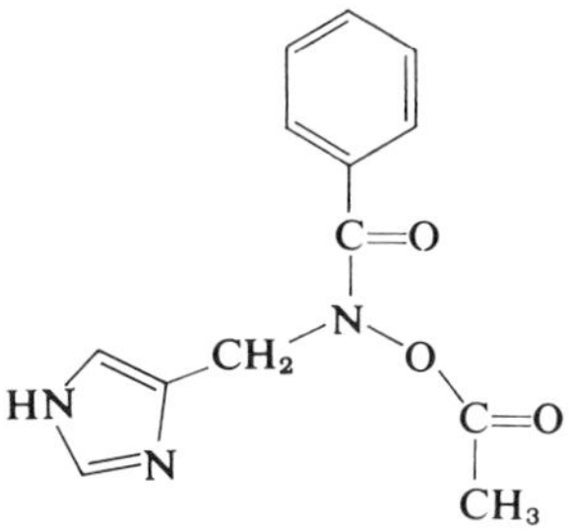

$k_H/k_D = 2.0$ [12] 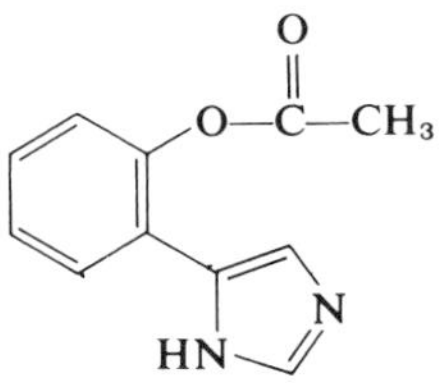 $k_H/k_D = 3.23$ [22] 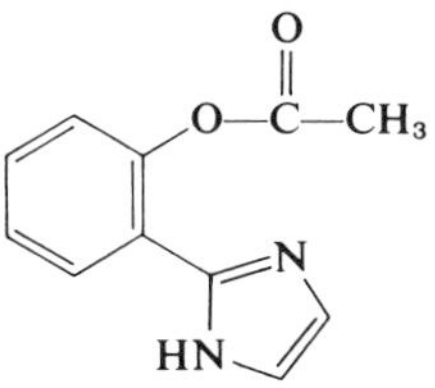 $k_H/k_D = 3.91$ [23]

MICROENVIRONMENTAL EFFECTS ON THE REACTIVITY OF ANIONIC NUCLEOPHILES: THE CONCEPT OF HYDROPHOBIC ION PAIR

It has been found that anionic nucleophiles such as alkoxide [25,26], oximate [27,28], hydroxamate [29,30], thiolate [31,32], and imidazole [33] anions are considerably activated in the presence of cationic micelles and cationic polysoaps. Table 3 gives some examples of the reactivity enhancement of a hydroxamate nucleophile [30]. The apparent second-order rate constant at pH 8.91 is $0.64 \text{ M}^{-1} \text{ sec}^{-1}$ in the presence of ethylated (85%) poly(4-vinylpyridine). The $k_{a,obs}$ value is enhanced to $390 \text{ M}^{-1} \text{ sec}^{-1}$ in the presence of laurylated (33%) poly(4-vinylpyridine) (4VP-L). The rate enhancement becomes even greater at a lower ionic strength. Poly(vinylpyridine) with only 3 mole % of the lauryl substituent was effective. Since the quaternized polymer is known to form a micellelike structure when the content of the lauryl group exceeds ca. 15 mole % [34], the micellar structure is not required for the rate enhancement. Therefore, the hydroxamate anion is activated, probably by the formation of the hydrophobic ion pair between the laurylpyridinium group and the long-chain hydroxamate.

Evidence supporting this concept is obtained from study of reactions in

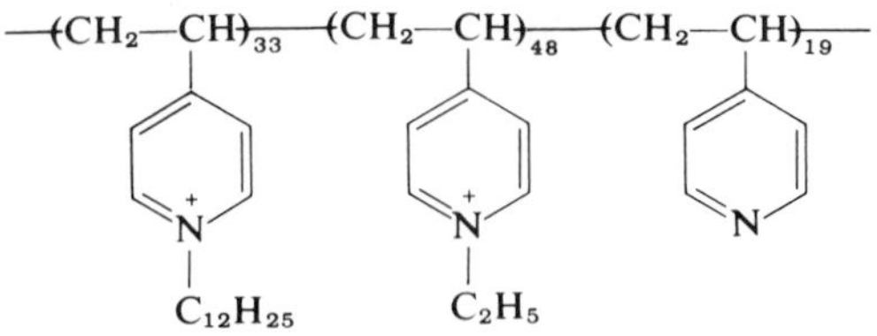

$$-(CH_2-CH)_{33}-(CH_2-CH)_{48}-(CH_2-CH)_{19}-$$

[Structure 4VP-L with pyridinium rings: $C_{12}H_{25}$, C_2H_5, and N]

4VP-L

(15)

aqueous nonionic micelles [35]. The nucleophilicity of the N-methylmyristo-hydroxamate (MMHA) (16) anion toward PNPA is enhanced in cationic and nonionic micelles relative to the nonmicellar system. Although the observed rate enhancement is much greater in the cationic (cetyltrimethylammonium bromide, CTAB) micelle than in nonionic micelles, the difference can be made smaller by adding a small amount of trioctylmethylammonium chloride (TMAC). Addition of less hydrophobic ammonium ions ($Et_4N^+Br^-$ and $Bu_4N^+Br^-$) does not cause a rate increase. Furthermore, a zwitterionic nucleophile, DHDB (17), shows much enhanced reactivity in nonionic micelles relative to the nonmicellar system (60-fold enhancement). The addition of the hydrophobic ammonium ion (TMAC) (19) to this system does not result in a further rate increase.

From these results, the rate enhancement of anionic nucleophiles in cationic micelles can be largely ascribed to the formation of hydrophobic ion pairs [intermolecular ($MMHA^-CTAB^+$) and intramolecular (DHDB) ion pairs] in the hydrophobic environment of the micellar phase. Characteristics peculiar to the cationic micelle, such as high charge density at the micellar surface, are not required for the rate enhancement. Then the concept of the hydrophobic

TABLE 3

Reaction of N-Methylmyristohydroxamate Anion with PNPA in the Presence of a Micelle and Quaternized Poly(vinylpyridine)

	Ionic strength	pH	Lauryl unit ($\times 10^4\ M$)	$k_{a,obs}$ ($M^{-1} sec^{-1}$)
Laurylpyridinium bromide	0.5	7.24	291	5.38
4VP-L (15)[a]	0.5	7.24	9.87	8.10
4VP-L (15)[a]	0.08	8.92	14.7	1100
4VP-L (15)[a]	0.5	8.92	14.7	390
4VP-E$_{85}$[b]	0.5	8.91	0	0.64

[a] L indicates laurylated.
[b] E$_{85}$ indicates 85% ethylated.

$$CH_3 \!-\!\!(CH_2)_{12}\!-\!\overset{\displaystyle \underset{\displaystyle O}{\|}}{C}\!-\!N\!\!\begin{array}{c} CH_3 \\ \diagdown OH \end{array}$$

MMHA
(16)

$$CH_3 \!-\!\!(CH_2)_{11}\!-\!\overset{\overset{\displaystyle CH_3}{\displaystyle |}}{\underset{\underset{\displaystyle CH_3}{\displaystyle |}}{\overset{+}{N}}}\!-\!CH_2\!-\!\overset{\overset{}{}}{\underset{\underset{\displaystyle -O}{}}{C}}\!\diagdown$$

DHDB
(17)

$$CH_3 \!-\!\!(CH_2)_{15}\!-\!\overset{\overset{\displaystyle Br^-}{+}}{N}\!-\!(CH_3)_3$$

CTAB
(18)

$$(C_8H_{17})_3\!-\!\overset{\overset{\displaystyle Cl^-}{+}}{N}\!-\!CH_3$$

TMAC
(19)

ion pair is not limited to the conventional cationic micelles and polysoaps but is generally applicable to the system of hydrophobic aggregates in the aqueous phase. For example, TMAC (19) does not form the conventional micellar structure in aqueous systems, although some physicochemical measurements clearly indicate the formation of relatively small aggregates [36]. In the presence of 7×10^{-5} M TMAC (19) at pH 9, the reactivity of N-laurylbenzohydroxamic acid toward PNPA was remarkably enhanced: $k_{a,obs} = 6700$ M^{-1} sec^{-1}. This value is about 450 times larger than that of the nonmicellar hydroxamate (15 M^{-1} sec^{-1}).

The high reactivity of hydrophobic ion pairs may be attributed to (1) pK_a lowering of the nucleophilic function and (2) the enhanced reactivity of anionic nucleophiles. The first factor must be particularly important when the nucleophilic function is situated in close proximity to the positive charge, as in zwitterionic nucleophiles and at the micellar surface. As for the second factor, desolvation may be operating. It is usually difficult to separate the desolvation effect from other effects in aqueous aggregates.

The importance of desolvation can be unambiguously proved by experiments using organic solvents [37]. Table 4 summarizes the kinetic results of the reaction of tetraethylammonium N-methylmyristohydroxamate (MMHA$^-$NEt$_4$$^+$) with PNPA. The reaction is very fast in dry, aprotic solvents (dimethylformamide, acetonitrile, and benzene) and, in contrast, very slow in protic solvents (water, ethanol, and formamide). Addition of very small amounts of water to aprotic solvents brings large rate decreases. Therefore, the nucleophilic reactivity of the hydroxamate anion (or probably oxyanions in general) as desolvated, loose ion pairs is remarkably enhanced. The desolvation effect may be similarly important for the rate enhancement of hydrophobic ion pairs in aqueous systems.

Kirsh et al. [38] found that an oxime group bound to the pyridine ring by quaternization, PPyOX (20), possesses a much lowered pK_a value, yet retains

TABLE 4
Reaction of MMHA$^-$NEt$_4$$^+$ and PNPA at 30°C

Solvent	[H$_2$O] (mM)	k_a (M^{-1} sec^{-1})
Dimethylformamide	2.2–3.0	1130
CH$_3$CN	3.3–5.3	845
	570	12.6
Benzene	4.5–6.1	350
Ethanol	27	0.63
Formamide	—	$< 10^{-4}$
Water	—	~ 30

high nucleophilicity. The pK_a value of the polymer-bound oxime was 8.5 (25°C, $\mu = 0.01$) and not dependent on the extent of quaternization (β). The corresponding small-molecule oxime [PyOX (21)] had a pK_a of 9.6. The reactivity of the polymer-bound oxime with PNPA increases with increasing pH due to dissociation and with the decreasing extent of quaternization. The reactivity difference between the polymeric and nonpolymeric oximes is very large (~ 100 times at pH 9), and the second-order rate constant of the polymer-bound oximate anion ($\beta = 12$) is 1000 M^{-1} sec^{-1}. This enhanced reactivity is again ascribed to the lowered pK_a value and to the high reactivity of the oximate ion pair in the hydrophobic microenvironment.

PPyOX
(20)

PyOX
(21)

$$\beta = \frac{m}{m + n} \times 100$$

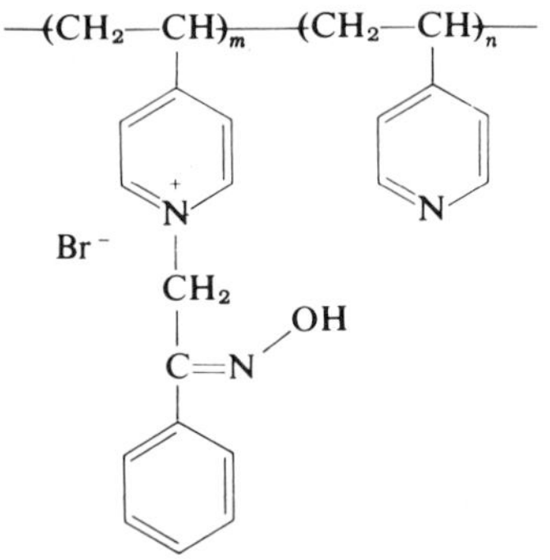

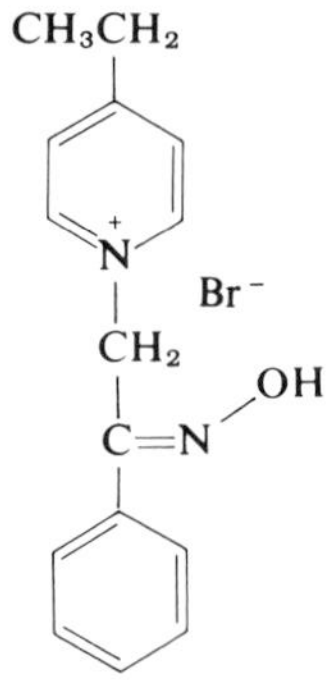

It is useful to point out the analogy of the concept of the hydrophobic ion pair with the mechanism of phase-transfer catalysis. Phase-transfer catalysts extract anionic reagents from the aqueous phase and place them in the organic phase [39]. The reaction takes place in the organic phase with anionic reagents activated by desolvation. Thus, the rate enhancement due to formation of the hydrophobic ion pair may be considered as arising from the microscopic phase-transfer catalysis.

CATALYTIC EFFICIENCY

In the preceding sections, we have discussed two factors that lead to efficient nucleophilic catalysis. They are the complementary combination of two functional groups and the enhanced nucleophilicity of hydrophobic ion pairs. Truly efficient catalysis is attained by the appropriate use of these two factors. Two examples along this line are given below. One of the examples is a micellar bifunctional catalyst, LImHA (**22**), and the other is partly quaternized poly(vinylpyridine) (PVP$^+$-HA) (**24**) and poly(vinylimidazole) (EIm$^+$-HA) (**25**), which can form zwitterionic nucleophiles.

Table 5 [10,14,38,40–44] lists apparent rate constants of acylation, deacylation, and turnover available at near pH 8 in the hydrolysis of PNPA. The

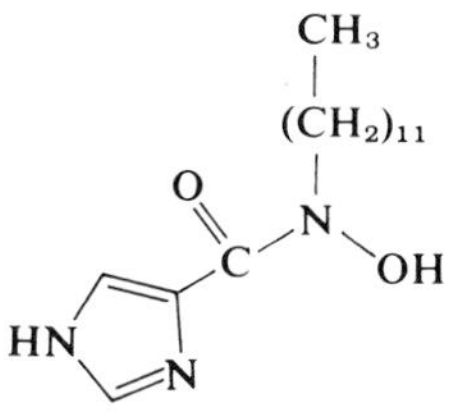

LImHA
(**22**)

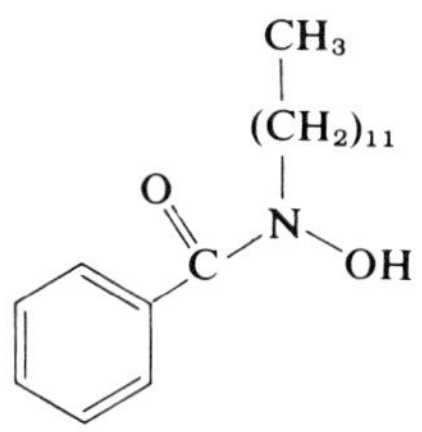

LBHA
(**23**)

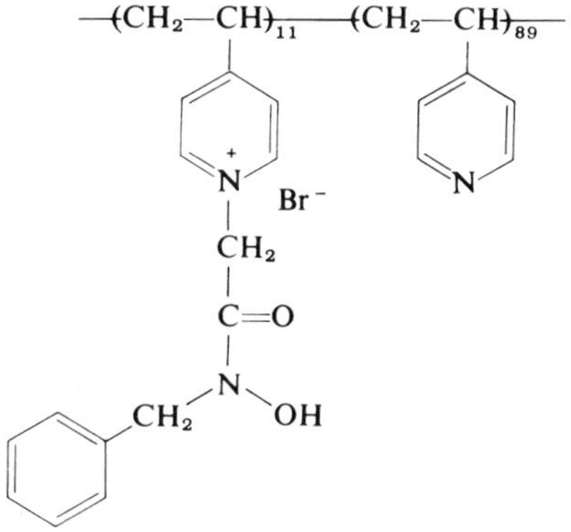

PVP-HA
(**24**)

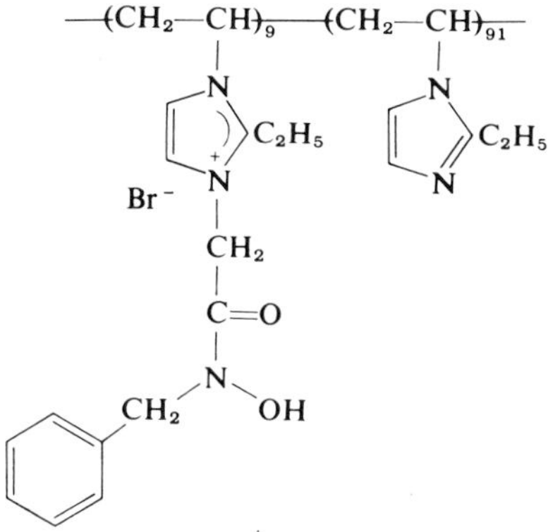

EIm-HA
(**25**)

value k_{turnover} is the rate constant of catalyst regeneration given by Eq. (14).

$$k_{\text{turnover}} = \frac{k_{\text{a,obs}}k_{\text{d,obs}}[\text{PNPA}]}{k_{\text{a,obs}}[\text{PNPA}] + k_{\text{d,obs}}} \tag{14}$$

It is immediately apparent that efficient catalysis is possible by the optimal combination of $k_{\text{a,obs}}$ and $k_{\text{d,obs}}$. For example, the $k_{\text{a,obs}}$ value is very large for PPyOX (20) and LBHA(N-laurylbenzohydroxamic acid)(23)–CTAB catalysts; however, k_{turnover} is small because of small $k_{\text{d,obs}}$ values.

In contrast, $k_{\text{d,obs}}$ values are large for polymer-bound hydroxamates, EIm$^+$-HA (25) and PVP$^+$-HA (24) [42], and for a micellar bifunctional catalyst, LImHA(22)–CTAB [43]. The deacylation reaction was not detected in the case of PPyOX (20). Therefore, it is indicated that the polymer-bound hydroxamate undergoes deacylation more efficiently than the corresponding oximate. The fast deacylation of LImHA(22)–CTAB is derived from the intramolecular imidazole catalysis, as discussed earlier. Its efficiency is much greater than that of the corresponding nonmicellar catalyst MImHA (6). Apparently, the highly reactive imidazole anion is formed in the micellar system.

The LImHA(22)–CTAB system is the most efficient catalyst in Table 5. Its k_{turnover} value is more than two times greater than that for α-chymotrypsin under comparable conditions.

TABLE 5

Catalytic Efficiency in the Hydrolysis of PNPA[a]

Catalyst	$k_{\text{a,obs}}$ (M^{-1} sec^{-1})	$10^4 \times k_{\text{d,obs}}$ (sec^{-1})	$10^4 \times k_{\text{turnover}}$ (sec^{-1})	Reference
α-Chymotrypsin[b]	400[c]	250	160	40
D(10%)-PEI-Im(15%)[d]	45	40	21	41
PPyOX ($\beta = 12$) (20)	200	—	—	38
PHA-VIm-AAm (11)	0.79	96	0.78	14
EIm$^+$-HA (25)	23	87	18	42
PVP$^+$-HA (24)	13	130	12	42
LImHA(22)–CTAB ($\mu = 0.01$)	930	650	380	43
LBHA(23)–CTAB ($\mu = 0.01$)	490	0.13	0.13	43
MImHA (6)	1.9	4	1.3	10
Imidazole	0.33	3.0	0.32	44

[a] Conditions: pH 8.0, [PNPA] $= 1 \times 10^{-4}$ M, 30°C, $\mu = 0.1$.
[b] Conditions: 25°C, 20 v/v% isopropanol–H$_2$O.
[c] $k_{\text{a,obs}} = k_{\text{cat}}/(K_{\text{m}} + [\text{PNPA}])$.
[d] PEI, poly(ethylenimine) with 10% dodecyl group and 15% imidazolylmethyl group.

CONCLUSION

It is now clear that the efficiency of the nucleophilic catalysis of hydrolysis can be remarkably enhanced by appropriate use of multifunctionality and microenvironments. The turnover rate constant of a micellar, bifunctional catalyst, LImHA(22)–CTAB, is more than 1000 times greater than that of imidazole, the first model of hydrolytic enzymes. Only the hydrolysis of phenyl esters is discussed in this chapter. However, the concepts developed here should apply, with some modifications, to hydrolyses of aliphatic esters and amides and also to other classes of nucleophilic catalysis. Thus, hydrophobic ion pairs of the hydroxamate anion were shown to abstract proton from carbon acids very efficiently [45].

The active site of enzymes is generally considered to be highly hydrophobic. Therefore, the concept of hydrophobic ion pair should be applicable to a wide variety of enzyme-catalyzed reactions.

ACKNOWLEDGMENT

The author deeply appreciates the support of Dr. S. Shinkai, Y. Okahata, and other co-workers as well as financial assistance from the Ministry of Education.

REFERENCES

1. D. M. Blow, J. J. Birktoft, and B. S. Hartley, *Nature (London)* **221**, 337 (1969).
2. G. R. Schonbaum and M. L. Bender, *J. Am. Chem. Soc.* **82**, 1900 (1960).
3. D. C. Williams and J. R. Whitaker, *Biochemistry* **7**, 2562 (1968).
4. F. Schneider, *Hoppe-Seyler's Z. Physiol. Chem.* **348**, 1034 (1967).
5. F. Schneider and H. Wenck, *Hoppe-Seyler's Z. Physiol. Chem.* **350**, 1653 (1969).
6. J. Schoenleben and P. Lochon, *C.R. Hebd. Seances Acad. Sci., Ser. C* **278**, 1293 and 1381 (1974).
7. C. G. Overberger, T. J. Pacansky, J. Lee, T. St. Pierre, and S. Yaroslavsky, *J. Polym. Sci., Polym. Symp.* **46**, 209 (1974).
8. E.g., M. Dessolin, M. Laloi-Diard, and M. Vilkas, *Bull. Soc. Chim. Fr.* p. 2573 (1970).
9. W. B. Gruhn and M. L. Bender, *J. Am. Chem. Soc.* **91**, 5883 (1969); *Bioorg. Chem.* **4**, 219 (1975).
10. T. Kunitake and S. Horie, *Bull. Chem. Soc. Jpn.* **48**, 1304 (1975).
11. H. Tokunaga, Y. Okahata, and T. Kunitake, *Annu. Meet. Chem. Soc. Jpn., 1975* p. 1304 (1975).
12. T. Kunitake, Y. Okahata, and T. Tahara, *Bioorg. Chem.* **5**, 155 (1976).
13. T. Kunitake, Y. Okahata, and R. Ando, *Macromolecules* **7**, 140 (1974); T. Kunitake and Y. Okahata, *Bioorg. Chem.* **4**, 136 (1975).
14. T. Kunitake and Y. Okahata, *Chem. Lett.* p. 1057 (1974); *Macromolecules* **9**, 15 (1976).
15. Y. Kitaura and M. L. Bender, *Bioorg. Chem.* **4**, 237 (1975).

16. T. Kunitake, Y. Okahata, and R. Ando, *Bull. Chem. Soc. Jpn.* **47**, 1509 (1974).
17. M. L. Bender and T. H. Marshall, *J. Am. Chem. Soc.* **90**, 201 (1968).
18. T. Kunitake and Y. Okahata, *J. Am. Chem. Soc.* **98**, 7793 (1976).
19. T. Kunitake, Y. Okahata, R. Ando, and S. Hirotsu, *Bull. Chem. Soc. Jpn.* **49**, 2547 (1976).
20. C. J. Belke, S. C. Su, and J. A. Shafer, *J. Am. Chem. Soc.* **93**, 4552 (1971).
21. A. J. Kirby and G. J. Lloyd, *J. Chem. Soc., Perkin Trans. 2* p. 637 (1974).
22. S. M. Felton and T. C. Bruice, *J. Am. Chem. Soc.* **91**, 6721 (1969).
23. G. A. Rogers and T. C. Bruice, *J. Am. Chem. Soc.* **96**, 2463 (1974).
24. A. J. Kirby and A. R. Fersht, *Prog. Bioorg. Chem.* **1**, 6 (1969).
25. C. A. Bunton and L. G. Ionescu, *J. Am. Chem. Soc.* **95**, 2912 (1973).
26. K. Martinek, A. V. Levashov, and I. V. Berezin, *Tetrahedron Lett.* p. 1275 (1975).
27. A. K. Yatsimiriski, K. Martinek, and I. V. Berezin, *Tetrahedron* **27**, 2855 (1971).
28. W. Tagaki, I. Takahara, and D. Fukushima, *Pap., 32nd Annu. Meet. Chem. Soc. Jpn. 1975* p. 1308 (1975).
29. I. Tabushi, Y. Kuroda, and S. Kita, *Tetrahedron Lett.* p. 643 (1974); I. Tabushi and Y. Kuroda, *ibid.* p. 3613.
30. T. Kunitake, S. Shinkai, and S. Hirotsu, *J. Polym. Sci., Polym. Lett. Ed.* **13**, 377 (1975).
31. W. Tagaki, T. Amada, Y. Yamashita, and Y. Yano, *Chem. Commun.* p. 1131 (1972).
32. H. Chaimovich, A. Blancho, L. Chayet, L. M. Costa, P. M. Monteiro, C. A. Bunton, and C. Paik, *Tetrahedron* **31**, 1139 (1975).
33. K. Martinek, A. P. Osipov, A. K. Yatsimirski, V. A. Dadolt, and I. V. Berezin, *Tetrahedron Lett.* p. 1279 (1975).
34. U. P. Strauss, N. L. Gershfeld, and E. H. Crook, *J. Phys. Chem.* **60**, 577 (1956).
35. T. Kunitake, S. Shinkai, and Y. Okahata, *Bull. Chem. Soc. Jpn.* **49**, 540 (1976).
36. Y. Okahata and T. Kunitake, *J. Am. Chem. Soc.* (in press).
37. S. Shinkai and T. Kunitake, *Chem. Lett.* p. 109 (1976).
38. Yu. E. Kirsh, A. A. Rahnanskaya, G. M. Lukovkin, and V. A. Kabanov, *Eur. Polym. J.* **10**, 393 (1974).
39. A. W. Herriott and D. Picker, *J. Am. Chem. Soc.* **97**, 2345 (1975).
40. H. Gutfreund and J. M. Sturtevant, *Biochem. J.* **63**, 656 (1956).
41. I. M. Klotz, G. P. Royer, and I. A. Scarpa, *Proc. Natl. Acad. Sci. U.S.A.* **68**, 263 (1971).
42. Y. Okahata and T. Kunitake, *J. Polym. Sci., Polym. Chem. Ed.* (in press).
43. T. Kunitake, Y. Okahata, and T. Sakamoto, *Chem. Lett.* p. 459 (1975); *J. Am. Chem. Soc.* **98**, 7799 (1976).
44. T. C. Bruice and J. L. Herz, *J. Am. Chem. Soc.* **86**, 4109 (1964).
45. S. Shinkai and T. Kunitake, *J. Chem. Soc., Perkin Trans. 2* p. 880 (1976).

9

The Interplay of Theory and Experiment in Bioorganic Chemistry: Three Case Histories

G. M. Maggiora and R. L. Schowen

INTRODUCTION

Although good theorists and good experimentalists in every field always keep a weather eye on each other, it is fair to say that in bioorganic chemistry a strong and complementary relationship of theory and experiment is a sign of recent times. A status of friendly divorcement existed only a few years ago in which even the most admirable theoretical work seldom gave rise directly to an experiment, and theoretical calculations were rarely based on particular experimental findings. This situation had an unarguable basis in technical pragmatism. The theory of that period lacked the range to treat problems in a way that could directly influence experimental design, while at the same time the experimental formulations (particularly in bioorganic mechanisms) lacked the definition required to orientate and inspire a theoretical contribution.

These points are readily brought home by illustrations from two famous books of 1963, one on biochemical theory and one dealing in part with enzyme mechanisms. Figure 1a is a typical illustration from the Pullmans' "Quantum Biochemistry" [1]. It exemplifies then-current theoretical approaches to acyl-transfer mechanisms (a problem we consider below). It consists of a simple Hückel molecular orbital (MO) calculation of the charge

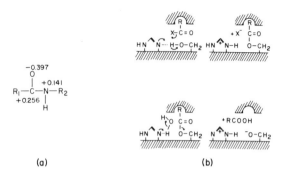

Fig. 1 (a) An illustration from Pullman and Pullman, "Quantum Biochemistry," a theoretical text published in 1963 [1, p. 659]. A simple Hückel MO calculation giving the charge distribution for the amide function is shown. (b) The mechanism of chymotrypsin-catalyzed hydrolysis of acyl derivatives as portrayed in Haurowitz, "Chemistry and Function of Proteins," also published in 1963 [2, p. 301]. No data were available at the time to allow any specification of enzyme structure beyond the shaded regions shown in the diagram.

distribution in the amide function. While valuable in itself and as groundwork for later developments, such a calculation was likely to be little used by experimentalists. Only very small systems could be examined, only a few calculations on each system were feasible, many experimentalists lacked confidence in the reliability of the results, and there was no generally accepted relationship linking the numerical results of the calculations to experimental observables. The theorist who surveyed the experimental scene found an equally uncomfortable view. Consider Fig. 1b, which shows the mechanism of enzymatic acyl transfer, as presented in Haurowitz's "Chemistry and Function of Proteins" [2]. The scheme is an accurate summary of the existing knowledge of the time, nearly all of which has survived intact to the present. The shaded "brick walls" representing the then-unknown enzyme structure, however, left the theorist with an infinitude of unmanageable degrees of freedom. It is not surprising that workers of both persuasions, experimental and theoretical, proceeded along their separate routes, each group largely innocent of the other's efforts.

   The situation has changed. *Ab initio* and semiempirical quantum mechanics is now capable of generating detailed information in quantity on large molecules [3]. Crystallographic data on protein structure have shown us the apparatus of enzyme catalysis [4] and are proceeding to ever greater levels of refinement [5,6]. Mechanistic thought at all levels is becoming daily more precise [7,8]. The result has been that, in a variety of important advances in bioorganic thought, a vigorous interactive mode of theory and experiment has emerged. It is our purpose in this chapter to illustrate this phenomenon by several case histories and to give a brief analysis of the prerequisites for such an advantageous reinforcement. Our case histories involve two questions of enzyme catalytic power ("charge-relay" catalysis and acyl-transfer transition states) and a problem in ligand–protein interactions involving the visual pigment rhodopsin.

These three examples involve studies in which the authors have themselves participated, and they admittedly cover only the narrowest selection of possible cases from the field of bioorganic chemistry. Even within this severely constricted range, the portrayal of the interplay between theory and experiment is entirely formed by the subjective perceptions of the authors and therefore large numbers of important contributions of both theoretical and experimental character have been omitted. Hence, we offer in advance our apologies to those whose work has not been adequately dealt with and to the reader for our failure to acquaint him with every aspect of each problem. An attempt to survey every contribution of significance, and to give it its proper place, would have consumed far more time than the authors had available and far more space than the editor would have allowed. Sad to say, the result would have been no less subjective and little more complete than the present chapter.

CHARGE-RELAY CATALYSIS

Enzymatic General Catalysis

It has long been thought that enzyme catalysis may originate in part with acidic or basic groups attached to the enzyme. Indeed, the idea may be traced at least to Isaac Newton [9], who noted in 1692 that "in all Fermentation there is an Acid latent or suppress'd. . . ." The pH dependence of the hydrolytic action of the serine proteases (ionization of an acid of pK_a near 7 required for catalytic effect) and observations on imidazole catalysis in model systems made natural the assumption of a general base catalytic entity derived from an enzymatic histidine [10]. The rate of the enzymatic reaction is typically depressed 2- to 4-fold in deuterium oxide solvent [11], just as is the case for nonenzymatic general catalyzed hydrolysis of similar substrates [12]. However, the experience in small-molecule systems is that general catalysis leads to small accelerations, suggesting a doubtful role for simple general catalysis in the major accelerations produced by hydrolytic enzymes [7].

Crystallographic Discoveries

A possible new form of general catalysis was, however, proposed on the basis of the crystallographic structure for α-chymotrypsin. This "charge-relay" catalysis was supposed to originate with an aligned group (Fig. 2)

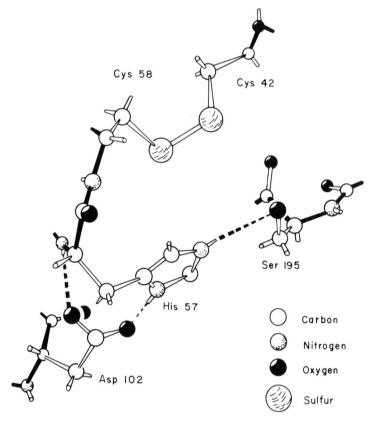

Fig. 2   The hydrogen bond chain formed from Asp-102, His-57, and Ser-195 in the active site of α-chymotrypsin, as derived by Blow, Birktoft, and Hartley [13] from X-ray crystallographic information. Note the contrast between the structural detail in this figure and in Fig. 1b.

composed of Asp-102, His-57, and Ser-195, linked by hydrogen bonds [13]. Such an alignment would be unusually difficult to achieve in small-molecule chemistry because of the large entropic cost, and the "charge-relay" system and any associated catalytic power thus probably constitute an original chemical invention of molecular evolution. The special significance of the system arises from its potential for coupled motion of the protons in its two component hydrogen bonds (Fig. 3), relaying negative charge from the buried, and therefore unusually basic, carboxylate center to the serine oxygen in acylation or to a water oxygen in deacylation. Support for the mechanism was derived from nuclear magnetic resonance titrations of the active site [14], which showed no change in positive charge at the histidine

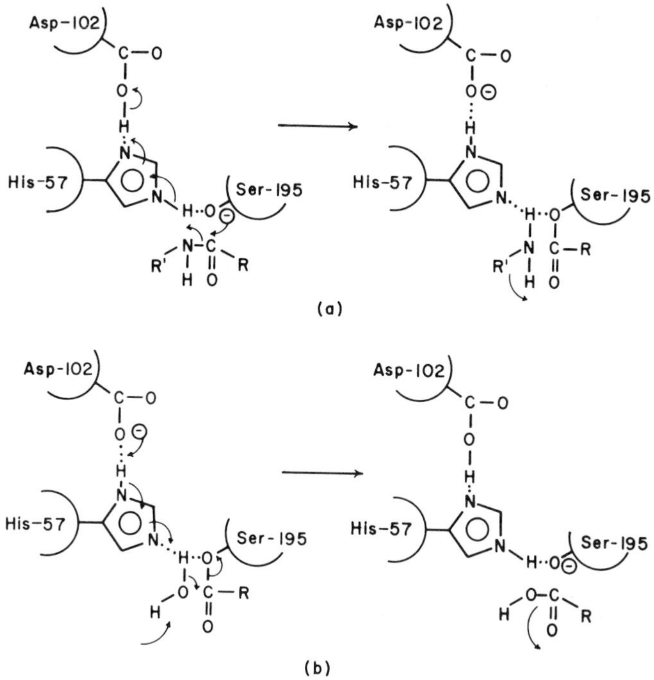

Fig. 3 The function of the "charge-relay" system, as depicted by Blow, Birktoft, and Hartley in their 1969 paper [13]. Part (a) shows relay of a proton to the departing amine. Part (b) shows relay of a proton from an entering water.

center on protonation or deprotonation, and from the greatly reduced reactivity when one nitrogen of His-57 was methylated [15]. However, the titrations do not directly probe the catalytic events and the methylation represents a particularly severe perturbation so that neither form of evidence is conclusive. Model studies [16] indicated by contrast that charge-relay catalysis was an unlikely source of large acceleration.

Proton Inventories: Theory

A different probe, capable of direct reporting from the catalytic transition state, was offered by the "proton inventory" method. This technique employs rates in mixtures of protium and deuterium oxides [17] to measure the number of protons contributing to the solvent kinetic isotope effect and the magnitude of the contribution from each. The basis of the approach is given in Eqs. (1)–(4), in which k_n is the rate constant in a mixture of H_2O and D_2O of atom fraction n of deuterium. Equation (1) shows k_n to be given by the rate constant in H_2O, k_0, multiplied by a series of correction factors J_i, one for each proton

$$k_n = k_0 \prod_i^v J_i \tag{1}$$

$$J_i = (1 - n + n\phi_i^T)/(1 - n + n\phi_i^R) \tag{2}$$

$$k_n = k_0 \prod_i^v (1 - n + n\phi_i^T) \tag{3}$$

$$k_n = k_0 + c_1 n + c_2 n^2 + \cdots + c_v n^v \tag{4}$$

that contributes to the isotope effect. These factors have the form shown in Eq. (2), depending on the atom fractions of protium $(1 - n)$, deuterium (n), and the isotopic fractionation factors for the reactant state (ϕ_i^R) and transition state (ϕ_i^T). It is frequently a useful approximation to consider those reactant-state fractionation factors that change on activation (so that $J_i \neq 1$) to be essentially unity [18]. This produces Eq. (3) and, as Eq. (4) emphasizes, k_n becomes a polynomial in n, and the order v of this polynomial measures the number of "active" protons (those that produce an isotope effect) required by the data for rates in isotopic mixtures.

In the case of charge-relay catalysis, the proton inventory probe assumes a simple form. If the charge-relay system functions in a truly coupled sense, with both protons changing their binding states simultaneously, then k_n should depend on isotope effect contributions from both protons and should therefore be *quadratic* in n [Eq. (5)]. If, on the other hand, the observed solvent isotope effects [11] arise from a single catalytic proton, with the function of the active-site hydrogen bond chain being, for example, to orient His-57 or to stabilize its positive charge electrostatically, then k_n should receive only the single-proton isotope effect contributions and [as in Eq. (6)] should depend *linearly* on n.

$$k_n = k_0(1 - n + n\phi_1^T)(1 - n + n\phi_2^T) \tag{5}$$

$$k_n = k_0(1 - n + n\phi_1^T) \tag{6}$$

Proton Inventories for Serine Protease Action

Experiments to test these alternative hypotheses were conducted by Pollock and Hogg, Venkatasubban, and Elrod [19–19b]. Pollock *et al.* [19] first reported that a plot of V_n vs. n for deacetylation of acetyl-α-chymotrypsin was linear, indicating that the enzyme was functioning as a one-proton catalyst in its own deacetylation and thus that the charge-relay system was not behaving in a coupled manner. Now it was conceivable that this linear plot arose from a coincidental cancellation between a quadratic dependence of V_n on n and a number of smaller inverse isotope effects neglected in the formulation of Eq. (3) [20]. This is not the case, as is shown by the similar linearity of $V_n(n)$ for deacetylation of such other acetyl enzymes as acetyltrypsin, acetylthrombin, and acetylelastase [19a,b] and for acylation of α-chymotrypsin by N-acetyl-L-tryptophanamide [19a]. Furthermore, various other reactions of serine proteases including the deacylation reactions of α-N-benzoyl-L-arginyltrypsin and α-N-benzoyl-L-arginylthrombin and the acylation of elastase by α-N-carbobenzyloxy-L-alanine p-nitrophenyl ester all exhibit linear dependences of V_n on n. Figure 4 shows typical examples of these results. It is quite clear that the serine proteases are fully capable of exhibiting strong catalytic effect *without* coupled motion of the protons of the hydrogen bond chain linking Ser-195 or nucleophilic water through His-57 to Asp-102.

It was at this point that the modern experimentalist could turn with a reasonable degree of confidence to chemical theory. Experiments showed the charge-relay chain to be uncoupled. Of theory it might be asked, Why? Is the chemical constitution of the serine protease hydrogen bond chain ($\mathrm{ROH}\cdots$ $\mathrm{ImH}\cdots\bar{\mathrm{O}}_2\mathrm{CR}$) such as to render it unfit for coupled proton shifts? Or are there perhaps certain geometrical circumstances of the active site which make it unfavorable for coupling?

Theory of Coupling in Hydrogen Bond Chains

To examine these two questions, Gandour *et al.* [21] employed the INDO semiempirical quantum mechanical technique to generate potential-energy surfaces for proton motions in hydrogen bond chains. The studies have since been extended by Gandour and Rodgers [22]. The fundamental strategy involved the process of Eq. (7). In this scheme, the identities of A, B, and C

$$\underset{\underset{R_1}{\underbrace{}}}{\overset{\overset{r_a}{\overbrace{}}}{{}^+\mathrm{A}\!-\!\mathrm{H}}}\quad \underset{\underset{R_2}{\underbrace{}}}{\overset{\overset{r_b}{\overbrace{}}}{\mathrm{B}\!-\!\mathrm{H}}}\quad :\mathrm{C} \longrightarrow \begin{bmatrix} \mathrm{A}: & \mathrm{H}\!-\!\overset{+}{\mathrm{B}}\!-\!\mathrm{H} & :\mathrm{C} \\ \mathrm{A}\cdots\mathrm{H}\cdots\mathrm{B}\cdots\mathrm{H}\cdots\mathrm{C} \\ {}^+\mathrm{A}\!-\!\mathrm{H} & \bar{\mathrm{B}}: & \mathrm{H}\!-\!\mathrm{C}^+ \end{bmatrix} \longrightarrow \mathrm{A}: \mathrm{H}\!-\!\mathrm{B}\ \mathrm{H}\!-\!\mathrm{C}^+ \tag{7}$$

can be varied to test the effect of chemical constitution, and the values of R_1 and R_2 can be varied to test the effect of overall geometry. For any given system, a potential-energy surface can be calculated and presented either in three-dimensional form (as a "transect diagram"; Fig. 5) or as a contour map. The minimum-energy path (MEP) on such a surface may represent either uncoupled motion of the two protons (routes around the circumference of the map in which r_a and r_b change sequentially) or coupled motion (a route passing across the center of the map such that r_a and r_b change in concert). Uncoupled motion generates intermediate structures such as a cationic intermediate [top

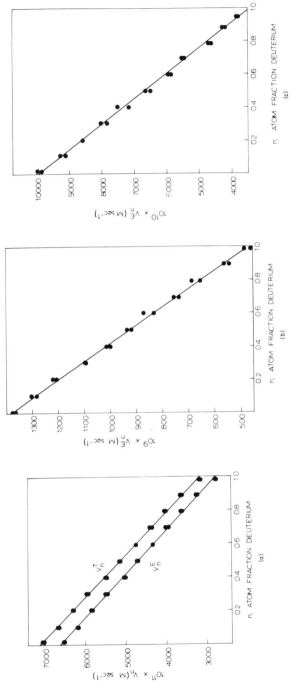

Fig. 4 (a) Variation of the zero-order rate constants for the deacetylation of acetylelastase as a function of the atom fraction n of deuterium in binary mixtures of protium oxide and deuterium oxide. The rates were determined for the elastase-catalyzed hydrolysis of p-nitrophenyl acetate. The term V_n^E represents the rates of the enzyme-catalyzed reaction, which were obtained by correction of the total hydrolysis rate V_n^T for the small component of buffer-catalyzed reaction. The solid lines are linear least-squares fits to the data (statistically significant with $P < 0.001$). A quadratic fit, although significant at $P < 0.01$, yields an isotope effect of 1.00 for the second proton. (b) A similar plot for deacylation of α-N-benzoyl-L-arginyltrypsin. Here the rates are for trypsin-catalyzed hydrolysis of α-N-benzoyl-L-arginyl ethyl ester, and the background reaction is negligible. Again, the solid line represents a least-squares fit (statistically significant with $P < 0.001$). In an attempted quadratic fit, the term in n^2 is not significant at $P < 0.2$. (c) The same experiment as in (b) with thrombin as catalyst. The linear fit again has $P < 0.001$, while the quadratic term is not significant at $P < 0.1$.

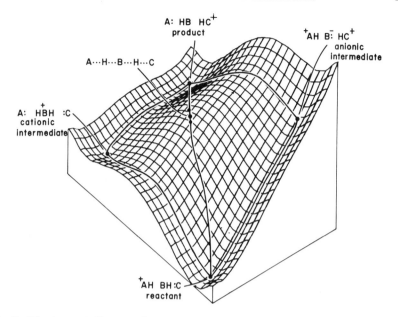

Fig. 5 The transect diagram of a potential-energy surface for a hydrogen bond chain undergoing proton transfer as in Eq. (7). The lines connecting the dots illustrate possible reaction paths for motion of the two protons. The path passing through the cationic intermediate is the MEP for this surface. The other uncoupled route, through the anionic intermediate, is a higher-energy path. The coupled route, passing across the center of the diagram, does not include a true transition state on this surface, since the region near the center does not form a saddle point (as required for an actual transition state), but rather represents a local maximum with respect to both coordinates.

structure on the arrow of Eq. (7)] or an anionic intermediate [bottom structure on the arrow of Eq. (7)]. Coupled motion generates only a transition state such as the center structure on the arrow of Eq. (7). In the calculations, the overall distances R_1 and R_2 are held constant, simulating the constraint of enzyme structural features on the hydrogen bond distances in the charge-relay system.

The results can be inferred from the transect diagrams of Fig. 6. Figure 6a shows surfaces for four quite different chemical constitutions, all with the distances R_1 and R_2 chosen as quite long, that is, 3.0 Å. The surfaces all appear rather similar and all have MEP's only of the uncoupled variety. A contrast is offered by the set of surfaces in Fig. 6b. Here the chemical constitutions cover the same range as before and again all the surfaces are similar in character to each other. Now R_1 and R_2 have been shortened to 2.75 Å, and it is apparent that coupled pathways have become far more favorable.

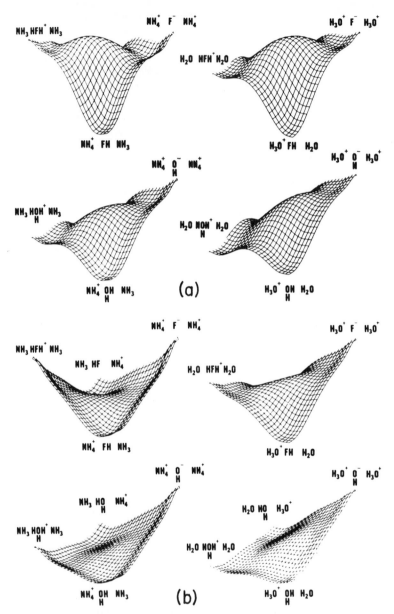

Fig. 6 (a) Transect diagrams for hydrogen bond chains of various chemical constitutions, shown on the diagrams ($R_1 = R_2 = 3.0$ Å for all surfaces). The minimum-energy paths on all these surfaces are uncoupled. (b) Transect diagrams for the same systems as in (a) but with R_1 and R_2 now reduced to 2.75 Å. This compression greatly favors coupled bonding changes for the two protons. Details of these calculations are available in Gandour *et al.* [21]. Reprinted with permission from *J. Am. Chem. Soc.* **96**, 6967 (1974). Copyright by the American Chemical Society.

Indeed, if R_1 and R_2 are shortened to 2.5 Å, the central region of these diagrams becomes a global energy minimum [21]!

The conclusions that theory presents at this junction to the experimentalist are as follows (see Gandour *et al.* [21] for details):

1. The chemical constitution of a hydrogen bond chain is relatively unimportant (compared to geometrical factors) in determining the degree of coupling in its protonic motions. Thus, there is no reason to believe that the chemical structures constituting the charge-relay system of the serine proteases are responsible for its uncoupled behavior in the reactions discussed above.

2. The distances across the component hydrogen bonds of the charge-relay system are likely to be critical to its coupling; short distances favor coupling, and longer distances favor uncoupled behavior. Relatively small changes, of the order of tenths of an angstrom unit, may couple or uncouple the system.

It is important to note that these conclusions do not depend on any of the detailed features of the method of theoretical calculation nor on the exact choice of models nor on the precision of the energies obtained. For example, Fig. 7 shows that exactly similar results are obtained when the system is unsymmetrical. Figure 8 represents a calculation with imidazole as the bridging moiety, just as in the enzymatic charge-relay chain; the surface is extremely similar to those of Fig. 6a. The conclusions are of a general and qualitative character and are, in fact, completely consistent with simple considerations based on a building-up principle from hydrogen bond potential functions [21].

Subsequent and Consequent Experimental Developments

The theoretical treatment thus indicated that relatively minor changes in the distance across the charge-relay system in the catalytic transition state might make the difference between coupled and uncoupled bonding changes at the two protonic centers. Two lines of experimental evidence suggested that just such changes might arise from a lengthening of the substrate peptide chain beyond that already examined (a specific peptide unit plus an *N*-acyl function). These lines of evidence were (1) the findings of Huber and his colleagues that a structural compression in the charge-relay system accompanied the binding of the long-chain trypsin inhibitors (which may simulate the substrate part of the catalytic transition state) [23] and (2) the discovery by Thompson and Blout [24] that residues at the third and more remote peptide positions from that attacked catalytically are important in elastase action.

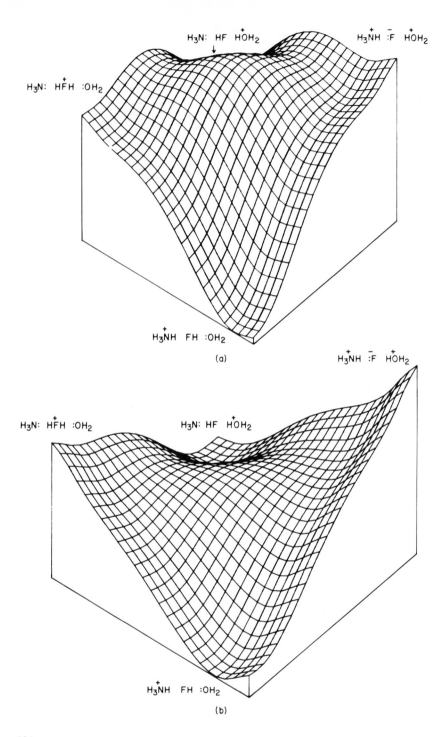

$H_3N: HF \overset{+}{H}OH_2$

$H_3\overset{+}{N}H :\overline{F} \overset{+}{H}OH_2$

$H_3N: H\overset{+}{F}H :OH_2$

$H_3\overset{+}{N}H FH :OH_2$

(a)

$H_3\overset{+}{N}H :\overline{F} \overset{+}{H}OH_2$

$H_3N: H\overset{+}{F}H :OH_2$

$H_3N: HF \overset{+}{H}OH_2$

$H_3\overset{+}{N}H FH :OH_2$

(b)

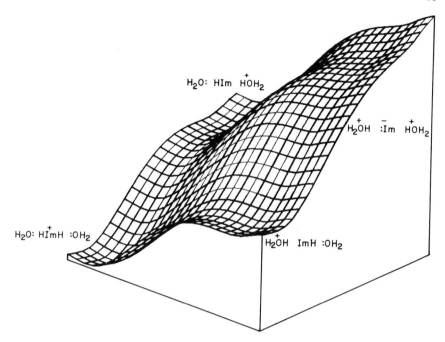

Fig. 8 An uncoupled minimum-energy pathway is observed when imidazole forms the bridge in a system with $R_1 = R_2 = 3.0$ Å. In fact, the surface is nearly identical in character to the surface for $H_2OH^+ \cdots OH_2 \cdots OH_2$ in Fig. 6a.

Therefore, the possibility did not seem remote that a longer-chain substrate might lead to coupling of the charge-relay system in the catalytic transition state. Indeed, this is found to be true. In Fig. 9 we present the proton inventory results for three substrates of trypsin, all examined under deacylation conditions. The very small acetyl group generates one-proton catalysis and a quite small isotope effect ($k_H/k_D = 1.4$). The N-benzoyl-L-arginyl group, simulating a dipeptide, continues to be removed with one-proton catalysis, although the isotope effect is considerably increased. However, the pseudo-tetrapeptide group N-benzoyl-L-phenylalanyl-L-valyl-L-arginyl couples the charge-relay system. Here, $V_n(n)$ is quadratic with each transition-state proton generating a kinetic isotope effect $k_H/k_D \sim 2$. The prediction that arises from a rather complex mixture of theoretical and experimental findings, that a sufficiently long polypeptide substrate of a serine protease will bring

---

Fig. 7 (a) Potential-energy surface for the $NH_4^+ \cdots FH \cdots OH_2$ system with $R_1 = R_2 = 3.0$ Å. Note the similarity to Fig. 6a. (b) A surface for the same system with $R_1 = R_2 = 2.75$ Å, showing how the shorter distance favors the coupled route as in Fig. 6b.

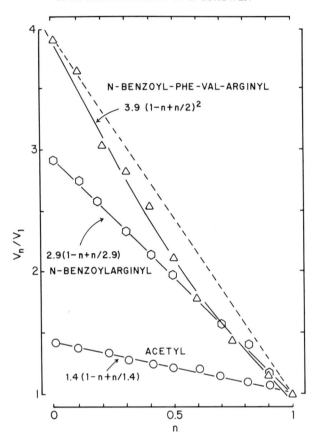

Fig. 9   A series of proton inventory results for three examples of trypsin catalysis. With a very small substrate-derived function (acetyl, lowest line), the enzyme is a one-proton catalyst with a small isotope effect. Increasing the size of this function to that of a dipeptide analog (*N*-benzoylarginyl, middle line) produces a larger isotope effect, but the enzyme remains a one-proton catalyst. Finally, a tetrapeptide analog (*N*-benzoylphenyl-alanylvalylarginyl) couples the charge-relay chain, and the enzyme becomes a two-proton catalyst.

about coupling of the bonding changes in the charge-relay system of the enzyme in the catalytic transition state, appears to be confirmed.

Summary

The historical development we have just recounted is summarized in Fig. 10. A notable feature of this history is the "great collaborative expertise of science" which so often daunts the practitioners of other arts (see the

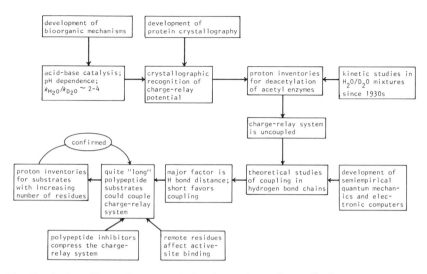

Fig. 10 A chart illustrating theoretical and experimental contributions to the development of current concepts in charge-relay catalysis.

remarks of J. Holloway in "A Reply to Sir Peter Medawar" [25]). The mechanistic background of physical and organic chemistry first led to the postulation of acid–base catalysis in the serine proteases, and this prepared the minds of Blow and his collaborators, themselves building on a long heritage of crystallographic development, to realize the possible significance of the charge-relay system when they observed it. From another direction, the mechanistic applications of H_2O–D_2O mixtures ever since the discovery of deuterium suggested this technique as a natural approach to testing for charge-relay catalysis. The development of large-scale computational quantum chemistry was an absolute prerequisite to calculation of potential-energy surfaces for hydrogen bond chains. The importance of geometrical factors in charge-relay coupling emerged from these calculations almost coincidently with the crystallographic and enzymological data that suggested the importance of polypeptide character in the substrate for coupling of the charge-relay system.

In future work, it remains to be established whether coupling of the charge-relay system gives rise to any substantial catalytic acceleration or if perhaps it is a mere structural and dynamic coincidence that it is coupled with physiological substrates but not with "small" substrates. The simplest conclusion at the moment is that no large acceleration is associated with coupling, but the problem merits further attention.

ENZYME CATALYSIS AND TRANSITION-STATE STRUCTURE

Fundamental Concepts

A major task, perhaps the most pressing one, facing researchers in bio-organic mechanisms today is the elucidation of the origins of enzyme catalytic power. Several decades of work have led to a wealth of general concepts and a wide spectrum of opinion, but a full accounting of the catalytic capability of any enzyme remains absent.

An approach we have taken to this task is based on the postulate that we can understand the chemical features of enzyme catalysis if we can answer the following questions:

1. How do the structure and energy of the substrate-derived portion ("core") of the transition state for enzyme catalysis compare to the corresponding structures for uncatalyzed, equivalent reactions?

2. Given this core structure for the enzymatic reaction, how does it interact with the enzyme-derived portion of the catalytic transition state?

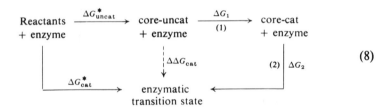

$$(8)$$

As the scheme of Eq. (8) shows, the difference in free energies of activation between the nonenzymatic reaction ($\Delta G^*_{\text{uncat}}$) and the enzymatic reaction (ΔG^*_{cat}) is given by the negative quantity $\Delta\Delta G_{\text{cat}}$, which measures the catalytic acceleration produced by the enzyme.* The larger the absolute magnitude of $\Delta\Delta G_{\text{cat}}$, the more efficacious the enzyme. If $\Delta\Delta G_{\text{cat}} = -13.6$ kcal/mole at 298 K, the enzyme effects 10^{10}-fold acceleration; if $\Delta\Delta G_{\text{cat}} = -27.2$ kcal/mole, a 10^{20}-fold acceleration, etc. Our postulate above envisions the hypothetical dissection of $\Delta\Delta G$ into two contributions. The first, ΔG_1, measures the free energy required to alter the uncatalyzed-reaction transition-state core structure [core-uncat in Eq. (8)] to the structure it will possess in combination

---

* The formulation of Eq. (8) applies to enzyme catalysis when substrate concentration is well below the dissociation constant of the enzyme–substrate complex. The concept must be reformulated to include inhibition by substrate stabilization when saturation becomes appreciable. It is also important to remember that different transition states (or sets of transition states) may determine the rate under different conditions.

with the enzyme [core-cat of Eq. (8)]. This is the quantity relevant to question 1 above. The second component, ΔG_2, represents the free energy released on development of the interactions between enzyme and core-cat. It is relevant to question 2 above.

An important role that theory will eventually play in this field is in estimating the energies associated with the distortion of core-uncat to core-cat (i.e., ΔG_1). However, at this point, the major contribution has been in development of experimental methodology for determining the structural differences between core-uncat and core-cat. It is this interaction of theory and experiment that we shall discuss here.

Kinetic Isotope Effects: Model Calculations

The most promising experimental approach to transition-state structures in both enzymatic and nonenzymatic reactions lies in the field of kinetic isotope effects. These measurements directly probe the transition state. Inasmuch as they do not perturb the charge distributions but only nuclear masses, they leave the potential-energy surface of the reaction unchanged. Isotopic substitution being a minimal perturbation, it is ideal for investigation of sensitive enzymatic systems. The theoretical apparatus that is interposed between a measured kinetic isotope effect and its structural interpretation is much more reliable than the theoretical transformations that connect all other mechanistic probes to their interpretations.

A protocol for using kinetic isotope effects to investigate enzyme catalytic power is outlined in Fig. 11. As shown at the top of the figure, the desired end is the interpretation of experimentally observed enzymatic accelerations in terms of enzymatic transition-state structure and interactions as deduced from kinetic isotope effects. To relate the experimental isotope effects to possible transition-state structures, we use the Bigeleisen equation [26] [Eq. (9)] in the context of the computational procedures worked out by Wolfsberg and Stern [27]. Equation (9) exhibits the fact that the isotopic rate ratio, k/k', depends only on the set of ν_{Ti} and ν'_{Ti} (the vibration frequencies of the two isotopic transition states) and the set of ν_{Rj} and ν'_{Rj} (the vibration frequencies of the two isotopic reactant molecules). These frequencies, or equivalent information such as bond force constants and molecular structures, are usually

$$k/k' = \frac{\displaystyle\prod_{i}^{3N_T-6} (\nu_{Ti}/\nu'_{Ti}) \prod^{3N_T-7} e^{-h(\nu_{Ti}-\nu'_{Ti})/2kT}(1-e^{-h\nu'_{Ti}/kT})/(1-e^{-h\nu_{Ti}/kT})}{\displaystyle\prod_{j}^{3N_R-6} (\nu_{Rj}/\nu'_{Rj}) \, e^{-h(\nu_{Rj}-\nu'_{Rj})/2kT}(1-e^{-h\nu'_{Rj}/kT})/(1-e^{-h\nu_{Rj}/kT})} \tag{9}$$

available for reactant molecules from infrared and Raman spectroscopy. If this is so, and k/k' is measured, then the ν_{Ti} and ν'_{Ti} for the transition state are

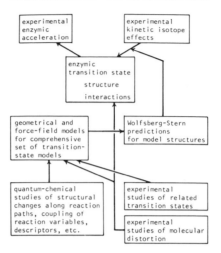

Fig. 11 A program for explaining enzyme catalytic accelerations in terms of enzymatic transition-state structures through the use of kinetic isotope effects for establishment of these structures. The structural interpretation of the kinetic isotope effects requires a complex interplay of theoretical and experimental methods and results.

the only unknown quantities in Eq. (9). Although Eq. (9) cannot be "solved" for the ν_{Ti} and ν'_{Ti} (for a transition state containing five atoms, the ν_{Ti} alone are nine in number while a maximum of five different k/k' measurements can be made at any one temperature), it is possible to choose a model structure (consisting of bond lengths, angles, and force constants) for the transition state and to generate a value of k/k' from it. Comparison with an experimental value of k/k' tests the chosen model for consistency with experiment. The technique is exactly analogous to the fitting of structural models for stable molecules to vibrational spectroscopic data.

Indeed, the Schachtschneider–Snyder vibrational analysis program (based on the Wilson **FG** method [28]) is incorporated in the program developed by Wolfsberg and Stern for isotope effect calculation.* Its mode of operation is summarized in Fig. 12. The input data for the reactant state are readily obtained, and indeed observed reactant-state frequencies may be used directly if it is desired.

Development of Model Transition-State Structures

The trouble comes, of course, with the transition state. We are not so fortunate as the vibrational spectroscopist when he sits down to consider the

---

* A computationally more efficient procedure has now been programmed by Professor L. B. Sims, Department of Chemistry, University of Arkansas, Fayetteville, Arkansas. This version avoids the **FG** formalism.

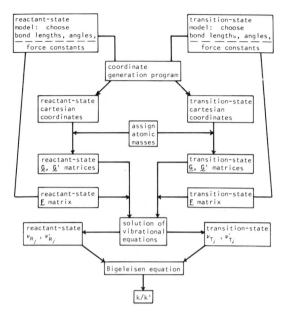

Fig. 12 Calculation of kinetic isotope effects (k/k') from transition-state model structures by the Wolfsberg–Stern procedure. Geometrical and force-field models are used to generate vibration frequencies for the isotopic species by the Wilson **FG**-matrix method. The Bigeleisen equation then provides the corresponding isotope effects.

range of possible models that may fit his spectrum. Behind him he has the intellectual legacy of more than a century of structural chemistry. Guided by the rules of this field he quickly eliminates all of the large number of "chemically unreasonable" structures consistent by logic alone with the molecular formula. There remain only a few possibilities, in general, which must be tested against the spectrum.

We have essentially no chemical history in the field of transition-state structures, but in its place information from three other sources can come to our aid. These sources are BEBO (bond energy, bond order [29]) approaches to gas-phase transition states, crystallographic results on "frozen transition states" or molecular distortions in the solid phase [30], and exploration of structures along reaction paths by molecular quantum mechanics.

Acyl-Transfer Transition States

The nature of the problem can be formulated for acyl-transfer transition states. The acyl-transfer process [Eq. (10)] has been the subject of extensive mechanistic study in both enzymatic and nonenzymatic systems [31] and thus

$$
\underset{\substack{\text{R}^{C}\diagdown\text{X}\\ +\ \text{Y}^{(-)}}}{\overset{\text{O}}{\parallel}} \rightleftharpoons \left[\underset{\substack{\text{R}^{C}\diagup\text{X}\\ \text{Y}}}{\overset{\text{O}}{\parallel}} \right] \rightarrow \underset{\substack{\text{R}^{C}\diagup\text{X}\\ \text{Y}}}{\overset{\text{O}^{-}}{\mid}} \rightleftharpoons \left[\underset{\substack{\text{R}^{C}\diagdown\text{X}\\ \text{Y}}}{\overset{\text{O}}{\parallel}} \right] \rightarrow \underset{\substack{\text{R}^{C}\diagdown\text{Y}\\ +\ \text{X}^{(-)}}}{\overset{\text{O}}{\parallel}}
$$

$$(10)$$

forms an excellent candidate for study by the techniques described above. Acyl-transfer enzymes are biologically important in events as disparate as digestion, reproduction, and function of the central nervous system. At the same time, the heavy structural and mechanistic efforts that have focused on such acyl-transfer enzymes as the serine proteases put them in the class of the best understood biological catalysts. Work in model systems has supported the formulation of Eq. (10), in which formation of the bond to the acyl acceptor occurs in a first step, generating a "tetrahedral" intermediate (in which the central carbon achieves sp^3 hybridization), and fission of the bond from the acyl donor occurs in a second step, regenerating the trigonal (sp^2-C) structure that characterizes both reactant and product. It is reasonable to assume that the structure of either transition state lies between the limiting extremes of trigonal and tetrahedral, and we can therefore in principle characterize the effective transition state for an acyl-transfer reaction (which-ever step determines the rate) by its "tetrahedrality." The principal structural features that differ between the trigonal and tetrahedral structures along the reaction path are as follows [see Eq. (10)]:

1. The bond length to the nucleophile (X), which decreases in the first step of Eq. (10), and the bond length to the leaving group (Y), which increases in the second step of Eq. (10)

2. The carbonyl bond C—O, which lengthens as the formal double bond of the trigonal structure is converted to the formal single bond of the adduct

3. The planar trigonal structure, which undergoes out-of-plane distor-tion to the pyramidal form it possesses in the adduct

Figure 13 shows a structural representation of these changes for the nucleo-philic attack of NH_3 on $CH_2{=}O$. The question we need to answer in order to write down a comprehensive set of transition-state model structures spanning the range from completely planar to completely tetrahedral is, How do the C—X or C—Y bond length changes correlate with the C—O bond length changes and the out-of-plane distortion? When the C—X bond is half-formed, to what extent have the other changes proceeded?

BEBO Contributions

Our first requirement is to define the meaning of the phrase "the C—X bond is half-formed." Here BEBO theory [29] helps us in two ways. Its

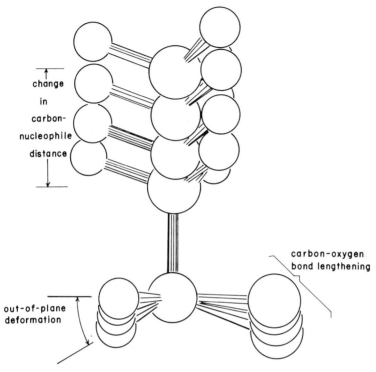

change
in
carbon-
nucleophile
distance

carbon-oxygen
bond lengthening

out-of-plane
deformation

Fig. 13 A "ball-and-stick" representation of structural changes along the reaction path for nucleophilic attack by NH_3 (approaching from above) on $CH_2{=}O$. As the C—N bond forms, the formaldehyde moiety experiences an out-of-plane deformation and the C—O bond (pointing to the right) becomes longer.

successful treatment of gas-phase transition states for simple reactions such as hydrogen atom transfer suggests that an appropriate descriptor for the degree of bond formation is the Pauling bond order, B, given [32] by Eq. (11), in which r_B is the length of a partial bond of order B and r_1 is the length of a full bond. The coefficient is obtained from a plot of $\ln B$ vs. bond distance for triple ($B = 3$), double ($B = 2$), and single ($B = 1$) bonds between first-row elements. A choice of C—X or C—Y bond distance thus provides a bond

$$B = e^{(r_1 - r_B)/0.31} \tag{11}$$

order B for this bond, which varies from $B \sim 0$ for very long bonds to $B = 1.0$ for $r_B = r_1$. A further, very useful result of the BEBO method is that the force constants of partial bonds are proportional to the Pauling bond order of the partial bond [Eq. (12)]. This will be of obvious use in generating

$$f_B = Bf_1 \tag{12}$$

force fields from geometrical model structures in order to calculate isotope effects.

Crystallographic Contributions

The next step comes from parallel studies in crystallography and quantum chemistry, which yield similar and complementary results. The crystallographic findings are those of Bürgi and his collaborators [33]. These workers have examined crystal structures for compound (1) and seven other compounds, all of which contain tertiary amino groups and carbonyl groups. As

(1)

emphasized in (1), interaction between these functions may occur in the solid phase. The molecules examined had C—N distances ranging from 1.49 Å ($B \sim 1$) to 2.91 Å ($B \sim 0$) and thus provided, in Burgi's words [30], "a series of snapshots along a continuous reaction path describing nucleophilic addition of amine nitrogen to carbonyl carbon." The development of pyramidal character about the carbonyl carbon was measured by the distance Δ by which the carbonyl carbon departed from the plane defined by its three substituents. This quantity varied from 0.064 Å when $B \sim 0$ to 0.42 Å when $B \sim 1$. Most interestingly, $(\Delta/0.42)^2$ (the fractional "pyramidalization" of the carbonyl group [30]) is about equal to B for the C—N bond, as calculated from Eq. (11)! Similarly, if it is assumed that B for the carbonyl bond, B_{CO}, is just equal to $2 - B_{CN}$ (conservation of bond order at carbon), then Eq. (10) correctly accounts for the observed C—O bond lengths in the structures examined. The conclusion is that the Pauling bond order of the carbonyl–nucleophile bond, B_{CN}, seems to be a universal structural descriptor of the transition state; structural alterations such as increase in C—O bond length and out-of-plane distortion will have proceeded to a fractional extent measured by B_{CN}.

Such a conclusion, if valid, would permit the immediate generation of a comprehensive set of transition-state model structures between $B_{CN} = 0$ and $B_{CN} = 1$. Structural features other than C—N bond length can be calculated from B_{CN} and known values for trigonal and tetrahedral states, and force constants can be calculated from Eq. (12) or similar relationships. However, as Bürgi [30] has noted, "the structural parameters [in the treatment above] are obtained from molecules and crystal lattices at equilibrium and do not yield energetic information." Thus, the hypothesis that the correlation observed "represents a model for the structural course of the reaction or, in other words, coincides with a valley of the corresponding energy surface... has to be justified, at least in a qualitative sense...."

Quantum Chemical Contributions

To provide such a justification, Bürgi *et al.* [34] carried out Hartree–Fock SCF–LCGO–MO calculations on the reaction of hydride ion, H^-, with formaldehyde, $CH_2{=}O$, to yield $CH_3{-}O^-$. The results strongly support the view that the crystallographic data accurately reproduce structural correlations along the reaction pathway for carbonyl addition or elimination. In our group, Hogan, Gandour, and Kessler have examined many features of the reaction pathways for addition of NH_3 [35] and HO^- [36] to formaldehyde, using the INDO semiempirical method. We shall describe some of these findings here.

Figure 14 shows the calculated changes in energy and structure for addition of ammonia to formaldehyde. As has been observed in studies such as this,* the reaction path does not display an energy maximum. Further, the energy of formation of the adduct (-122 kcal/mole) is unrealistically exothermic, suggesting an underestimation of the stability of the free reactants. Nevertheless, it is reasonable to hope that the representation of structural correlations along the reaction path is accurate. Wherever comparisons can be made (Fig. 15) between these calculations and either the *ab initio* calculations or the crystallographic findings, the agreement is such as to increase confidence in this hope.

Reaction Variables

In Fig. 16, we see a representation of the structural variables to be considered along the reaction path. Some of these are fundamental to the description of the transition state itself:

r_{CN} or B_{CN}: The carbon-to-nucleophile distance or bond order
r_{CO} or B_{CO}: The carbonyl bond distance or bond order

---

* *Ab initio* quantum mechanical calculations on a variety of S_N2 reactions [A. Dedieu and A. Veillard, *J. Am. Chem. Soc.* **94**, 6730 (1972)] show that the transition-state configuration corresponds to a global energy minimum if a minimal basis set is employed. Only when extensive basis sets are used is the expected reaction barrier generated. Minimum basis set *ab initio* calculations on amide hydrolysis also show no reaction barrier [G. Alagona, E. Scrocco, and J. Tomasi, *J. Am. Chem. Soc.* **97**, 6976 (1975)].

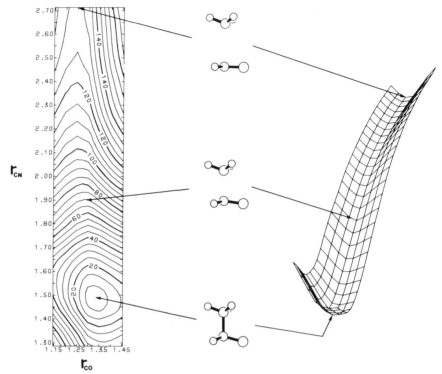

Fig. 14 Energy changes along the minimum-energy path for reaction of NH_3 with $CH_2{=}O$, as calculated by the INDO method. "Ball-and-stick" structures (drawn to scale) are shown for "reactantlike," medial, and adduct structures. The medial structure corresponds to the point at which one-half of the energy of reaction has been released [35].

β: An angle that measures the out-of-plane distortion of the carbonyl function

θ: The angle between the carbonyl substituents

If these variables change in some degree in a correlated manner, then to that extent it is sensible to speak of a single descriptor of transition-state structure (e.g., a "20% tetrahedral transition state"). If their changes are instead absolutely uncorrelated, then it will make sense only to perform sufficient experiments to elucidate the magnitude of each variable independently and to specify, therefore, the structure of the transition state in detailed terms: a C—N bond length of x Å, a C—O bond length of y Å, a β angle of z degrees, etc.

It is also necessary, of course, to relate the geometrical structures (whatever approach is used) to variables that are directly probed by particular experiments. Isotope effects, for example, arise from changes in force constants

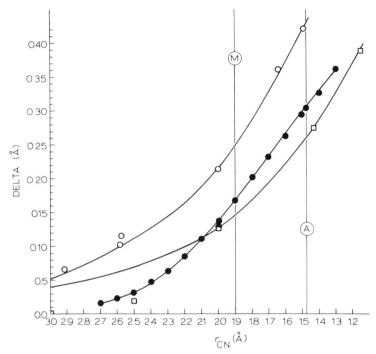

Fig. 15 The progress of "tetrahedralization" as a function of the C—N bond length along the reaction path for nucleophilic attack on the carbonyl function, as derived from crystallographic measurements [33] (○), *ab initio* quantum mechanical calculations for hydride addition to formaldehyde [34] (□), and INDO quantum mechanical calculations for ammonia addition to formaldehyde [35] (●). The quantity Δ, which measures the degree of tetrahedral character, is the perpendicular distance of the carbonyl carbon atom from a plane defined by the carbonyl oxygen atom and the other two carbonyl substituents. The points labeled M and A correspond to the medial and adduct structures of Fig. 14. The agreement between the experimental and theoretical results is quite satisfactory.

along the reaction path, and substituent effects arise from changes in charge distribution. In this connection we have examined the following properties along the reaction path which relate to possible experiments:

K_β: The bending force constant for out-of-plane distortion; related to the α-deuterium secondary isotope effect

K_{CO}: The carbonyl stretching force constant; related to a $C^{18}O$ isotope effect and to a ^{13}CO or ^{14}CO isotope effect

ΔQ_N: The charge alteration on the nucleophilic center; related to a substituent effect experiment in which nucleophile structure is varied

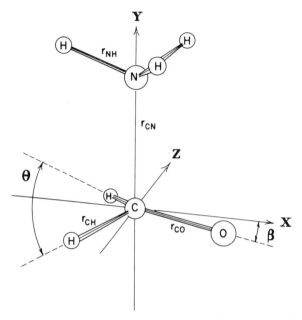

Fig. 16 Definitions of the reaction variables used to describe structures along the minimum-energy path for ammonia addition to formaldehyde.

ΔQ_R: The charge alteration at the carbonyl substituent; related to substituent effects in which these structures are varied

Normalized Reaction Variables and Coupling of Variables

Now the correlation of any one of these variables with any other, in a general sense, presents a primitive problem in dimensions: The distances are in angstrom units, the angles in degrees, etc. This problem is most readily disposed of through the use of *normalized reaction variables*. Such a variable is defined [35] by Eq. (13), where $\Delta \xi_i^{tot}$ represents the entire change suffered by the ith variable ξ_i during complete reaction, $\Delta \xi_i(\lambda)$ is the change experienced up to point λ along the reaction path, and $\hat{\xi}_i(\lambda)$ is the value of the normalized variable at point λ. Any $\hat{\xi}_i$ will thus vary between zero (reactant

$$\hat{\xi}_i(\lambda) = |\Delta \xi_i(\lambda)/\Delta \xi_i^{tot}| \tag{13}$$

state) and unity (product state). A plot of $\hat{\xi}_1$ vs. $\hat{\xi}_2$ (Fig. 17) will then produce a line that displays the degree of correlation of the two reaction variables.*

* In case of perfect correlation, this line will pass diagonally across the diagram and have a dimensionless length of $\sqrt{2}$. In case of perfect anticorrelation, $\hat{\xi}_1$ will change completely independently of $\hat{\xi}_2$, and the line will pass around the edge of the diagram (by either a "northwest" or "southeast" route), with a dimensionless length of 2.

Level of Resolution of Transition-State Structure

Plots of $\hat{\xi}_1$ vs. $\hat{\xi}_2$ for addition of ammonia to formaldehyde are shown for several reaction variables in Fig. 18. Since none of the lines representing the relationship of $\hat{\xi}_1$ and $\hat{\xi}_2$ lies precisely along the diagonal, no combination of the variables changes in precise concert. However, they all change in a reasonably concerted manner, and an interesting generalization emerges about the small departures from exact coupling: If a given change is energetically "easy" then its fractional progress will be large at an early point on the reaction path. Thus, as the nucleophile–carbon bond forms, both the out-of-plane angle β and the carbonyl C—O distance increase. The former distortion (a bending motion) is "easier" to effect than the latter stretching motion and should therefore be fractionally more advanced early on the reaction path. As Fig. 18a shows, this is the case.

The question, to what degree, for example, can the C—N bond order $\hat{B}_{CN}$ be employed as a universal descriptor of transition-state structure, can be considered with the aid of Fig. 19. It will be recalled that the crystallographic studies indicated that this would be a strong possibility. The question is examined in more detail here. In this figure, the fractional progress of a number of important reaction variables is compared at points along the reaction path, as measured by the C—N distance r_{CN}. One variable shown is the average of three bond order values for the C—N bond (the Wiberg bond index, Ehrenson–Seltzer bond parameter, and Pauling bond order). The Pauling bond order $\hat{B}_{CN}$ changes most "slowly" of these and, if it had been plotted, the corresponding point on the figure would have appeared to the right of the average point. The conclusion one must draw from Fig. 19 is that no universal descriptor is likely to be generally valid if sufficiently precise experimental probes can be brought to bear on transition states. For example, the points at which the different variables reach 50% completion of their

---

One measure [19a] of the degree of correlation in intermediate cases is the length L of the line $\hat{\xi}_1(\hat{\xi}_2)$. The quantity ω defined in Eq. (14), based on L, varies from zero (perfect

$$\omega = (2 - L)/(2 - \sqrt{2}) \qquad (14)$$

anticorrelation) to unity (perfect correlation) and has been used in consideration of hydrogen bond chains [19a]. This definition is disadvantageous, however, because ω drops from unity only rather slowly as L departs from $\sqrt{2}$. A better measure of correlation would seem to be the function Γ of Eq. (15). This also varies from zero for anti-

$$\Gamma = 1 - 2\left| \int_0^1 \hat{\xi}_1(\hat{\xi}_2)\, d\hat{\xi}_2 - (1/2) \right| \qquad (15)$$

correlation to unity for correlation but employs the area between the line $\hat{\xi}_1(\hat{\xi}_2)$ and the diagonal as a measure of departure from correlation. The integral can be determined analytically, e.g., by fitting a polynomial to $\hat{\xi}_1(\hat{\xi}_2)$ by regression analysis, or graphically by using a planimeter or weighing of paper, as in gas chromatography.

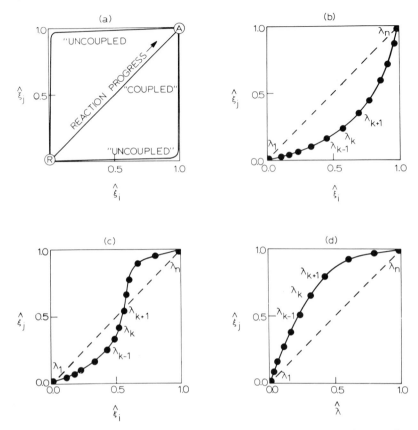

Fig. 17 Diagrams of normalized reaction variables which illustrate various modes of coupling. (a) A coupling diagram of $\hat{\xi}_i$ vs. $\hat{\xi}_j$ produces a diagonal straight line connecting the reactant point R to the product point A if the $\hat{\xi}_i$ and $\hat{\xi}_j$ are perfectly coupled while, if they are perfectly uncoupled, the line passes around one edge of the diagram (both possibilities are shown). (b) A coupling diagram in which $\hat{\xi}_i$ and $\hat{\xi}_j$ exhibit an intermediate degree of coupling; the filled circles might represent quantum mechanical calculations at points λ_{k-1}, λ_k, λ_{k+1}, etc., along the reaction path. (c) A coupling diagram illustrating a more complex relationship of $\hat{\xi}_i$ and $\hat{\xi}_j$. (d) A coupling diagram in which one reaction variable is chosen to be the normalized length $\hat{\lambda}$ along the reaction path.

overall changes occur over a range of r_{CN} from 2.0 to 1.85 Å. As Fig. 19 shows, some of the reaction variables change by as much as 25% over this range, and thus this defines an error necessarily inherent in describing a transition state as "50% tetrahedral."

It is, however, true that in many cases a large error (such as 25%) in a normalized variable does not correspond to a large error in absolute structural terms. The total change in bond distance experienced by the C—O bond, for

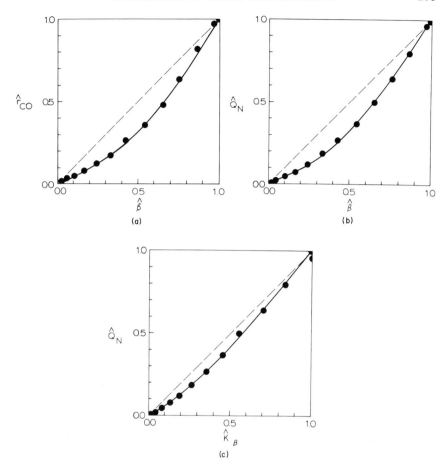

Fig. 18 Coupling diagrams illustrating the degree of coupling among various normalized reaction variables for ammonia addition to formaldehyde, calculated by the INDO method. (a) The normalized C—O bond length $\hat{r}_{CO}$ is not precisely coupled to the normalized out-of-plane angle $\hat{\beta}$, but rather the angular distortion precedes the more difficult bond stretching along the reaction path. (b) The normalized charge development on nitrogen $\hat{Q}_N$, compared to the normalized out-of-plane distortion angle $\hat{\beta}$. This plot suggests a relationship between the experimentally measured substituent effect in the nucleophile (which is determined by $\hat{Q}_N$) and the transition-state structural variable $\hat{\beta}$. (c) The normalized charge development on nitrogen $\hat{Q}_N$, compared to the normalized out-of-plane bending force constant $\hat{K}_\beta$. Here a connection is suggested between two experimentally measurable quantities, the nucleophile substituent effect (related to $\hat{Q}_N$) and the α-deuterium isotope effect (related to $\hat{K}_\beta$). In all three plots, the dashed line represents the expected dependence for perfect coupling. Note that the degree of coupling is greatest in (c) and least in (a).

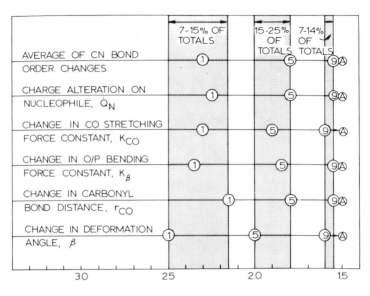

Fig. 19 Changes in normalized reaction variables as a function of the nitrogen–carbon distance r_{CN} (plotted on the abscissa) along the reaction path for ammonia addition to formaldehyde. The circles encompass the points at which each normalized variable reaches values of 0.1, 0.5 and 0.9, corresponding to 10, 50, and 90% of their total changes over the reaction path. The angle β reaches its point of 10% change earliest, while the carbonyl bond distance r_{CO} reaches its point of 10% change latest; all other variables achieve a 10% change between these limits. The range from $r_{CN} = 2.5$ Å (at which $\hat{\beta} = 0.1$) to $r_{CN} = 2.15$ Å (at which $\hat{r}_{CO} = 0.1$) thus defines a "band" (shaded in the figure) within which all reaction variables will have passed the point $\hat{\xi}_i = 0.1$. Examination of the changes in the individual $\hat{\xi}_i$ across this band shows the smallest change to be 7% of its total change during reaction (i.e., $\Delta\hat{\xi}_i = 0.07$) and the largest change to be 15% of the total (i.e., $\Delta\hat{\xi}_i = 0.15$). Thus, the "width" of this band corresponds to a 7–15% change in the reaction variables. Similar considerations generate bands with widths of 15–25% for $\hat{\xi}_i = 0.5$ and 7–14% for $\hat{\xi}_i = 0.9$, also shown in the figure. The "average of CN bond orders" shown at the top of the figure includes the Wiberg bond index and the Ehrenson–Seltzer bond parameter as well as $\hat{B}_{CN}$ (O/P denotes out-of-plane).

example, is only 0.2 Å so that a 25% error is about 0.05 Å. The total change in β is around 15° so that a 25% error represents about 4°. Eventually we want to define transition-state structures to better limits than these, but at the moment we cannot do so. In the models described below, we have for the sake of simplicity employed $\hat{B}_{CN}$ as a universal descriptor. The calculations require us, in doing this, to keep in mind that (1) this procedure will not be correct in fine detail, and (2) since $\hat{B}_{CN}$ "trails" most other reaction variables, transition states are likely to be "more tetrahedral" than our models show.

Construction of Transition-State Models

We now wish to construct, on the basis of the crystallographic and quantum chemical findings, as comprehensive a set of transition-state model structures as we can. In doing so, we make many simplifications and compromises which we are confident later results will require us to repair. However, we are wary of building into our initial models any degree of complexity which is not absolutely demanding in its character. Our procedure has been to use the "recipe" below to generate highly simplified transition-state model structures and then to make isotope effect predictions for several different isotope effect probes from these models. Measurements of these probes will then simultaneously yield information about transition-state structures and force improvements in the formulation of the models. In this iterative manner, we can expect eventually to arrive at acceptable transformations for relating the experimental measurements to detailed transition-state structures.

The simplified recipe for generation of acyl-transfer transition-state models is based in the BEBO method and involves the following steps:

1. We select $\hat{B}_{CN}$ as the fundamental structural descriptor, and every other variable is related to $\hat{B}_{CN}$. Of course, r_{CN} might have been used (and is a more graphic structural variable), but $\hat{B}_{CN}$ has the advantages of being a normalized variable and of having an important function in the BEBO formulation.

2. All reaction variables are now assumed to change in concert with $\hat{B}_{CN}$. For bond angles and force constants, we take $\hat{\xi}_i = \hat{B}_{CN}$ or $\xi_i = \xi_0 + \hat{B}_{CN} \Delta \xi_i^{tot}$. For bond lengths, we take $\Delta \hat{B} = \hat{B}_{CN}$ and calculate the bond length from Pauling's rule. We also use the BEBO principle of "conservation of bond order" so that, e.g., $\hat{B}_{CO} = 2 - \hat{B}_{CN}$.

In this procedure, we are clearly ignoring the quantum mechanical finding that the angle β and the bond length r_{CO} do not change in precise concert. However, as pointed out above, they do change in approximate concert and our treatment is currently of insufficient sophistication to permit us to try to formulate and include the observed departures from concerted change of these and other variables.

Predicted Kinetic Isotope Effects

The Wolfsberg–Stern procedure (see Fig. 12) can at this point be employed to generate a set of kinetic isotope effects corresponding to each transition-state model [37]. Three particularly useful isotope effects are those shown in (2): the α-deuterium effect ("α-D probe"), the β-deuterium effect ("β-D probe"), and the *carbonyl*-^{18}O effect ("$C^{18}O$ probe"). Figure 20 shows the

$$\overset{18}{O}$$

(diagram 2: structure with ^{18}O double bonded to C, connected to another C bearing N and $D(\alpha)$, $D(\beta)$)

(2)

results of the Wolfsberg–Stern calculations for each of these three isotopic probes. The quantities k_H/k_D or k_{16}/k_{18} are plotted as ordinates vs. $\hat{B}_{CN}$, as an independent reaction variable on the abscissa. Within the context of the highly simplified models employed for the calculations, the conclusions given below should be valid.

Relatively simple relationships seem to exist between all the probes and transition-state structure. As $\hat{B}_{CN}$ increases from zero to unity, both the α-D

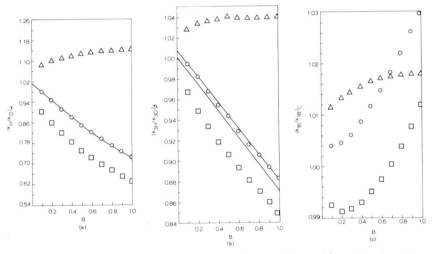

Fig. 20 Predictions of kinetic isotope effects for model transition-state structures, calculated by the Wolfsberg–Stern procedure as outlined in the text. (a) The α-D effect as a function of $\hat{B}_{CN}$. The solid line connecting the calculated isotope effects (circles) is nearly linear, and the maximum predicted effect (for $\hat{B}_{CN} = 1.00$) agrees with the measured equilibrium effect for hydration of acetaldehyde. The squares give the zero-point energy contribution and the triangles the contributions from excited vibration states and changes in moments of inertia and molecular weight. (b) the β-D effect (for 3H vs. 3D) as a function of $\hat{B}_{CN}$. The solid line connecting the calculated effects (circles) is again nearly linear, and the effect for $\hat{B}_{CN}$ agrees with expectation from equilibrium measurements. The lower solid line (which is too close to the calculations to be distinguished by currently feasible measurements) is simply the straight line connecting $k_{3H}/k_{3D} = 1$ at $\hat{B}_{CN} = 0$ to $k_{3H}/k_{3D} = 0.87$ (the measured equilibrium effect) at $\hat{B}_{CN} = 1.0$. The squares and triangles have the same significance as in (a). (c) The $C^{18}O$ isotope effect (circles) as a function of $\hat{B}_{CN}$. The squares and triangles have the same significance as in (a) and (b).

and β-D effects become steadily more inverse. Their magnitudes at $\hat{B}_{CN} = 1$ should be approximated by equilibrium isotope effects for carbonyl hydration, since this represents the complete conversion of a trigonal reactant state to a tetrahedral ($\hat{B}_{CN} = 1$) final state. Indeed, this is the case, $(k_{3H}/k_{3D})_\beta$ being 0.87 for hydration of 1,3-dichloroacetone [37] (corrected to the value for three deuteriums), while $(k_H/k_D)_\alpha$ for hydration of acetaldehyde is 0.74 [38]. These values can be compared to 0.89 in Fig. 20b and 0.76 in Fig. 20a. The agreement is very good. It should also be noted—and this is a finding to be used below—that the isotope effects fall off with $\hat{B}_{CN}$ in an essentially *linear* fashion.

The $C^{18}O$ probe (Fig. 20c), which relates to the weakening of the carbonyl π bond as addition takes place, increases to larger, normal values (as expected) as $\hat{B}_{CN}$ increases. The relationship is not linear. Furthermore, the isotopic oxygen atom participates in reaction-coordinate motion so that the effect is not really a secondary isotope effect. This means that it cannot easily be calibrated against equilibrium isotope effects, which in any case are not available. One study [39] of kinetic isotope effects of this kind (for methoxide attack on aryl acetate esters) gave values of 1.018 and 1.024, which are in the calculated range.

Figure 21 shows scaled structural representations of three transition states for acyl transfer (or carbonyl addition), together with the expected corresponding ranges of experimental isotope effects.

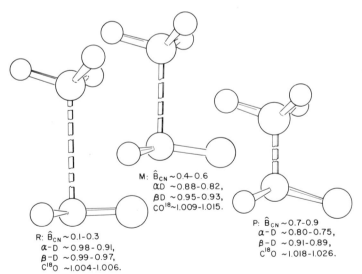

M: $\hat{B}_{CN} \sim 0.4-0.6$
$\alpha D \sim 0.88-0.82$,
$\beta D \sim 0.95-0.93$,
$CO^{18} \sim 1.009-1.015$.

R: $\hat{B}_{CN} \sim 0.1-0.3$
α-D $\sim 0.98-0.91$,
β-D $\sim 0.99-0.97$,
$C^{18}O \sim 1.004-1.006$.

P: $\hat{B}_{CN} \sim 0.7-0.9$
α-D $\sim 0.80-0.75$,
β-D $\sim 0.91-0.89$,
$C^{18}O \sim 1.018-1.026$.

Fig. 21 Approximate structural representations of reactantlike (R), medial (M), and productlike (P) transition states for carbonyl addition and predictions of the corresponding kinetic isotope effects.

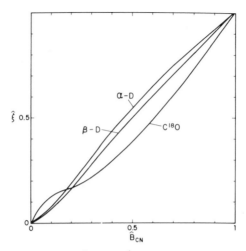

Fig. 22 Coupling diagram of $\hat{B}_{CN}$ with $\hat{\xi}_{\alpha\text{-D}}$, the normalized α-deuterium isotope effect, $\hat{\xi}_{\beta\text{-D}}$, the normalized β-deuterium effect, and $\hat{\xi}_{C^{18}O}$, the normalized *carbonyl*-^{18}O effect. These three probes are reasonably coupled to each other and to $\hat{B}_{CN}$.

A final useful question is whether each of these isotopic probes (the α-D, β-D, and $C^{18}O$ probes) gives essentially equivalent information about the transition state or whether they yield any independent data. Our model and our conclusions to this point would suggest that the probes should be highly coupled and thus yield equivalent measures of transition-state structure. This point may be tested by use of the normalized-variable approach already described. In Fig. 22, the three probes are compared, in normalized form, as functions of $\hat{B}_{CN}$. As is apparent, they are strongly coupled. We may conclude that, other things being equal, all such measurements should agree in their structural indications if properly interpreted.

Measurement of Kinetic Isotope Effects: Technology

Isotope effect measurements, to yield data of sufficient precision to resolve transition-state structures into individual classes, must obviously be reasonably careful. For example, the universe of the β-D effect (on which we will concentrate here) is only around 15% for the entire range from limiting-trigonal to limiting-tetrahedral transition states. To resolve even a 5-fold division of structures would demand isotope effects to approximately $\pm 1.5\%$.

It is generally possible to determine isotope effects (which are a class of relative rates) by either a competitive method or a direct method. In the competitive method, a mixture of isotopic substrates is employed and their relative reaction rates are found by analysis of the isotopic composition of either reactants or products at various times during the reaction. Accumulation of the heavy isotope in the reactant or the light isotope in the product (for example) shows the light isotope to be undergoing faster reaction; quantitative analysis of the degree of accumulation with time yields k/k'. This technique has been used with great effect by Kirsch [40] and O'Leary [41] to determine nitrogen, oxygen, and carbon isotope effects in enzymatic reactions. The competitive method can be difficult to apply in enzymatic reactions because it requires the isolation, purification, and isotopic assay of materials from the usually very dilute reaction medium.

To apply the direct method, on the other hand, for measurement of secondary hydrogen isotope effects or any heavy-atom isotope effects places very strong demands of precision on rate constant measurement. If a ratio k/k' is, as noted above, to be known to $\pm 1.5\%$, then both k and k' must be known to about ± 0.7–0.8%. One approach that we have taken to this problem is outlined in Fig. 23. Although illustrated for spectrophotometric kinetics obtained using instruments such as the Cary-16 or Cary-118 spectrophotometers, related techniques have been used for fast-reaction studies with

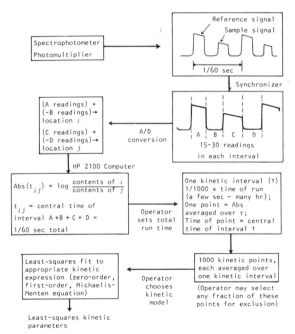

Fig. 23 Technique used for precise rate measurements in the direct determination of small kinetic isotope effects.

the Durrum stopped-flow, T-jump system and for slow rates measured using ion-selective electrodes. Rate constants from this system are usually precise to at least a few parts per thousand, and frequently much better, within a single kinetic run. Reproducibility from run to run with enzymatic reactions has been brought to a level just under $\pm 1\%$ currently, and further improvements can be expected to reduce this level by a factor of 2–5.

Measured Isotope Effects and Transition-State Structures

We present here, for illustrative purposes, a few β-D isotope effects which have been obtained for both enzymatic and nonenzymatic reactions. Figure 24 shows a plot in which a limit of k_{3H}/k_{3D} (i.e., the isotope effect for CH_3— vs. CD_3— substrates) of 0.85 is adopted for the limit of a fully tetrahedral transition state. This is approximately the effect for equilibrium hydration of ketones and, although subject to some uncertainty, it is sufficiently well

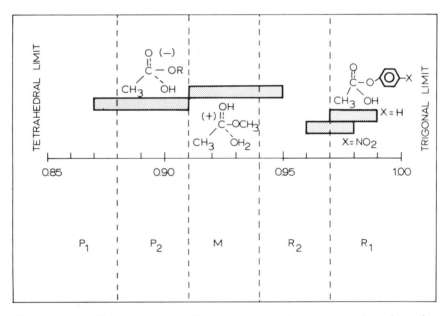

Fig. 24 Some β-D kinetic isotope effect measurements for nonenzymatic acyl-transfer reactions. An approximate limit of $k_{3H}/k_{3D} = 0.85$ is adopted for the "tetrahedral limit" of transition-state structure, and 1.00 is taken for the trigonal limit. All effects are at 25°C. The entries, from left to right, are for hydroxide-promoted alkyl ester hydrolysis, acid-catalyzed alkyl ester hydrolysis, and hydroxide-catalyzed hydrolysis of two aryl esters. In this diagram, it is imagined that the measurements can be approximately dissected into about five structural regions resembling reactants (R_1 and R_2), medial configurations (M), and products (P_2 and P_1).

known to use in this illustration of the concept. The limit of 1.00 corresponds to no change in force constants at the β centers on forming the transition state from the trigonal reactants. It therefore represents a still-trigonal transition state. In agreement with the calculations shown in Fig. 20, a linear relation between the isotope effect and the bond order of the forming bond is adopted. With the cautions noted above, we divide the total range of isotope effects into five parts, corresponding to five structural classes for the transition state. Finally, four examples of β-D isotope effects for nucleophilic reactions at ester centers are entered on the diagram, with limits that encompass the range of the experimental measurements.

The acidic and basic hydrolysis reactions of alkyl esters fall into the M and P_2 (medial and rather productlike) classes, respectively, although a true distinction between these would really require more careful study from both theoretical and experimental sides. In any case, the general expectation [42] that the "tetrahedralization" of the carbonyl center would be relatively advanced in these transition states (since formation of the tetrahedral intermediate is endothermic) seems to be met.

The basic hydrolysis of aryl esters presents a contrast. Here the quite small isotope effects show that the effective transition state for these reactions is in the (R_2, R_1) (rather to very reactantlike) class with a relatively small degree of "tetrahedral character." We say "effective transition state" advisedly, because it is not certain that a single elementary rate process (such as formation of the carbon–nucleophile bond) governs the rates of these reactions (or of other acyl transfers). If more than a single process contributes to determination of the rate, then the isotope effect we observe will be a weighted-average quantity. If competing (parallel) processes are involved, the lowest free-energy transition states will be the heaviest contributors. If successive steps (in series) are similar in rate, then the highest free-energy transition states will be most heavily weighted.

In all such cases, the transition-state structure inferred from experimental data will be that for a virtual transition state. The structure of this virtual species is a meaningful construct because its properties are indeed those that determine the dynamic properties of the reacting system. In the present case, it is quite possible that some process such as desolvation of the nucleophile, in the presence of the carbonyl center, is partially rate limiting. If this is so, our data will refer to a virtual transition-state structure that is an average of the structures of the desolvation transition state and of the bond formation transition state. Since the desolvation transition state will still have a completely trigonal carbonyl center, the effect of its contribution to the virtual structure will be to move it toward the trigonal limit. Whether an effect of this sort influences the structural interpretation of the data shown in Fig. 23, we are at this time unable to say. However, it is correct to say that the virtual

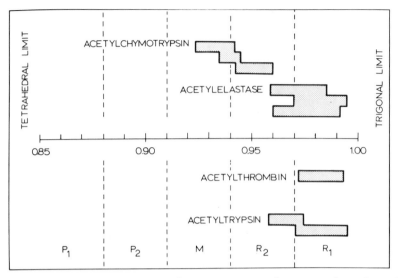

Fig. 25 Some β-D kinetic isotope effect measurements for enzymatic acyl-transfer reactions (at 25°C) plotted as in Fig. 24. The effects are for deacylation of the indicated species. The same structural dissection as in Fig. 24 is shown.

transition-state structures for the alkyl ester reactions resemble the (P_2, M) structures of Fig. 18, while those for the aryl esters resemble the (R_2, R_1) structures.

Exactly the same considerations come into the interpretation of the enzymatic data shown in Fig. 25. These findings strongly indicate that the virtual transition-state structures for deacetylation of four acetyl enzymes of the serine protease class fall in the reactantlike (R_2, R_1) class (with perhaps a bit of "leakage" toward M). Here again, we cannot be certain whether this is because the transition state for a single bond formation process determines the rate and has a quite trigonal structure, or whether the transition state of another process is also contributing. In these cases, of course, the potential for partial rate determination by protein reorganization processes is very important.

Summary

The course of events leading from the fusion of transition-state theoretical ideas with enzyme mechanism concepts up to our current primitive information on virtual transition-state structures is portrayed in Fig. 26. All of the features we noticed in our previous case history appear here, in particular the importance of accidental collaboration through the immediate and direct

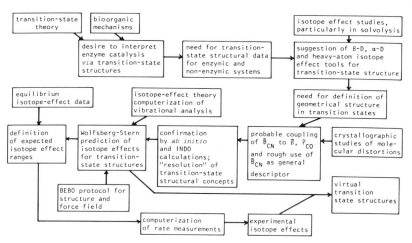

Fig. 26 A chart illustrating theoretical and experimental contributions to the development of current methods for establishing enzymatic transition-state structures from isotope effects.

incorporation of independent results into an ongoing program. For example, our work was well advanced when Bürgi and his colleagues published their crystallographic findings, but these immediately formed an integral part of our approach. Further, the enormous utility of techniques developed for other reasons underscores once again the unpredictable course of scientific work and the hopelessness of attempting to "target" the solution of difficult scientific problems. The isotope effect theory, which was so critical here, actually grew in part from the atomic bomb project of World War II [43], and the development of secondary isotope effects as mechanistic tools emerged almost wholly from the research of the 1950's and 1960's on solvolysis reactions [44].

MICROENVIRONMENTAL POLARIZABILITY EFFECTS IN VISION

The Retinal Chromophore

11-*cis*-Retinal (Fig. 27) is the primary visual chromophore of the vertebrate eye. Its absorption of light energy is the first of a chain of events that ultimately leads to the transduction of this energy into the visual sensory information employed by the brain in constructing our mental images of the world. An understanding of the spectroscopic behavior of this molecule is therefore

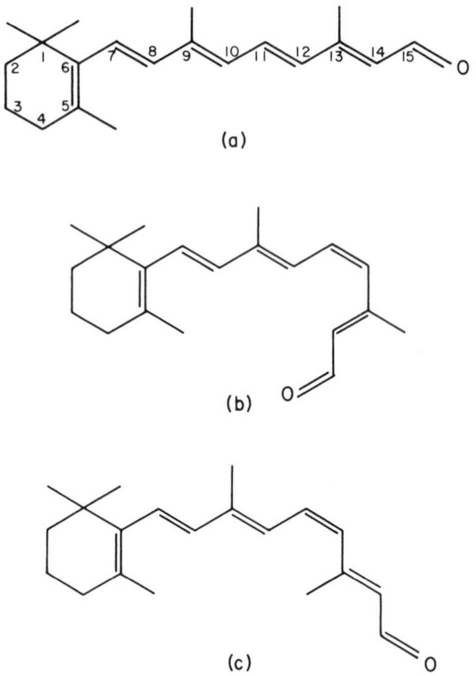

(a)

(b)

(c)

Fig. 27   Structural representation of three conformers of the visual chromophore retinal: (a) *all-trans*; (b) 11-*cis*, 12-*s-cis*; (c) 11-*cis*,12-*s-trans*.

fundamental not only to neurophysiology but also to experimental psychology and, one can reasonably maintain, to any detailed comprehension of the acquisition and character of human knowledge.

In the last decade extensive studies of both 11-*cis*- and *all-trans*-retinal have provided a reasonably satisfactory account of chemical and photochemical properties of these important molecules in solution. Indeed, an examination of the manner in which the various theoretical and experimental studies have been interwoven into the fabric of our understanding of these molecules exemplifies a most productive interplay between theory and experiment [45–50]. Unfortunately, the status of our knowledge of the physiochemical properties of retinals *in situ*, i.e., in conjunction with the apoprotein opsin, is not nearly as advanced. This is due in large measure to the lack of detailed information on the complex and unusual protein microenvironment surrounding the chromophore.

Experimental approaches have not yet succeeded in determining the structure or even the composition of this environment because of the extraordinarily refractory character of the protein. Even when this becomes known, the elucidation of the interactions of this environment with excited states of the chromophore, which govern its behavior in light absorption, will remain a very difficult problem. In both of these areas, theoretical approaches are currently producing an effective combination with experimental spectroscopy.

Characteristics of the Bathochromic Shift

A particularly good example of the importance of the protein microenvironment to the *in situ* properties of the chromophore is found in the rhodopsin bathochromic shift illustrated in Fig. 28. The "shift" occurs when the visual chromophore 11-*cis*-retinal binds to the apoprotein opsin, giving rise to the visual pigment rhodopsin, which is largely responsible for vertebrate vision.

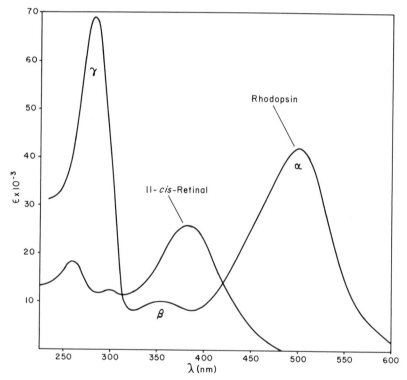

Fig. 28 Absorption spectra of 11-*cis*-retinal and rhodopsin illustrating the bathochromic shift that takes place when 11-*cis*-retinal ($\lambda_{max} \sim 380$ nm) binds to the apoprotein opsin to yield rhodopsin ($\lambda_{max} \sim 500$ nm). The α and β bands of rhodopsin represent chromophore transitions, while the γ band represents primarily transitions in the aromatic amino acid residues of the apoprotein. (Redrawn from Ebrey and Honig [50] with permission from *Q. Rev. Biophys.*, Cambridge University Press.)

The magnitude of the bathochromic shift is considerable, spanning the spectral region from the intense $\pi \rightarrow \pi^*$ transition in 11-*cis*-retinal (~ 380 nm) to the main absorption band in rhodopsin (~ 500 nm). An understanding of the molecular basis of this effect is important since it may play a fundamental role in the wavelength regulation necessary for color vision.

Due to the highly hydrophobic nature of rhodopsin it is unlikely that X-ray diffraction techniques (which have contributed so substantially to our knowledge of the microenvironments of enzyme active sites) will be able to provide, in the near future, similar detailed structural information on the chromophore binding site of rhodopsin. However, this does not mean that progress on the elucidation of the microenvironment of the chromophore binding site has been or will be nonexistent until the crystallographers are able to crystallize rhodopsin. As was and still is the case for many enzyme systems, the structure of the chromophore binding site can effectively be probed through the clever application of a variety of chemical and spectroscopic techniques. And it is at this point that the necessity for a strong interplay between theory and experiment is clearly manifest.

Essentially three questions need to be answered in order to elucidate the mechanism of the bathochromic shift:

1. What are the structure and composition of the protein microenvironment surrounding the chromophore?

2. What is the structure of the chromophore *in situ*?

3. How does the protein microenvironment interact with the chromophore to produce the observed bathochromic shift?

Since resolution of the third question involves excited electronic states the need for quantum as well as classical and semiclassical chemical theoretical explanations becomes evident. In this regard, the present example represents a new situation in which the application of quantum theoretical analyses to the problem is not only desirable but also required.

Spectral Consequences of the Chromophore–Protein Linkage

A number of models have been proposed to explain the bathochromic shift, each model being synthesized from a varied mixture of experimental and theoretical data on related systems. Essentially all models assume that the chromophore is bound to opsin in a protonated Schiff base linkage. While such a linkage is now generally accepted as correctly representing the mode of chromophore–protein binding [50], it was not until the recent resonance Raman studies of Lewis *et al.* [51] and of Oseroff and Callender [52] that the issue was considered resolved.

As shown in Fig. 29 protonation of the retinylidene Schiff base (RSB, **3**) (portrayed below in its *all-trans* conformation) to protonated retinylidene

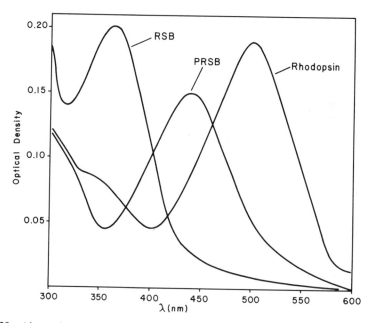

RSB
(3)

PRSB
(4)

Schiff base (PRSB, **4**) shifts the main absorption base from ~ 360 nm (which is close to that of free 11-*cis*-retinal; see Fig. 27) to ~ 440 nm, respectively. Hence, a major part of the observed bathochromic shift can be accounted for simply by protonation of the free base RSB (**3**) [53]. Although protonation produces an appreciable shift (~ 80 nm), an additional shift of ~ 60 nm or

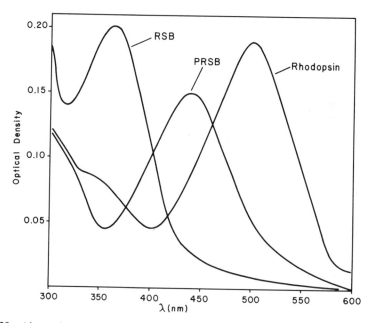

Fig. 29 Absorption spectra of the retinylidene Schiff base (RSB) (**3**), protonated retinylidene Schiff base (PRSB) (**4**), and rhodopsin. Note that λ_{max} of RSB occurs at a slightly shorter wavelength than that of 11-*cis*-retinal. Protonation to give PRSB produces approximately one-half of the opsin-induced bathochromic shift. (Redrawn from Dartnall [46, p. 41].)

about half of the total effect remains to be explained.* It is this additional shift that has led to the variety of models proposed.

Basically these models can be divided into two classes, both of which involve some form of protein–chromophore interaction:

1. Structural perturbation of the chromophore (e.g., double-bond rotations)

2. Electronic perturbation of the chromophore (e.g., point charge perturbations from nearby ionized carbonyl groups)

Descriptions and analyses of the different models can be found in a number of excellent reviews [48–50,54]. An examination of these models clearly shows that a completely satisfactory explanation of the bathochromic shift is not at present available. However, recent experimental work by Irving, Byers, and Leermakers [55–57] has led to the development of a model that may provide a consistent interpretation of many of the spectroscopic features of rhodopsin.

Microenvironmental Polarizability Model

Irving, Byers, and Leermakers [55–57] have proposed a microenvironmental polarizability model based on the observation that solvents of high *polarizability* (as measured, for example, by their indices of refraction), essentially independently of their *polarity* (as measured by their dielectric constants), were able to induce substantial bathochromic shifts in an analog of *all-trans-*PRSB (5) and related molecules. Their argument rests on the ability of highly

(5)

polarizable solvent molecules to undergo effectively instantaneous changes in polarity in response to induced dipole changes in the chromophore brought about by the absorption of light. This results in a differential stabilization of the chromophore excited state over its ground state due to the larger dipole moment of the latter.

Furthermore, they argue that the protein microenvironment surrounding the chromophore in rhodopsin not only provides the required highly polar-

---

* On an energy and not wavelength scale the remaining ~60 nm shift represents about 35% of the total observed shift from RSB to rhodopsin. Furthermore, it is important to note that shifts up to ~580 nm have been observed, thus greatly increasing the magnitude of the shift due to the protein microenvironment.

izable groups in the form of aromatic amino acid residues, but also places these groups in a proper, specific orientation to the chromophore such that large local polarization effects, larger than would be expected in bulk solvent, can be obtained.

Three other experimental results provide further evidence for the plausibility of such chromophore–protein interactions:

1. Recent resonance Raman experiments on bovine rhodopsin [51,52] show the existence of a low-frequency C C stretching vibration, lower than that observed in the free PRSB. The frequency lowering was interpreted by Oseroff and Callender [52] to indicate an increased delocalization of the polyene π electrons through some type of chromophore–protein interaction. Lewis et al. [51] have proposed that specific chromophore–aromatic amino acid interactions, resulting in a type of "π complex," are responsible for the frequency lowering.

2. An electron spin resonance (esr) study on frog rhodopsin irradiated at its α-band wavelength (see Fig. 27) also implicated the possibility of protein–aromatic amino acid interactions. By examining the weak esr signal of the forbidden $\Delta m = \pm 2$ transition, Shirane [58] was able to determine the two zero field splitting parameters, D and E. The values obtained were consistent with those expected for the triplet state of a tryptophan molecule, and hence Shirane proposed that the tryptophan triplet state was created by some type of charge-transfer interaction between the chromophore and the tryptophan residue.

3. Finally, Ishigami et al. [59] and Mendelsohn [60] demonstrated that weak complexes, of unknown structure, formed between retinal and both tryptophan and indole. These complexes may simulate the interaction between protein and chromophore in rhodopsin. Furthermore, these complexes are capable of absorbing light well into the visible range, lending further credence to the hypothesis that they may simulate the situation in rhodopsin.

Quantum Chemical Investigation of Microenvironmental Polarizability Effects

Quantum chemistry is currently in a position to take up the burden of investigation here: to pursue the question of whether polarization effects of aromatic amino acid residues, aligned in a reasonable fashion, can account quantitatively for the observed bathochromic shift. Hence, a series of theoretical all valence-electron CNDO/S calculations were carried out on model PRSB–benzene complexes [61,62]. Computational details can be found in an earlier publication dealing with spectral properties of retinals [63]. Since the

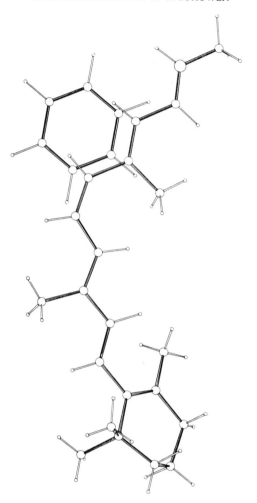

Fig. 30 An ORTEP drawing of the planar 11-*cis*,12-*s-trans*-PRSB–benzene complex. The alicyclic ring of PRSB is shown at the left and the Schiff base linkage at the right (the nitrogen is shaded). The polyene chain lies in the plane of the paper with the benzene ring behind. The benzene moiety is oriented such that maximum π–π overlap with the polyene chain (at the C-12—C-13 and C-13—C-14 bonds) is achieved.

molecular structure of the chromophore binding site is unknown, the development of a reasonable structural model for the PRSB–benzene complex was guided by results of the previously cited experiments mixed liberally with "chemical intuition." Particular emphasis was given to the structures of the general type illustrated in Fig. 30. Such structures represent maximum π–π "overlap" configurations and were suggested by the results of the resonance

Raman experiments [51,52]. The polyene chain in all structures examined was maintained in the 11-*cis* conformation. However, due to the low barrier to rotation about the C-12—C-13 single bond observed in 11-*cis*-retinal [47,48,64–66], several "distorted" 12-*s* conformations were considered.*

The results of several calculations are summarized in Fig. 31, along with the spectra of bovine rhodopsin (in digitonin) and an 11-*cis*-PRSB analog (in 1,2-dichloroethane). The similarity between the two experimental spectra over the whole spectral region is remarkable and clearly indicates the ability of a highly polarizable solvent (i.e., 1,2-dichloroethane) to emulate the effect of the protein microenvironment surrounding the chromophore in rhodopsin.†

A comparison of the spectra in Fig. 31 shows excellent agreement between the experimental and calculated results, especially in the case of the planar 12-*s*-*trans* conformer. Of particular note is the prediction of the presence of a second transition in the region of the β band. Although it has previously been assumed that the β band represents a single transition, the "flatness" of the spectrum would make resolution of multiple bands quite difficult. In fact, the slight asymmetry observed in the β-band circular dichroism (CD) spectrum of rhodopsin [67] is quite consistent with the presence of two bands, both with rotational strengths of the same sign. Whether this *a priori* prediction of the quantum chemical investigation will be borne out by more detailed experimentation is a fascinating and suspenseful question.

On the basis of a comparison of the relative magnitudes of the calculated oscillator strengths, in particular those in the α- and β-band regions, with the experimental spectra, the present model predicts a 12-*s trans* conformation as the most likely chromophore conformation both in rhodopsin and in PRSB in a highly polarizable solvent.‡ The prediction for rhodopsin is consistent with the recent proposal of Doukas and Callender [68], based on flow resonance Raman experiments, that the conformation of the chromophore in rhodopsin is distorted 12-*s*-*trans*.

Such agreement between theory and experiment does not, of course, prove the correctness of the theoretical model ("agreement" between theory and

---

* In particular, three conformations were considered: (1) 11-*cis*,12-*s*-*cis* (38.7°); (2) 11-*cis*,12-*s*-*trans* (141.3°); and (3) 11-*cis*,12-*s*-*trans* (planar). The distorted geometries correspond to those observed in the crystal by X-ray diffraction techniques (see Honig and Ebrey [48] for a review).

† The difference in the optical density between rhodopsin and the 11-*cis*-PRSB analog is attributable to the absorption of aromatic amino acid residues on the protein but not necessarily those in close proximity to the chromophore.

‡ As in the case of 11-*cis*-retinal it is expected that an equilibrium exists between the 12-*s*-*cis* and 12-*s*-*trans* conformers. However, in the present case, i.e., PRSB in a highly polarizable solvent, the 12-*s*-*trans* conformer would become more stable. Temperature-dependent spectral studies would help to clarify this point.

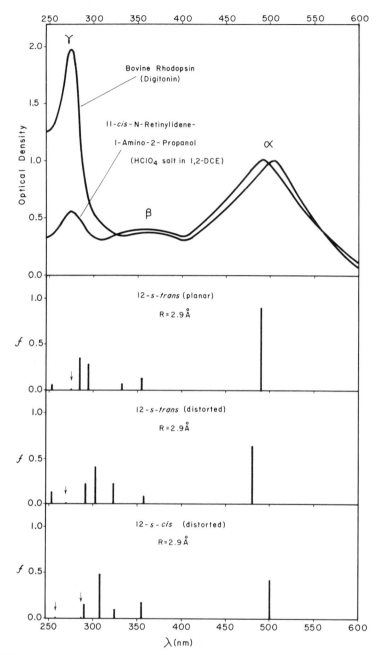

Fig. 31 Comparison of experimental and calculated absorption spectra. Experimental spectra are shown at the top for bovine rhodopsin in digitonin (a detergent) solution and for an 11-*cis*-PRSB in 1,2-dichloroethane (1,2-DCE) solution. The lower three boxes contain calculated spectra (both λ_{max} and oscillator strength, f, are shown) for 11-*cis*-PRSB–benzene complexes with a 2.9 Å distance (R) separating the benzene from the

experiment is a necessary but not a sufficient condition for the correctness of a theoretical model), but it does add a measure of confidence in the suitability of the model as an analytical and predictive tool for the investigation of chromophore–protein interactions in rhodopsin.

In order to further assess the characteristics and implications of the theoretical model, a more detailed examination of the effect of varying the benzene–polyene distance (R) is illustrated in Fig. 32 for the planar 12-*s-trans* conformer. Similar behavior was also observed for the distorted 12-*s-cis* and 12-*s-trans* conformers. Figure 32 illustrates the direct relationship between R and both the transition energy (i.e., bathochromic shift) and calculated oscillator strength of the α band. Thus, it is clear that the present theoretical model also provides a reasonable mechanism for wavelength regulation. And, although such comparisons have not to date been made, it would be of interest to examine the relationship between the magnitude of the batho-chromic shift and the shape of the α band [69,70] in light of the present calculated transition energies and oscillator strengths.

Table 1 presents a summary of calculated results for the planar 11-*cis*,12-*s-trans*-PRSB–benzene complex. Values obtained for corresponding properties of the other complexes investigated show the same general trends. Of particu-lar interest are the results describing the benzene → PRSB charge-transfer character of the various transitions.

From Table 1 it is quite clear that the α-band transition has very little charge-transfer character, and thus the calculated bathochromic shift can reasonably be attributed to mutual polarization interactions between the PRSB and benzene moieties. This is further substantiated by the calculated polarization of the transition, which shows it to be effectively in-plane polarized along the long axis of the polyene chain. The lack of out-of-plane polarization shows that no significant transfer of charge occurs in the transi-tion. The current calculations also indicate that formation of the PRSB–benzene complex results in less than 0.01 of an electronic charge being transferred from the benzene to PRSB moiety (see footnote b of Table 1).

The first calculated β-band transition (354 nm) shows some charge-transfer character, although, as in the α-band transition, the lack of significant out-of-plane polarization would argue against such an assignment. The in-plane polarization angle shown in Table 1 indicates that this transition should be assigned as the "*cis*" band by analogy to the similar band observed in retinals [63]. This assignment is also in agreement with previous assignments [50].

The second calculated β-band transition (332 nm) represents the only clear

---

polyene chain. Spectra are given for the 12-*s-trans* planar structure and for the 12-*s*-trans and 12-*s*-cis distorted conformers in which an out-of-plane rotation of 38.7° about the C-12—C-13 bond has been effected. This distortion is observed in the crystal structure of 11-*cis*,12-*s-cis*-retinal [48]. [Experimental spectra redrawn from J. O. Erickson and P. E. Blatz, *Vision Res.* **8**, 1357 (1968).]

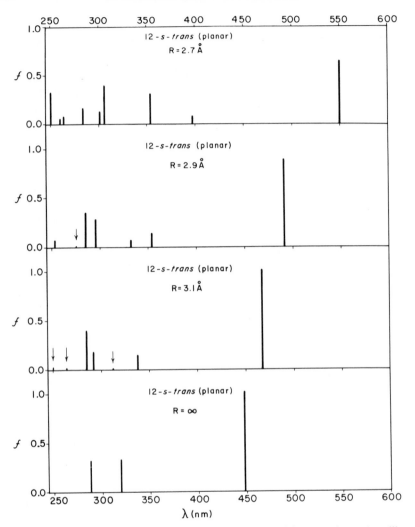

Fig. 32   An illustration of the effect on the calculated transition energies and oscillator strengths, *f*, of varying the PRSB–benzene distance, *R*. Calculations shown are for the planar 11-*cis*,12-*s-trans*-PRSB–benzene complex.

example of a charge-transfer type of transition. The amount of formal charge transfer (0.58) and the polarization direction substantiate this assignment. Careful examination of the in-plane and out-of-plane polarizations shows that the transition dipole moment oscillates approximately along a line connecting the center of the benzene ring with the C-13 carbon of the polyene chain (see Fig. 27). This is certainly consistent with the presence of substantial

TABLE 1

Summary of Calculations for the Planar 11-*cis*,12-*s-trans*-PRSB–Benzene Complex ($R = 2.9$ Å)

Band assignment[a]	Wavelength (nm)	Charge transfer[b]	Oscillator strength	Polarization[c] In-plane (deg)	Polarization[c] Out-of-plane (deg)
α	493	0.09	0.09	7	2
β	354	0.36	0.14	64	8
β	332	0.58	0.07	137	47
γ	296	0.10	0.29	139	8
γ	286	0.33	0.36	14	0

[a] See Fig. 31 for a summary of the band assignments.

[b] Amount of charge transferred from benzene to PRSB during the indicated transition. Less than 0.01 charge transfer occurs on formation of the ground-state PRSB-benzene complex.

[c] In-plane polarization is measured relative to the line connecting the C-6 carbon with the Schiff base nitrogen (see Figs. 27 and 30). Out-of-plane polarization is measured relative to the plane formed by the line connecting the C-6 and C-12 carbons and the line connecting the C-6 carbon and Schiff base nitrogen (see Figs. 27 and 30).

charge-transfer character to the transition. Both calculated γ-band transitions are essentially in-plane polarized, arising primarily from the PRSB moiety.

From the above data it is clear that the proposed theoretical model provides a consistent interpretation of a number of experimental facts and also makes a number of predictions that can in principle be tested. Additional calculations on different complex geometries and ligands (e.g., indole to mimic tryptophan) are currently in progress. A theoretical evaluation of the effect of the counterion on the spectra of PRSB's and PRSB–benzene complexes is also underway.

However, much more testing of the model is necessary before it can be considered to provide a reasonable model of the actual chromophore–protein interaction present in rhodopsin. For example, calculation of the rotatory strengths of the different transitions would provide an additional source of comparison with experimental spectroscopic data. Furthermore, magnetic circular dichroism studies may help establish the presence of two transitions as predicted in the β band.

Summary

In a molecular system as complex as rhodopsin, when detailed structural information such as that obtained by X-ray diffraction techniques is unavail-

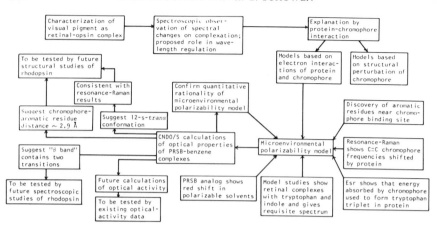

Fig. 33 Chart illustrating theoretical and experimental contributions to the current view of microenvironmental polarizability effects in vision.

able, it becomes necessary to study the chromophore binding site with a variety of experimental and theoretical probes. And, as each probe usually provides only limited information, it is necessary to interpret, evaluate, and correlate a number of different data in order to develop a detailed picture of the molecular structure and interactions in the chromophore binding site.

Figure 33 summarizes the relationship of the various theoretical and experimental studies on the bathochromic shift. Through the interplay of theory and experiment, as indicated in Fig. 33, a clearer picture of the underlying mechanism of the bathochromic shift is beginning to emerge.

In particular, it is clear that the microenvironmental polarizability model provides a plausible and consistent explanation of a number of features of the bathochromic shift. And, although other factors such as "double-bond twist" may contribute to the total shift, it is difficult to eliminate microenvironmental polarizability effects from playing a significant role. Thus, by examining the theoretical predictions that arise out of PRSB–benzene complex calculations, information, which bears on the three questions mentioned earlier, is obtainable. For example, the calculations are consistent with an 11-*cis*,12-*s-trans* chromophore structure in rhodopsin, the nearby placement of at least one aromatic amino acid residue in the chromophore binding site, and induction of the bathochromic shift via mutual polarization, not charge transfer, between the chromophore and aromatic amino acid residue. Hence, in the present case, theory provides an important bridge between the elegant experiments carried out by a number of investigators and molecular structural information—a necessary bridge if a detailed molecular explanation of the bathochromic shift and wavelength regulation is to be forthcoming.

THE REINFORCEMENT OF THEORY AND EXPERIMENT
IN BIOORGANIC CHEMISTRY

In the sense that the theoretical and experimental approaches to scientific problems exemplify, respectively, the contemplative and active ways of life, the differences between them go deep into the history of thought [71] and probably into the biology of the human mind. The Platonic division of form from substance, the Cartesian dualism of mind and matter, and Nietzsche's Apollonian and Dionysian modes all point up a concern throughout Western philosophy with a separation between ideas and materials, thought and action.* Even the Eastern dualisms of the Taoist and Zoroastrian traditions have a similar effect.† Of course, the monistic tenets of Buddhist and quasi-Buddhist systems present a contrast.‡ The rather optimistic view that both approaches to science are ideally combined in a single individual has been expressed by the aphorism, "The complete physical chemist blows his own apparatus and solves his own equations."§

In practice, there are certain forms of theory and certain types of experimentation that are currently capable of complementary interaction in an optimally beneficial way. Some features of the cases we have described in this chapter were important in promoting this kind of interaction. We believe the chief among these were the following:

1. The function of the theory in these studies was to illuminate general principles which could then be used in interpretation of experiments and design of further experiments. The object was not merely to make direct numerical comparisons of theoretical and experimental quantities. Such direct comparisons should be most meaningful with very simple molecular systems rather than with those of the complexity encountered in bioorganic chemistry. In the study of charge-relay catalysis, theory first suggested the

---

* On Plato and Descartes, see B. Russell, "A History of Western Philosophy," pp. 119ff. and 567–568, Simon & Schuster, New York, 1945. For Nietzsche, see F. Nietzsche, "Gesammelte Werke," Vol. 3, pp. 21ff, Musarion Verlag, Munich, 1920.

† For this point of Taoism, see "The Texts of Taoism" (transl. by J. Legge), pp. 12–44, Dover, New York, 1962, and on Zoroastrianism, E. Herzfeld, "Zoroaster and His World," Vol. 1, pp. 309ff, Princeton Univ. Press, Princeton, New Jersey, 1947. On the fascinating idea that Eastern dualism, through Manichaeism, Gnosticism, and Neoplatonism, may have underlain the beginnings of modern scientific thought, see F. A. Yates, "The Rosicrucian Enlightenment," Routledge & Kegan Paul, London, 1972.

‡ Traditional Buddhist thought on this subject is discussed enjoyably by A. W. Watts in "The Way of Zen," Vintage Books, New York, 1957, and a broader swath is cut by R. M. Pirsig in "Zen and the Art of Motorcycle Maintenance," Bantam Books, New York, 1974.

§ Pronounced in a discussion of experimental and theoretical approaches by E. A. Moelwyn-Hughes, "Physical Chemistry," 3rd ed., Pergamon, Oxford, 1961.

importance of the distance across a charge-relay chain for its coupling and confirmed that the compression of the chain expected from the binding of polypeptide substrates should indeed produce coupling. This was then shown by proton inventory experiments. The magnitudes of the quantities involved were not a part of theoretical–experimental interplay. Similarly, the theoretical finding of a rough (not exact) coupling of carbon–nucleophile bond order to other structural features of acyl-transfer transition states, combined with crystallographic indications of the same thing, provided an underpinning for the interpretation of the experimental isotope effects. Again, there was no collision of magnitudes. Finally, in the rhodopsin study, the theoretical observations were that a nearby polarizable molecule could produce spectral shifts in the correct direction and with a proper distribution of transition energies and oscillator strengths and further that the latter were susceptible of regulation by relatively small geometrical changes. These were the critical features, rather than the exact values of the calculated properties as compared with experiment.

2. The quantities that were explored theoretically were associated with properties of individual molecules rather than with bulk properties that would have required statistical extensions of the quantum theoretical results. The potential surfaces examined for hydrogen bond chains, the structures, charge distributions, force constants, etc., of the acyl-transfer transition states, and the optical properties of the PRSB–benzene complexes all share this characteristic.

3. There was no reason to expect that gross solute–solvent interactions or other bulk environmental influences would alter the major character of any of these results. This is important because quantum mechanical calculations of the present day necessarily reflect the situation in the gas phase. This is likely to be true in some sense for a period into the future, although a certain amelioration can be expected from the considerable current effort directed toward introducing general environmental effects into quantum chemical calculations [72].

4. These studies were based on experimental methods in which a reliable relationship exists between properties accessible to theory and properties measurable by experiment. The experimental spectroscopic transitions of the rhodopsin case are directly susceptible of theoretical calculation. In the isotope effect cases, the structures that are reached by theory are related by the well-established procedures of vibrational analysis and the Bigeleisen formulation to experimental isotope effects.

5. It was characteristic of the experimental contributions that they reduced the degrees of freedom for the theoretical studies and that these in turn narrowed the potential scope of further experimental investigations. Thus, the crystallographic results on chymotrypsin focused the theory on

hydrogen bond chains within certain structural limits, while the theoretically discovered relation of coupling and distance across the chain centered experimental effort on this relationship. In the acyl-transfer study, experimental findings limited the theoretical exploration of transition states to a small volume of the 18-dimensional energy hyperspace, while the theory confined experimental interpretations (at least at present) to the set of transition-state structures with nearly coupled reaction variables. Experimental information on the rhodopsin problem dictated the general geometrical features of the calculational model. The results of the calculation now show the most fruitful routes for further optical and structural studies.

ACKNOWLEDGMENTS

We are happy to thank the National Institutes of Health and the National Science Foundation for support of the research described in this chapter and to express deep gratitude to our colleagues who have assisted in the work and in the development of our ideas about the subject. An invitation to speak at the Mardi Gras Symposium, 1976, issued by the Department of Chemistry of Louisiana State University, provided the initial stimulus for working out much of the content of the chapter. R.L.S. is pleased to thank the department for its kind invitation and generous hospitality.

REFERENCES

1. B. Pullman and A. Pullman, "Quantum Biochemistry." Wiley (Interscience), New York, 1963.
2. F. Haurowitz, "The Chemistry and Function of Proteins." Academic Press, New York, 1963.
3. R. E. Christoffersen, *Adv. Quantum Chem.* **6**, 333 (1972); G. G. Hall, *Chem. Soc. Rev.* **2**, 21 (1973).
4. "Cold Spring Harbor Symposia on Quantitative Biology," Vol. 36. Cold Spring Harbor Lab., Cold Spring Harbor, New York, 1972.
5. H. Fritz, H. Tschesche, L. J. Greene, and E. Truscheit, eds., "Proteinase Inhibitors, Proceedings of the 2nd International Research Conference." Springer-Verlag, Berlin and New York, 1974.
6. D. M. Collins, F. A. Cotton, E. E. Hazen, Jr., E. F. Meyer, Jr., and C. N. Morimoto, *Science* **190**, 1047 (1975).
7. W. P. Jencks, *Adv. Enzymol.* **43**, 219 (1975).
7a. T. C. Bruice, *Annu. Rev. Biochem.* **45**, 331 (1976).
7b. T. H. Fife, *Adv. Phys. Org. Chem.* **11**, 1 (1975).
8. R. D. Gandour and R. L. Schowen, eds., "Transition States of Biochemical Processes." Plenum, New York, 1977.
9. Isaac Newton, "Some Thoughts about the Nature of Acids," 1692.
10. M. L. Bender and F. J. Kézdy, "Proton Transfer Reactions" (E. Caldin and V. Gold, eds.), Chapter 12. Chapman & Hall, London, 1975.
11. M. L. Bender and F. J. Kézdy, *Annu. Rev. Biochem.* **34**, 49 (1965).

12. S. L. Johnson, *Adv. Phys. Org. Chem.* **5**, 275 (1967).
13. D. M. Blow, J. J. Birktoft, and B. S. Hartley, *Nature (London)* **221**, 337 (1969).
14. M. W. Hunkapiller, S. H. Smallcombe, D. R. Whitaker, and J. H. Richards, *Biochemistry* **12**, 4732 (1973).
15. R. Henderson, *Biochem. J.* **124**, 13 (1971).
16. G. A. Rogers and T. C. Bruice, *J. Am. Chem. Soc.* **96**, 2473 (1974); cf. Bruice [7a].
17. A. J. Kresge, *Pure Appl. Chem.* **8**, 243 (1964); V. Gold, *Adv. Phys. Org. Chem.* **7**, 259 (1969); W. J. Albery, *in* "Proton Transfer Reactions" (E. Caldin and V. Gold, eds.), Chapter 9. Chapman & Hall, London, 1975; K. B. Schowen, *in* "Transition States of Biochemical Processes" (R. D. Gandour and R. L. Schowen, eds.), Chapter 6. Plenum, New York, 1977.
18. M.-S. Wang, R. D. Gandour, J. Rodgers, J. L. Haslam, and R. L. Schowen, *Bioorg. Chem.* **4**, 392 (1975).
19. E. Pollock, J. L. Hogg, and R. L. Schowen, *J. Am. Chem. Soc.* **95**, 968 (1973).
19a. J. P. Elrod, R. D. Gandour, J. L. Hogg, M. Kise, G. M. Maggiora, R. L. Schowen, and K. S. Venkatasubban, *Faraday Symp. Chem. Soc.* **10**, 145 (1975).
19b. J. P. Elrod, Ph.D. Thesis in Chemistry, University of Kansas, Lawrence (1975).
20. A. J. Kresge, *J. Am. Chem. Soc.* **95**, 3065 (1973).
21. R. D. Gandour, G. M. Maggiora, and R. L. Schowen, *J. Am. Chem. Soc.* **96**, 6967 (1974).
22. R. D. Gandour, J. Rodgers, G. M. Maggiora, and R. L. Schowen, unpublished results.
23. R. Huber, D. Kukla, W. Steigemann, J. Deisenhofer, and A. Jones, *Proc. Int. Res. Conf. Proteinase Inhibitors, 2nd,* 1973 pp. 497ff (1974).
24. R. C. Thompson, *Biochemistry* **12**, 47 (1973); R. C. Thompson and E. R. Blout, *ibid.* p. 57.
25. P. B. Medawar, "The Hope of Progress," p. 36. Anchor Press, Garden City, New York, 1973.
26. J. Bigeleisen and M. Wolfsberg, *Adv. Chem. Phys.* **1**, 15 (1959).
27. M. Wolfsberg and M. J. Stern, *Pure Appl. Chem.* **8**, 225 (1964); M. J. Stern and M. Wolfsberg, *J. Pharm. Sci.* **54**, 849 (1965); R. E. Weston, Jr., *Science* **158**, 332 (1967).
28. E. B. Wilson, J. C. Decius, and P. C. Cross, "Molecular Vibrations." McGraw-Hill, New York, 1955.
29. H. S. Johnston, "Gas Phase Reaction Rate Theory." Ronald Press, New York, 1966.
30. H. B. Bürgi, *Angew. Chem., Int. Ed. Engl.* **14**, 460 (1975).
31. W. P. Jencks, "Catalysis in Chemistry and Enzymology." McGraw-Hill, New York, 1969; M. L. Bender, "Mechanisms of Homogeneous Catalysis from Protons to Proteins." Wiley (Interscience), New York, 1975.
32. L. Pauling, "The Nature of the Chemical Bond," 3rd ed., p. 239. Cornell Univ. Press, Ithaca, New York, 1960.
33. H. B. Bürgi, J. D. Dunitz, and E. Shefter, *J. Am. Chem. Soc.* **95**, 5065 (1973); *Acta Crystallogr., Sect. B* **30**, 1517 (1974); E. Shefter, *in* "Transition States of Biochemical Processes" (R. D. Gandour and R. L. Schowen, eds.), Chapter 9. Plenum, New York, 1977.
34. H. B. Bürgi, J. M. Lehn, and G. Wipff, *J. Am. Chem. Soc.* **96**, 1956 (1974).
35. P. Hogan, R. D. Gandour, G. M. Maggiora, and R. L. Schowen, to be published.
36. K. Kessler, G. M. Maggiora, and R. L. Schowen, to be published.
37. J. L. Hogg and R. L. Schowen, to be published.
38. C. A. Lewis and R. D. Wolfenden, personal communication.

39. C. G. Mitton and R. L. Schowen, *Tetrahedron Lett.* p. 5803 (1968).
40. C. B. Sawyer and J. F. Kirsch, *J. Am. Chem. Soc.* **97**, 1963 (1975).
41. M. H. O'Leary and M. D. Kluetz, *J. Am. Chem. Soc.* **94**, 665 (1972).
42. G. S. Hammond, *J. Am. Chem. Soc.* **74**, 334 (1955).
43. J. Bigeleisen, *in* "Isotopes and Chemical Principles" (P. A. Rock, ed.), p. 1. Am. Chem. Soc., Washington, D.C., 1975.
44. V. J. Shiner, *in* "Isotopes and Chemical Principles" (P. A. Rock, ed.), p. 163. Am. Chem. Soc., Washington, D.C., 1975.
45. E. W. Abrahamson and S. E. Ostroy, *Prog. Biophys. Mol. Biol.* **17**, 179 (1967).
46. H. J. A. Dartnall, ed., "Handbook of Sensory Physiology," Vol. 7, Part 1. Springer-Verlag, Berlin and New York, 1972.
47. E. L. Menger, ed., *Acc. Chem. Res.* **8**, 1 (1975).
48. B. Honig and T. G. Ebrey, *Annu. Rev. Biophys. Bioeng.* **3**, 151 (1974).
49. E. W. Abrahamson and R. S. Fager, *Curr. Top. Bioenerg.* **5**, 125 (1973).
50. T. G. Ebrey and B. Honig, *Q. Rev. Biophys.* **8**, 129 (1975).
51. A. Lewis, R. S. Fager, and E. W. Abrahamson, *J. Raman Spectrosc.* **1**, 465 (1973).
52. A. R. Oseroff and R. H. Callender, *Biochemistry* **13**, 4243 (1974).
53. G. A. J. Pitt, F. D. Collins, R. A. Morton, and P. Stok, *Biochem. J.* **59**, 122 (1955).
54. B. Honig, A. Warshel, and M. Karplus, *Acc. Chem. Res.* **8**, 92 (1975).
55. C. S. Irving and P. A. Leermakers, *Photochem. Photobiol.* **7**, 665 (1968).
56. C. S. Irving, G. W. Byers, and P. A. Leermakers, *J. Am. Chem. Soc.* **91**, 2141 (1969).
57. C. S. Irving, G. W. Byers, and P. A. Leermakers, *Biochemistry* **9**, 858 (1970).
58. K. Shirane, *Nature (London)* **254**, 722 (1975).
59. M. Ishigami, Y. Maeda, and K. Mishima, *Biochim. Biophys. Acta* **112**, 372 (1966).
60. R. Mendelsohn, *Nature* **243**, 22 (1973).
61. L. J. Weimann, M. Muthukumar, and G. M. Maggiora, to be published.
62. L. J. Weimann and G. M. Maggiora, to be published.
63. L. J. Weimann, G. M. Maggiora, and P. E. Blatz, *Int. J. Quantum Chem., Quantum Biol. Symp.* **2**, 9 (1975).
64. B. Honig and M. Karplus, *Nature (London)* **229**, 558 (1971).
65. R. Rowan, A. Warshel, B. D. Sykes, and M. Karplus, *Biochemistry* **13**, 970 (1974).
66. R. S. Becker, S. Berger, D. K. Dalling, D. M. Grant, and R. J. Pugmire, *J. Am. Chem. Soc.* **96**, 7008 (1974).
67. T. G. Ebrey and B. Honig, *Proc. Natl. Acad. Sci. U.S.A.* **69**, 1897 (1972).
68. A. Doukas and R. H. Callender, *Biophys. J.* **16**, 98a (1975).
69. H. J. A. Dartnall, *in* "The Eye" (H. Davson, ed.), 1st ed., Vol. 2, pp. 323–533. Academic Press, New York, 1962.
70. C. D. B. Bridges, *Vision Res.* **7**, 349 (1967).
71. G. R. Levy, "Religious Conceptions of the Stone Age and Their Influence upon European Thought," pp. 300ff. Harper, New York, 1963.
72. J. Hylton-McCreery, R. E. Christoffersen, and G. G. Hall, *J. Am. Chem. Soc.* **98**, 7191, 7198 (1976).

10

Enzymatic Olefin Alkylation Reactions

Robert M. McGrath

INTRODUCTION*

The enzymes involved in the utilization of IPP and DMAPP are varied and, for the purposes of this review, they are all deemed prenyltransferases in accordance with the terminology proposed by Holloway and Popják [1]. They are classified as follows:

    1. Prenyl-prenyltransferases: those enzymes that (a) transfer $(C_5)_n$ units to IPP in a head-to-tail fashion or (b) condense C_{15} or C_{20} prenyl pyrophosphates in a tail-to-tail fashion.

    2. Prenyl-aryltransferases: those enzymes that transfer (a) a monoprenyl group or (b) a polyprenyl group to an aryl moiety.

There are other enzymes that utilize prenyl pyrophosphates, e.g., phosphatases or sialyltransferase, an enzyme involved in sialyl polymer synthesis, but the reaction mechanisms involving these enzymes are so different that they cannot be considered prenyltransferases. The enzymes described here generally mediate electrophilic substitution reactions, in which a carbonium

---

* Abbreviations used: IPP, isopentenyl pyrophosphate (3-methylbut-3-enyl pyrophosphate); DMAPP, 3,3-dimethylallyl pyrophosphate (3-methylbut-2-enyl pyrophosphate); GPP, geranyl pyrophosphate; FPP, farnesyl pyrophosphate; GGPP, geranylgeranyl pyrophosphate; NPP, neryl pyrophosphate; MK-n, menaquinones; Q-n, ubiquinones. The integer n signifies the number of C_5 isoprenyl units attached to the aromatic moiety (see Fig. 6).

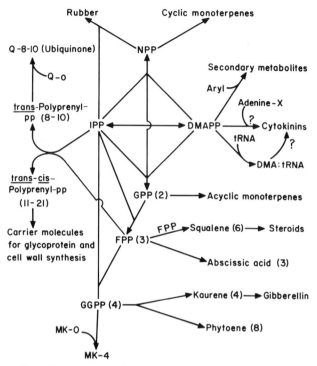

Fig. 1   An outline of isoprenoid pathways. Numbers in parentheses indicate numbers of isoprenoid (C_5) groups.

ion, generated by the pyrophosphate leaving group, bonds with an electron-rich atom of the cosubstrate.

Some of the types of compounds formed by the prenyltransferases are shown in Fig. 1, and it needs no eloquent exposition to underline the importance of these products. It is an interesting fact that a high percentage of such $(C_5)_n$ compounds, whether cyclized or acyclic, are involved in metabolic control in all forms of life. Thus, nature possesses the ability to synthesize, from DMAPP, IPP, and aryl molecules, groups of compounds, the stereochemistry and polarity of which are controlled by enzymes.

This review deals with the head-to-tail prenyl-prenyltransferases and the monoprenyl- and polyprenyl-aryltransferases. It lists the parameters of these enzymes and examines the mechanism and stereospecificity of such reactions in the belief that ultimately the differences and similarities will hold rewards in the study of metabolic control.

The regulation of the production of DMAPP and IPP has been reviewed by McNamara and Rodwell [2].

FORMATION OF PRENYL PYROPHOSPHATES

Mechanism

Cornforth [3] summarized his and his co-workers' conclusions, which are as follows:

1. The isomerization of IPP to DMAPP takes place in an *anti* fashion; i.e., it is a concerted mechanism, electrons being withdrawn and supplied from opposite sides of C-3 (Fig. 2a).

2. Elongation takes place by a *syn* mechanism; i.e., electrons are apparently withdrawn and supplied to the same side of C-3.

To rationalize this energetically unfavorable mechanism, suggested by the use of substrates stereospecifically labeled with isotopes of hydrogen, a two-step reaction was proposed. The allylic group and $X^{\ominus}$ add in an *anti* or *trans* fashion to IPP (Fig. 2b). This is followed by an *anti* elimination of $X^{\ominus}$ and H_R, which leads to a *trans** stereochemistry around the new double bond

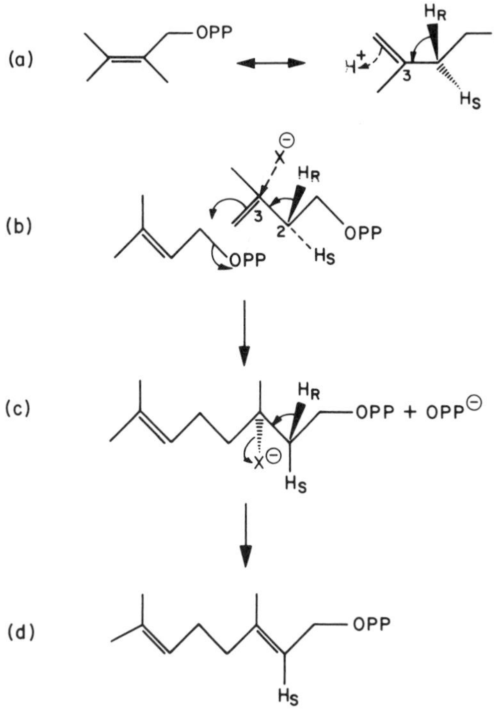

Fig. 2 The isomerization of IPP to DMAPP followed by elongation to GPP.

* The *cis–trans* notation is used throughout this chapter rather than the *Z–E* [6].

(Figs. 2c and 2d). Had H_S been eliminated, i.e., had the relative positions at H_S and H_R been interchanged, then a *cis* compound would have resulted as it does in rubber biosynthesis [4].

Recently Poulter, Satterwhite, and Rilling produced evidence that an alternative, and previously postulated, "ionization-condensation-elimination" mechanism, rather than the "displacement-elimination" mechanism presented here, was the more likely, at least when *trans*-3-trifluoromethylbut-2-enyl pyrophosphate was cosubstrate with IPP for the prenyltransferase from pig liver [5].

Most workers have tacitly accepted a general rule: Isoprenoids that lose the 2-pro-R hydrogen of IPP (equivalent to the 4-pro-S hydrogen of mevalonic acid) are biogenetically *trans* and vice versa. The change from S to R is merely one of nomenclature, brought about by observing the IUPAC rules [6]; the absolute configuration is unchanged.

However, doubts have been cast on this "rule" because reverse losses have been observed, in which the 4-pro-S hydrogen is lost and the product shows *cis* stereochemistry [7]. The doubts arose because no isomerase could be found which would have allowed the acceptance of a *trans* biosynthesis followed by isomerization [7].

Simultaneously, similar conclusions were reached, namely, that as well as an obvious isomerization reaction there may be (a) different prenyltransferases or (b) one prenyltransferase with different active sites that could position the substrates so that the pro-4-S hydrogen would always be lost irrespective of whether *cis* or *trans* bonds were being formed [8]. However, isomerization of a biogenetic *trans* arrangement via a redox system, for which there is some evidence [8,9] has been suggested. Since then, Shine and Loomis have shown the presence of an isomerase in carrot and peppermint which converts GPP to NPP in the presence of flavin, a thiol, and light. The reaction occurred in the dark if the flavin was first reduced [10].

Therefore, it seems likely that the general rule regarding the stereochemistry of products during enzymatic olefin alkylation will be reinstated *in toto*.

Holloway and Popják [1] investigated the reaction on isolated pig liver prenyltransferase and showed that it was an ordered sequential mechanism:

$$E + GPP \longrightarrow E^{GPP} + IPP \longrightarrow E^{GPP}_{IPP} - E^{FPP}_{PP} \longrightarrow$$
$$E_{PP} + FPP \longrightarrow E + PP$$

Three questions come to mind at this stage. What substrate stereochemistry is important in binding to the enzyme? How is chain length determined? What distinguishes a *cis*- from a *trans*-forming enzyme? Some answers are being found in substrate specificity studies.

Substrate Specificity of the Transferases

Popják's group prepared a series of analogs of GPP, two monophosphates and four pyrophosphates (i.e., four analogs in the alkyl group). They concluded that the polar pyrophosphate and relatively nonspecific apolar alkyl group are the important factors required in binding substrate to the pig liver enzyme [11]. They also demonstrated that the enzyme synthesized 10,11-dihydrofarnesyl pyrophosphate from 6,7-dihydrogeranyl pyrophosphate and IPP. Thus, the number of double bonds did not control chain length [12]. The simplest cosubstrate for IPP was DMAPP; replacement of either methyl group by hydrogen, a phenyl, or *tert*-butyl group caused loss of activity. This observation supports the idea of a narrow site for the apolar group [13].

The group of Ogura and Seto came to conclusions similar to that of Popják and his co-workers but used the transferase from pumpkin. They expanded the study and suggested a double site on the enzyme for GPP and DMAPP from their observations using *trans*-2-butenyl pyrophosphate and allyl pyrophosphate as inhibitors [14]. They also showed that *trans*-substituted compounds were bound more tightly than their *cis* analogs [15], and double bonds did not dictate chain length of the alkyl group [16].

The Japanese group prepared a series of analogs of DMAPP by extending

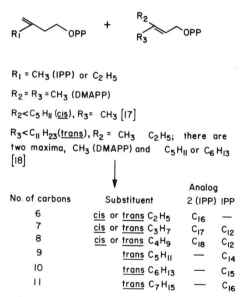

$R_1 = CH_3$ (IPP) or C_2H_5

$R_2 = R_3 = CH_3$ (DMAPP)

$R_2 < C_5H_{11}$ (cis), $R_3 = CH_3$ [17]

$R_3 < C_{11}H_{23}$(trans), $R_2 = CH_3$ C_2H_5; there are
two maxima, CH_3 (DMAPP) and C_5H_{11} or C_6H_{13}
[18]

No. of carbons	Substituent	Analog 2 (IPP)	IPP
6	cis or trans C_2H_5	C_{16}	—
7	cis or trans C_3H_7	C_{17}	C_{12}
8	cis or trans C_4H_9	C_{18}	C_{12}
9	trans C_5H_{11}	—	C_{14}
10	trans C_6H_{13}	—	C_{15}
11	trans C_7H_{15}	—	C_{16}

Fig. 3 Summary of some of the structural features that permit activity on pumpkin prenyltransferase [17,21]. The products are shown.

the chain length of either the *cis* or *trans* methyl groups. The effect of this elongation, summarized in Fig. 3, was to cause a peak activity with a *trans*-C_6H_{13} substituent. This compound resembles GPP in length and is readily acceptable to the enzyme at the GPP site; as a substrate it is nearly equivalent to GPP. Replacement by $—C_2H_5$ is not enough to take the substrate away from the DMAPP site. Further replacement by $—C_3H_7$ and $—C_4H_9$ caused partial recognition of both DMAPP and GPP sites; the analogs were not very active and gave mixed products [17,18]. It would appear from this work that spatial restrictions are important in binding the apolar moiety to the transferase.

The first changes in the IPP moiety were reported by Ogura, Koyama, and Seto when they showed that only the ethyl group could replace the methyl (Fig. 3) [19]. They then used the enzyme to produce an insect juvenile hormone skeleton using 3-ethylbut-3-enyl pyrophosphate and *cis*-3-methylpent-2-enyl pyrophosphate as substrates [20].

A comparison of enzymes from pig liver and pumpkin was made by Nishino, Ogura, and Seto. They found a broader specificity with some artificial substrates in liver enzyme, particularly for branched substituent groups [21].

When IPP was replaced by 4-methylpent-4-enyl pyrophosphate (i.e., an extra methylene group is interposed between the double bond and the pyrophosphate group) in the reaction with GPP using pig liver transferase, the result was a nonallylic homofarnesyl pyrophosphate, in which the analog had added in a *cis* fashion.

The authors explained this result by assuming that the binding site for IPP contains two subsites, P and M, which must be filled simultaneously with a pyrophosphate and methyl group, respectively. An extension of IPP by one

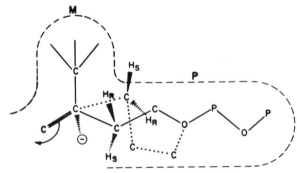

Fig. 4 Proposed position of IPP (solid bonds) filling its site on the enzyme. The dotted bonds show how a chain extension by one methylene group can also be accommodated on the subsites (P and M) although H_S and H_R are inverted. (From the work of Ogura, Saito, and Seto [22].)

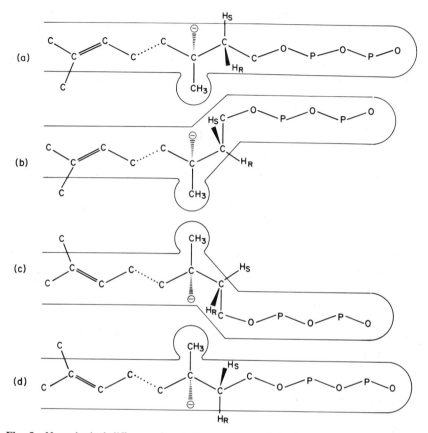

Fig. 5 Hypothetical difference in relative positions of the P and M sites showing the positional alteration of C-1, which in turn alters H_R and H_S with respect to a site on the enzyme. The elimination of H_R (a) and H_S (b) leads to *trans* and *cis* stereoisomers respectively; (c) and (d) are mirror images of (a) and (b) which lead to the same stereochemistry but with reverse proton losses. The dotted line shows the new bond formed. The stage in the reaction is equivalent to that shown in Fig. 2c.

methylene group only allows these conditions to be met. The new chain can be twisted so that stereochemistry allows subsites P and M to be filled; the pro-*S* and pro-*R* hydrogens are inverted, leading to an inversion of stereochemistry (Fig. 4) [22].

This experiment adds support to the generally accepted rule of biogenesis of *trans* or *cis* stereochemistry. Hypothetically, the same reasoning could be applied to *cis*- or *trans*-forming enzymes. Because there is no reason why the olefin alkylation reaction should not yield a mixture of *cis* and *trans* residues, one must accept that it is the enzyme which dictates the stereochemistry [23].

The enzyme difference could simply be one of the relative positions of the P and M subsites, which would dictate the position of the pro-S and pro-R hydrogens. This point is illustrated in Fig. 5, in which (a) and (d) are mirror images, as are (b) and (c): (a) and (b) yield *trans* and *cis* stereoisomers, with the observed loss of pro-R and pro-S hydrogens, respectively, whereas (c) and (d) give the reverse losses. Presumably, nature (with its usual bias for one chirality against another) favors enzymes (a) and (b), although there is no *a priori* reason for this being true.

The same authors studied the reaction between DMAPP and 4-methylpent-4-enyl pyrophosphate, which took place at a much lower rate than when GPP was substrate. The product with pig liver transferase as catalyst was a nonallylic homoneryl pyrophosphate, mimicking the previous work [24].

Shinka, Ogura, and Seto, after their extensive study of farnesylpyrophosphate synthetase, turned to geranylgeranylpyrophosphate synthetase from *Micrococcus lysodeikticus*. They found that neither *cis* nor *trans* derivatives of DMAPP, where R_2 or R_3 of Fig. 3 = C_2H_5–C_7H_9, would act as substrate. The longest chain produced was a C_{21} homolog brought about by the addition of two molecules of IPP to a C_7H_{15}-*trans* derivative (C_{11}) of DMAPP [25].

The same authors next compared the geranylgeranylpyrophosphate synthetase from *M. lysodeikticus* and pumpkin. They found that the activity profiles differed when IPP and some DMAPP derivatives (the *trans* methyl group was replaced by substituents increasing in length by a methylene group at a time) were incubated with the synthetase. Also, whereas GPP was the favored cosubstrate with IPP for the pumpkin enzyme, it was not preferred by the bacterial enzyme [26].

THE PRENYLTRANSFERASES

Some important parameters of these enzymes are shown in Table 1 [1,27–35,75]. Most are well known and all presumably have the same mechanism, with slightly variable substrate specificities. They are all sulfhydryl enzymes and are inhibited by sulfhydryl reagents such as iodoacetamide, N-ethylmaleimide, and p-mercurihydroxybenzoate, although some are more resistant than others. Possibly the most outstanding feature is the general requirement by geranylgeranylpyrophosphate synthetase for Mn^{2+} and of farnesylpyrophosphate synthetase for Mg^{2+}.

Recent work by Green and West indicates the existence of two species of FPP synthetase. Values for molecular weight, K_m, and activity of the enzymes are dependent on the protein concentrations, which suggests protein–protein interactions. Other workers may have ignored the nonlinearity between protein concentration and reaction rate (even when excess substrates are

Some Parameters of Prenyltransferases

Source	MW	Optimum pH	K_m (μM)	Purification	Cation	C_5 units	pI	Reference
Micrococcus lysodeikticus	>10⁵	DMAPP, 7.5; GPP, 6.5–8.0; FPP, 6.5–8.0	DMAPP, 29; GPP, 3.1; FPP, 3.8	20	Mg^{2+}	4	—	27
Pig liver	—	7.0	IPP, 6.6; FPP, 76.0	—	Mn^{2+}	4	—	28
Carrot	—	6.8	IPP, 10; FPP, 100	—	Mn^{2+}	4	—	28
Pig liver	—	7.0	IPP, 2.0; GPP, 4.0	10–15	Mn^{2+}	3	—	29
Pig liver	—	7.0–7.8	IPP, 1.25; DMAPP, 2.2	192	Mg^{2+}	3	—	30
Pumpkin	—	7.5	GPP, 1.3	—	Mg^{2+}	3	—	31
Pig liver	68,000	7.9	IPP, 1.45; GPP, 4.34	100	Mg^{2+}	3	—	
Pumpkin	—	6.8	—	—			—	1
Ricinus communis	(i) 72,500[a] 56,000[b] (ii) 72,500[a] 60,000[b]	6.8 6.8	IPP, 30–50[c]; 2–3[d] GPP, 30–50[c]; 4–6[d] DMAPP, 4–6[c]; 1–2[d]	970 645	Mn^{2+} Mg^{2+}	4 3	—	32 33
Avian liver	86,000 (2 × 43,000)	7.0	IPP, 0.5; GPP, 0.5	678	Mg^{2+}	3	5.72	34
Yeast	84,000 (2 × 43,000)	5.8–7.6	IPP, 4.0; DMAPP, 8.0; GPP, 14.0	341	Mg^{2+}	3	5.3	35
Penicillium cyclopium	64,000; approx. 50,000 with Mg^{2+}	8.0	—	—	Mg^{2+}	3	5.1	75

[a] Protein concentration, 25 mg/ml.
[b] Protein concentration, 20 μg/ml.
[c] Protein concentration, 42–56 μg/ml.
[d] Protein concentration, 12–20 μg/ml.
K_m values are approximated for both protein concentrations to include the two transferases (i) and (ii) in the range given.

present), which these results would imply [33]. The observation of Green and West is potentially important because protein–protein interactions and sub-unit assembly into oligomeric proteins, in which "quaternary structure might act principally to stabilize the correct conformation of the sub-unit" [36], is possibly of consequence in the prenyltransferases. Thus, the quaternary structure might organize the spatial relationships between the binding sites for DMAPP and GPP, FPP and IPP, or even between the P and M subsites.

The work of Reed and Rilling [34] and Eberhardt and Rilling [35] in purifying the enzymes to homogeneity, for the first time, removes any ambiguity as to product and substrate specificity. They have shown that the enzymes are dimers, and one wonders as to the possibility that each monomer might possess one active site.

THE POLYPRENYLTRANSFERASES

Hemming has pointed out that most polyprenoid material in nature is of mixed cis–trans stereochemistry. He lists the structures of castaprenol-10 to -13 (C_{50}–C_{65}) as all having the biogenetically first four isoprene units in the trans form and the rest cis while, in betulaprenol-6 to -9, the first three units are trans and the remainder are again cis [37]. Recently, the structures of cleomeprenol-9 to -11 have been shown to contain four trans residues and the remainder cis [38]. This suggests common starting materials of GGPP and FPP with cis additions thereafter. Although in plants materials such as the betulaprenols are not common, polyprenols from angiosperm leaves exhibit mixed stereochemistry of the castaprenol type [37].

In bacteria, undecaprenols are found also with the biogenetically first three residues trans and the remainder cis. They appear to be important in cell wall biosynthesis as a carrier of a carbohydrate moiety. In mammals, the longer-chained dolichols, consisting of 17 and more isoprene residues, also show a similar pattern of three biogenetic trans residues. They are probably associated with glycoprotein biosynthesis. Very little is known about the enzymes in Table 2 [4,39–43], molecular weights and binding constants are unknown, and even the pH quoted is not necessarily optimum. An exception is the all-cis-forming rubber synthetase. However, they all require Mg^{2+} as a cation. Purification is obviously a problem.

The isolation of a polyprenyltransferase by Keenan and Allen is the first example of a requirement for detergent (Triton X-100) to completely release the enzyme from a particulate fraction [43]. It is possible that, if Christenson, Scross, and Robbins [40] had used such a detergent, the enzyme from *Salmonella newington*, described in Table 2 as particulate, may have been solubilized. However, not only was the enzyme solubilized, but it was also

TABLE 2

Some Parameters of Polyprenyltransferases

Source	pH (assay)	Purification	Optimal cosubstrate	Cation	C_5 units	Comments	Reference
Micrococcus lysodeikticus	7.4	27	GPP FPP	Mg^{2+}	7–10 7 and 8[a]	All of the larger isoprene pyrophosphates bind strongly to protein and are not cleaved by alkaline phosphatase	39
Salmonella newington	7.4	—	FPP	Mg^{2+}	11	The enzyme is particulate	40
Salmonella newington	7.4	11	FPP	Mg^{2+}	About 8		40
Micrococcus lysodeikticus	7.4	27	FPP	Mg^{2+}	4–11, 8[a]	Monophosphate formed in crude extract incubation	41
Lactobacillus plantarum	7.4	Cell-free	—	—	11	The C_{55} isoprene is accompanied by an isomer containing one reduced residue	42
Lactobacillus plantarum	7.0	—	FPP	Mg^{2+}	11	Triton X-100 is needed for stability and extraction	43
Hevea brasiliensis	6.8–7.5	350	Rubber particles	Mg^{2+}	n	The stereochemistry is *all-cis*; MW = approx 60,000	4

[a] Predominates.

stabilized by Triton. This suggests that its native environment is apolar, e.g., membranous, and *in vivo* mevalonate feeding experiments using *Lactobacillus casei* indicated that bactoprenol (C_{55}) was biosynthesized in the mesosome and plasma membranes [44]. Keenan and Allen have also shown that, when Triton was removed, only cardiolipin out of four tested phospholipids could substitute for Triton [45].

The undecaprenol from *Lactobacillus plantarum* had been previously isolated and shown to be a prenol of the usual mixed stereochemistry [46], with which the product of the *L. plantarum* polyprenyltransferase might well be identical [43].

The polyprenyltransferases so far examined produce mixed *cis–trans* stereochemistry, but the ubiquinones [47] and menaquinones [48] are biosynthesized from a quinone and *all-trans*-polyprenyl pyrophosphate. Thus, at the moment, there is a gap in the isolation of enzymes involved in the synthesis not only of dolichols, but of the *all-trans*-polyprenyl pyrophosphates.

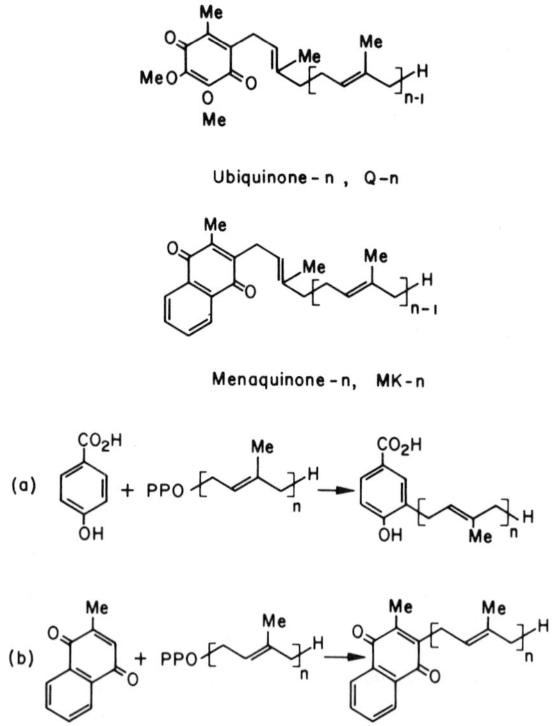

Ubiquinone - n , Q-n

Menaquinone - n,   MK - n

Fig. 6   Additions of polyprenols to aromatic compounds. Reactions mediated by enzymes *in vitro* and involved in ubiquinone and menaquinone (vitamin K) biosynthesis.

POLYPRENYL-ARYLTRANSFERASES

Products resulting from this class of enzyme include the ubiquinones (located in mitochondria in all aerobic organisms), plastoquinones (chloroplasts in plants), phylloquinones (chloroplasts in plants), and menaquinones (cell wall and plasma membrane in bacteria and higher animals). Of these, only the transferases involved in the biosynthesis of ubiquinones and menaquinones have been isolated.

Figure 6 shows the structures of the finished product resulting from the cell-free systems to be discussed. The nomenclature is that suggested by the IUPAC–IUB Commission [49]. No attempt will be made to trace the subsequent pathway to the ubiquinones because the authors quoted give complete references to this biosynthesis. Table 3 lists some features of the polyprenyl-aryltransferases [50,53,55,58,62].

p-Hydroxybenzoate:polyprenyltransferase
(*Ubiquinone Synthetase*)

This enzyme was found mostly in the inner mitochondrial membrane [50] and, as a matter of interest, so was *p*-hydroxybenzoate synthetase [51]. It carries out reaction (a) in Fig. 6. Earlier assays made use of *M. lysodeikticus* to preform radioactive polyprenyl pyrophosphate(s) which were then used as cosubstrate(s) with *p*-hydroxybenzoate. The product possesses an *all-trans* side chain [52] in spite of the fact that *M. lysodeikticus* is well known to produce the mixed *cis–trans* isomers. This implies enzyme specificity for the stereochemistry of the long-chain polyprenol and/or compartmentation of polyprenyl pyrophosphate.

Momose and Rudney [50] have shown that the transferase was unspecific with regard to the chain length of the polyprenyl substrate. Therefore, it is the availability of this substrate that determines the product. They not only demonstrated the presence of the enzyme in the inner mitochondrial membrane, but identified IPP-isomerase and prenyltransferase as being also present, implying lack of necessity for translocation of apolar polyprenyl materials. The enzyme required Mg^{2+} and had an optimum pH of 7.3; phosphate buffer, 0.1 M, was not inhibitory.

Young, Leppik, Hamilton, and Gibson using cell-free extracts of mutated *Escherichia coli* showed that the transferase and the precursor of the polyalkyl side chain were membrane bound and that Mg^{2+} was required for maximum activity [53].

Thomas and Threlfall demonstrated the presence of the enzyme in mitochondria of the broad bean and yeast. Unlike the liver transferase of Momose and Rudney [50], this mitochondrial system had to be supplied with preformed polyprenyl pyrophosphates [54].

TABLE 3
Some Parameters of Polyprenyl-aryltransferases

Enzyme	Source	MW	Optimum pH	Purification	Cation	C_5 units	Reference
p-Hydroxybenzoate: polyprenyltransferase	Rat liver mitochondria	Particulate	7.3	5 (approx)	$Mg^{2+} > Mn^{2+}$	9 and 10	50
	E. coli	Particulate	7.0	—	$Mg^{2+} > Mn^{2+}$	8	53
	Yeast mitochondria	Particulate	7.0	—	Mg^{2+}	3, 4, 5, 8; 9 and 10	55
	E. coli	Particulate	7.0	—	Mg^{2+}	3 > 4[a]; 8 > 9	58
Menaquinone-4 synthetase	Chick liver microsomes	Particulate	7.4	—	—	3 > 2 > 4	72

[a] Phytol.

The same authors showed that the transferase from yeast was activated by Mg^{2+}, had a pH optimum of 7.0, and was inhibited by phosphate buffer (unlike the enzyme from liver [50]). Further, the enzyme system could use FPP quite well as a cosubstrate with p-hydroxybenzoate to yield a C_{15} side chain [55].

Hachimi, Samuel, and Azerad have also used mutants of E. coli as did Gibson and co-workers [53,56]. They used two mutants, one defective in the transferase, the other blocked in the aromatic pathway. If the quinone mutant was supplemented with o-succinylbenzoate, a precursor of menaquinone-8 [57], then MK-8 and desmethylmenaquinone-8 were synthesized. The ubiquinone precursor p-hydroxybenzoate competed with MK-0 for the polyprenyl pyrophosphate, which suggests a common pool of isoprenoid in E. coli for Q and MK biosynthesis. On the other hand, if compartmentation is not a factor and if the polyprenoid is not rate limiting, then the same enzyme could possibly be involved.

These workers also showed the FPP was utilized better than solanesyl pyrophosphate (a C_{45} prenoid). However, the decarboxylating enzyme was found to be very specific for the 4-hydroxy-3-octoprenylbenzoic acid, which may explain the preponderance of Q-8 [58].

Hamilton and Cox [48] had reported that, when an E. coli aromatic auxotroph was grown in the presence of p-aminobenzoate, the polyprenylated derivative appeared. Alam, Nambudiri, and Rudney [59] showed that mitochondria from rat liver and brain also polyprenylated p-aminobenzoate, which also strongly inhibited the normal reaction. The implication was that it bound to the mammalian transferase and, whereas this did not irrefutably prove that it was the same transferase that polyprenylated the two benzoates, it must be regarded as very likely. Another inhibitor, p-chlorobenzoate, was not prenylated, which implied the electronic or steric importance of the NH_2 or OH groups in the reaction. The enzyme, like all the prenyltransferases, was dependent on —SH groups for activity [59].

A note of caution at this stage is appropriate. The enzyme from mitochondrial systems has not yet been solubilized, and observations made on whole mitochondria must take membrane effects into account.

Schechter, Nishino, and Rudney have, in fact, looked at the effect of Ca^{2+}, Mg^{2+}, EDTA, and EGTA (ethyleneglycolbis(β-aminoethyl ether)-N,N'-tetraacetic acid) with this in mind. These authors stated that the main point of their work was to introduce the idea that polyprenyl phosphates might take their place with other phospholipids when structure and function of mitochondrial membranes were being considered [60]. It is interesting to note, in this context, the work of Shechter, who has isolated an enzyme from Gibberella fujikuroi which converted FPP to the triphosphate and back to FPP again in the presence of AMP and Mg^{2+} to produce ATP and ADP [61].

The location of this enzyme was not reported, although one would like to imagine that it is associated with the mitochondria.

Menaquinone-0:polyprenyltransferase
(Menaquinone-4 Synthetase)

Dialameh, Yekundi, and Olson have demonstrated the polyprenylation of menaquinone-0 in different particulate suspensions of chick liver. The richest fraction in activity was the microsomes and the best cosubstrate of MK-0 was FPP, followed by GPP and finally GGPP (the natural cosubstrate). Thus, as in ubiquinone biosynthesis, the analogous enzyme was not specific for a prenyl pyrophosphate of a given chain length. This must be regulated by availability of prenyl pyrophosphates [62] so, presumably, in the chick there is a microenvironment of GGPP.

Lee, Houser, and Olson have since located the enzyme in the light smooth membrane of the microsomes. Treatment of this section of microsomal membranes with phospholipase C destroyed 60% of the activity, which was partly restored by the addition of lecithin [63].

There has been no report on the effect of divalent cations.

MONOPRENYL-ARYLTRANSFERASES

These enzymes seem to be associated with secondary metabolism, and, because secondary metabolism is probably best described as a method of ridding the cell of accumulating primary metabolites [64], should more significance not be placed, even hypothetically, on an enzyme that disposes of a molecule such as DMAPP? The potential products have been outlined in Fig. 1, e.g., ubiquinone, essential in electron transport; sterols, important in growth and reproduction of fungi [65]; and polyisoprenol sugars, important in cell wall metabolism [66], to name a few. Thus, the removal of such a compound might afford some form of control on other, more distal primary processes.

Figure 7 summarizes the systems for which a cell-free system, at least, was reported. An interesting feature of all these systems is that the dimethylallyl group has always entered a position where cyclization with an active hydrogen can occur. This is not entirely surprising, considering that the active hydrogens are always associated with $-NH_2$ and $-OH$ groups, which increase the electron density in any position ortho to them. The resultant increase in nucleophilicity is, theoretically, conducive to a reaction between the potential carbonium ion generated by the pyrophosphate leaving group in DMAPP. However, this illustrates the inherent dangers of neglecting the steric and

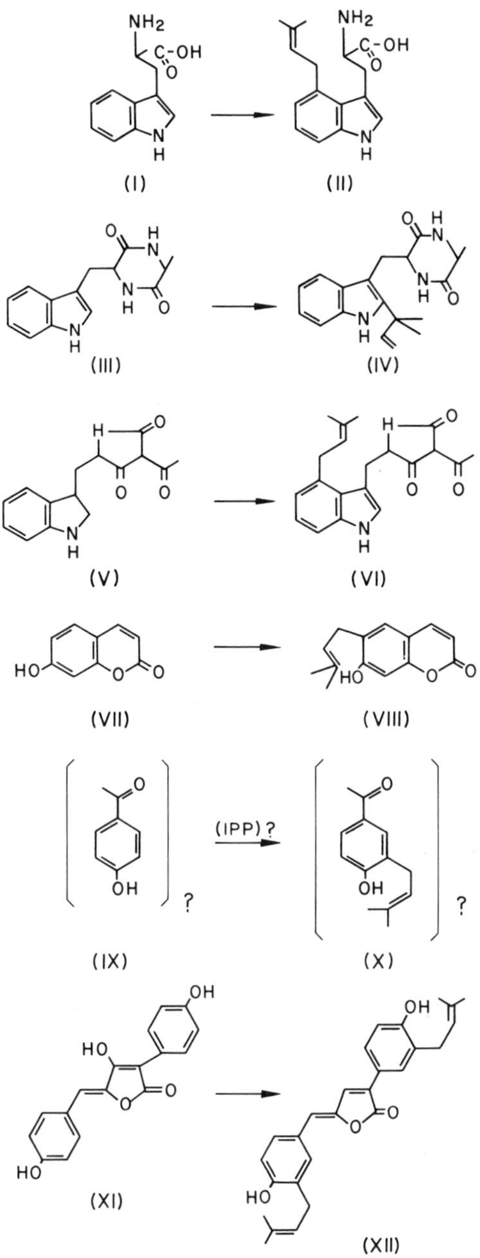

Fig. 7 *In vitro* reactions of aromatic compounds with DMAPP.

TABLE 4

Some Parameters of Monoprenyl-aryltransferases

Enzyme	Source	MW	Optimum pH	$K (\mu M)$	Purification	Cation	C_5 units	pI	Reference
4-Dimethylallyl-tryptophan synthetase	Claviceps sp.	73,000	7–8	DMAPP, 200 Trp, 67	63	$Ca^{2+} > Mg^{2+} > Mn^{2+}$	1	5.8	68
Cyclo-L-Ala-2(1,1-Dimethylallyl)-L-Trp synthetase	Aspergillus amstelodami	—	7 (assay)	—	—	Mg^{2+}	1	—	69
β-Cyclopiazonic acid synthetase	Penicillium cyclopium	96,000; 75,000 with Mg^{2+}	6–8	DMAPP, 2 cAATrp[a], 6	31	None	1	5.3	75
6-Dimethylallyl-umbelliferone synthetase	Ruta graveolens	—	7.5 (assay)	—	—	Mn^{2+} (Mg^{2+} does not activate)	1	—	76
3-Dimethylallyl-4-hydroxyaceto-phenone synthetase	Eupatorium rugosum	—	7–8	—	—	Mg^{2+}	1 (IPP)	—	77
Prenylpulvinone synthetase	Aspergillus terreus	—	7.0 (assay)	—	—	Mg^{2+}	2×1	—	78
Dimethylallyl: tRNA synthetase	E. coli	55,000	8.0	peak I tRNA, 0.66	350	Mn^{2+}, Fe^{2+}	$n \times 1$	—	79
	E. coli	—	7.5	tRNA, 13 DMAPP, 3	565	Mg^{2+}	$n \times 1$	—	80
	Lactobacillus acidophilus	—	7.5–8		24	Mg^{2+}	$n \times 1$	—	81

[a] cAATrp, Cycloacetoacetyl-L-tryptophanyl.

other effects of the enzyme, because position 4 of Trp should not be preferred over 2, as it is in Fig. 7, (I) → (II).

Once again, no attempt will be made to trace further biosynthesis of compounds (II)–(XII) (Fig. 7) because the references quoted invariably give a full bibliography.

Following the example of Heinstein, Lee, and Floss, these enzymes have been called dimethylallylpyrophosphate:aryl dimethylallyltransferases [67], or the "product" synthetase. The main features of these enzymes are listed in Table 4 [68,69,75–81].

Dimethylallylpyrophosphate:tryptophan Dimethylallyltransferase
(4-Dimethylallyltryptophan Synthetase) [Fig. 7, (I) → (II)]

This enzyme was demonstrated as active in a cell-free incubation of *Claviceps* species which had just ended its phase of balanced growth during stationary or shake culture. The dependence on protein concentration was not linear; in fact, it showed a peak of optimum activity at 500 μg/ml. The induction of this enzyme in the fungus correlated well with alkaloid production [67].

Lee purified the enzyme to homogeneity and showed the absence of a subunit structure. However, aggregates are formed on aging, presumably by intermolecular disulfide bonding. The effect of divalent cations on this enzyme is different from the enzymes already discussed. Here Ca^{2+} is the preferred ion; although Mg^{2+} is more effective at lower concentrations, it inhibits at 5 mM concentrations. The enzyme was highly specific for Trp, although 7-methyl-Trp showed 40% activity of the control. The mechanism appears to be sequential, but Lee was not able to distinguish between a random and ordered type [68].

Dimethylallylpyrophosphate:cyclo-L-alanyl-L-tryptophanyl Dimethylallyltransferase (Cyclo-L-alanyl-2(1,1-dimethylallyl)-L-tryptophanyl Synthetase) [Fig. 7, (III) → (IV)]

This synthetase was isolated from *Aspergillus amstelodami* as an ammonium sulfate precipitate (40–70%) by Allen [69]; [³H]DMAPP and cyclo[¹⁴C]L-alanyl-L-tryptophanyl used as substrates produced (IV) (Fig. 7), with the substrates in a 1:1 ratio.

The enzyme was quite specific for the aromatic substrate; if the substrate was not cyclic, as L-Ala-L-Trp or L-Trp-L-Ala, then there was no reaction. The isomer of DMAPP, IPP, did not serve as a substrate.

A subsequent publication showed that (IV) was indeed an *in vivo* precursor of echinulin [70]. It is interesting that echinulin requires the addition of two further 3,3-dimethylallyl moieties, but the cell-free system does not do this.

Deyrup and Allen [71] showed that the enzyme also utilized cyclo-L-Pro-L-Trp almost as well as cyclo-L-Ala-L-Trp. The analog is found in *A. ustus* [72]. However, an analog of DMAPP, cyclopentylidine ethyl pyrophosphate, which is active as a substrate for pig liver transferase [15], is not active with this enzyme.

Dimethylallylpyrophosphate:cycloacetoacetyl-L-*tryptophanyl Dimethylallyltransferase* (*β-Cyclopiazonic Acid Synthetase*) [*Fig.* 7, (**V**) → (**VI**)]

This enzyme was used as a cell-free extract from *Penicillium cyclopium* to detect an unknown cosubstrate of DMAPP for β-cyclopiazonic acid biosynthesis. The unknown material was identified as cycloacetoacetyl-L-tryptophanyl [73] and was found in the cells and in the culture filtrate in about equal concentrations (80 μM), while at the same time no Trp pool was evident. Dimethylallyltryptophan, the precursor of the clavine alkaloids and a possible precursor of cyclopiazonic acids, was not utilized in *in vivo* experiments, nor was Trp a substrate or an inhibitor for the enzyme producing cyclopiazonic acid. It was concluded that dimethylallyltryptophan was not a precursor of the cyclopiazonic acids.

The cell-free extract utilized the substrates in a 1:1 stoichiometry, and the product was very strongly protein bound [74]. Once again, the enzyme activity as a function of protein concentration was not linear.

The synthetase, prenyltransferase, and isomerase from *P. cyclopium* have been separated and partially purified. The enzymes all showed variations in molecular size as measured by gel filtration. Further, these changes could readily be brought about *in vitro* by the addition or removal of either Mg^{2+} or Mn^{2+} to or from the gel column. Magnesium ion caused a diminution in size of the synthetase from 96,000 to 75,000 daltons, whereas Mn^{2+} caused the discreet peak to smear out. The prenyltransferase and isomerase both changed from 64,000 to 50,000 daltons but retained well-defined peaks in the presence of both cations.

The three enzymes were studied in an attempt to understand the control of the branch point set up by the competition of the synthetase and the primary prenyltransferase. It was concluded that the single most important event was the appearance of cycloacetoacetyl-L-tryptophanyl. During this study, it was found that the synthetase was not inhibited by EDTA but that it was inhibited by Mn^{2+}. The primary prenyltransferase, on the other hand, was severely inhibited by EDTA, requiring Mg^{2+} for activity. Either the synthetase binds a divalent cation very strongly, or it does not require one, which would be unusual.

The β-cyclopiazonic acid synthetase was specific for its substrates. Not only did it not prenylate or bind Trp, but it did not accept *N*-acetoacetyl-L-Trp, GPP, or IPP (which is a poor inhibitor) as alternative substrates [75].

Dimethylallylpyrophosphate:umbelliferone Dimethylallyltransferase
(6-Dimethylallylumbelliferone Synthetase) [*Fig.* 7, **(VII)** → **(VIII)**]

This enzyme was demonstrated in cell-free extracts of *Ruta graveolens* [76]. It is a particulate enzyme which requires DMAPP and which does not use GPP or methylumbelliferone (herniarin) as substrates. The enzyme is specific for the formation of the linear furanocoumarins. Presumably another enzyme is reponsible for the biosynthesis of 8-dimethylallylumbelliferone, which leads to the angular furanocoumarins. The title synthetase requires Mn^{2+} for activity, which is not replaceable by Mg^{2+}. Attempts at solubilization were only partially successful. With Tween 80, about 50% activity remained in the supernatant, but the enzyme proved to be very unstable.

Isopentenylpyrophosphate:4-hydroxyacetophenone Dimethylallyl-
transferase (3-Dimethylallyl-4-hydroxyacetophenone Synthetase)
[*Fig.* 7, **(IX)** → **(X)**]

At the moment, there is some question as to what reaction this enzyme actually mediates. It exists as a cell-free extract from the leaf of *Eupatorium rogusum* and may mediate a step in the biosynthesis of dehydrotremetone [77]. Some evidence exists which indicates that IPP rather than DMAPP is the alkylating agent. This would be difficult to explain mechanistically. However, the authors wisely caution against accepting, at face value, results obtained from a complex system.

There appears to be some doubt as to whether the product of the reaction is indeed **(X)** (Fig. 7), which would perhaps support the observation that IPP is a cosubstrate or could mean instead that **(IX)** is not the true substrate. However, from the results given in the publication, it is not clear if this is so.

It would seem that an aromatic material derived from [^{14}C]IPP is incorporated into dehydrotremetone some 180 times more efficiently than authentic 4-hydroxy-3-dimethylallylacetophenone. On the other hand, in a competition between this unknown ^{14}C material and ^{3}H authentic compound, there is a small difference in relative incorporation, representing a 37% loss of the original ^{3}H. But, without an incubation, a simple mixture of these two compounds (inseparable by thin layer chromatography) eluted from SiO_2 with methanol into a scintillation vial showed an average loss of 26.5% ^{3}H. This might cause one to question the belief that the unknown and the authentic compounds are different as is claimed by the authors. Unfortunately, insufficient quantities of the unknown material were available for analysis, which could have placed the issue beyond doubt, although infrared spectra of the two compounds are different. Also, the initial observation of large differences in incorporation is unexplained. Perhaps it represents differences in cell-free extracts with different levels of cosubstrates (IPP) or the absence of an enzyme capable of converting authentic 4-hydroxy-3-dimethylallyl-acetophenone into the closely related unknown intermediate.

The enzyme is inhibited by iodoacetamide and so, like all of these enzymes, it is dependent on sulfhydryl groups for activity.

Dimethylallylpyrophosphate:pulvinone Dimethylallyltransferase(s)
[Prenylpulvinone Synthetase(s)) [*Fig.* 7, (**XI**) → (**XII**)]

The enzyme(s) from *Aspergillus terreus*, as a crude ammonium sulfate fraction, catalyzes the addition of two dimethylallyl groups to pulvinone to give (**XII**) [78]. There has as yet been no study of the system other than the requirement of DMAPP, but one would immediately wonder if one or two enzymes are involved.

Dimethylallylpyrophosphate:tRNA Dimethylallyltransferase
(*Dimethylallyl:tRNA Synthetase*)

This enzyme is classified here with the monoprenyl-aryltransferases even though it differs from its predecessors in not substituting directly into an aromatic ring.

The reaction is shown in Fig. 8. There is still a question as to how compounds like the cytokinins (**XV**) arise. One would expect the reaction (**XVI**) → (**XV**) to occur between an adenine derivative and DMAPP (the product of which would be modified to the cytokinin) by simple analogy to the enzymatic reactions described here, but such a reaction has yet to be discovered. On the other hand, the enzyme carrying out the reaction (**XIII**) → (**XIV**) is well known [79,80,81], but the catabolism of (**XIV**) → (**XV**) is open to question. Hall, in a review of cytokinins, presents evidence that there are alternative biosynthetic routes other than turnover of tRNA [82].

In adenine, the amino group is present as such and not as the imino form. Chemical behavior is not consistent with that of an aromatic amine, and it is rather unreactive in nonenzymatic reactions [83]. In nature, an isomer is found in *Gleditsia triacanthos* which is substituted at N-3 by the dimethylallyl group [84]. If a similar reaction occurred at N-1 followed by a Dimroth rearrangement (as it does quite readily in 0.02 M aq. KOH) to the 6-amino group, one would obtain the cytokinin [85]. On the other hand, the enzymatic 6-amino alkylation may not have been observed because adenine, or adenosine, is not the cosubstrate with DMAPP, or the enzyme is compartmented away from exogeneous substrates.

The tRNA modifying transferase is known, as has been said, and the modification caused by it is important for proper binding of tRNA to ribosomes [82].

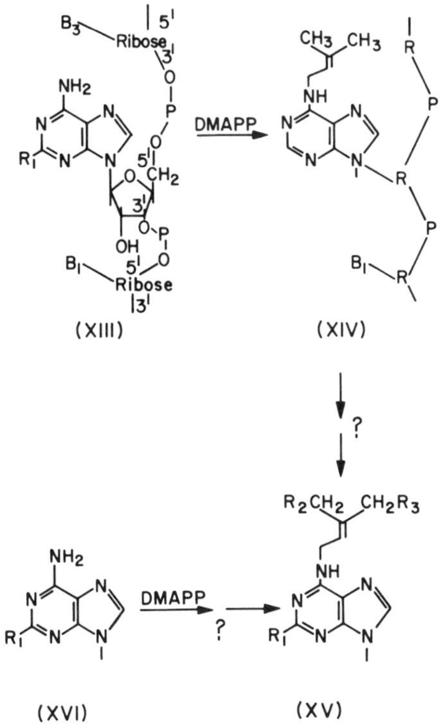

Fig. 8 Biosynthesis of cytokinins (**XV**), where R = ribose or H, R_1 = R_2 = R_3 = H, or R_3 = OH, or R_1 = CH_3S. Compound (**XIII**) is unmodified tRNA.

Bartz and Söll [79] used tRNA from *Mycoplasma* sp. (Kid), known to be deficient in modified tRNA, as a cosubstrate with DMAPP for the enzyme obtained from *E. coli*. They showed that it did not utilize small molecules in place of tRNA, e.g., adenosine or its 3'- or 5'-phosphate, or oligoadenylic acids (two to six chain length).

Rosenbaum and Gefter [80] also purified the enzyme from *E. coli*. Again, small cosubstrates were not prenylated; even a dodecyl nucleotide (obtained from the tRNA substrate as a breakdown product), which contained the adenosine base normally isoprenylated *in vivo*, was not alkylated. They demonstrated that it was indeed an adenosine moiety of the tRNA which was isoprenylated, confirming the work of Robins, Hall, and Thedford [86]. Further, the base was adjacent to the 3' end of the anticodon.

Recently, the enzyme was isolated from *Lactobacillus acidophilus*, and it appears, from the data given, to be little different from the *E. coli* enzyme [81].

In Table 4, the three enzyme preparations are summarized and a difference in cation requirement is evident, the Bartz and Söll preparation requiring Mn^{2+}, Fe^{2+}, or Ca^{2+} in preference to Mg^{2+}, which showed inhibitory effects. However, Rosenbaum and Gefter [80] reported that Mn^{2+} was only 60% as effective as Mg^{2+} at 3.3 mM, but they too reported inhibition by Mg^{2+}. Another difference lay in the effect of pyrophosphate; the latter workers' enzyme was much more inhibited. These differences may be artifactual or inherent, but in either case speculation would be worthless. A better understanding of the substrate specificities and the structure and control of the enzymes will no doubt clear up these minor points.

CONCLUDING REMARKS

The enzymes involved in olefin alkylation have been taken out of their various contexts—from ubiquinone and menaquinone biosynthesis, secondary metabolism, and tRNA modification—and juxtapositioned to highlight similarities or differences. They are all sulfhydryl proteins which, with at least one exception, require a divalent cation for maximum activity. Although they all generate a carbonium ion from either mono-, di-, tri-, or polyprenyl pyrophosphate, there is a wide variety of cosubstrates which can show remarkable changes in specificities, for example, 4-dimethylallyl-L-tryptophan and 4-dimethylallylcycloacetoacetyl-L-tryptophanyl synthetases, the former binding and utilizing tryptophan, the latter not binding tryptophan at all. Of importance is the probability of protein–protein interactions modifying the activity of these enzymes. Recent reports also tend to support the idea of subunit structures.

With the combined efforts of biochemists and organic chemists, a study of some of these enzymatic reactions, judiciously selected, should yield valuable information with regard to reaction control and the conformation of the enzyme itself.

The aim of this review unfortunately precluded a discussion of compartmentation (which has been reviewed recently by Davies [87]), an important control mechanism in the cell, which denies or permits access of substrates to enzymes. This is especially unfortunate, because it represents the antithesis of this work in which the cell-free system was sought. Nevertheless, the well-defined, minimal system is a necessary goal in the furtherance of bioorganic chemistry.

REFERENCES

1. P. W. Holloway and G. Popják, *Biochem. J.* **104**, 57 (1967).
2. D. J. McNamara and V. W. Rodwell, *in* "Biochemical Regulatory Mechanisms in Eukaryotic Cells" (E. Kun and S. Grisolia, eds.), pp. 205–243. Wiley (Interscience), New York, 1972.

3. J. W. Cornforth, *Angew. Chem., Int. Ed. Engl.* **7**, 903 (1968).
4. B. L. Archer and B. G. Audley, *Adv. Enzyml. Relat. Areas Mol. Biol.* **29**, 221 (1967).
5. E. D. Poulter, D. M. Satterwhite, and H. C. Rilling, *J. Am. Chem. Soc.* **98**, 3376 (1976).
6. IUPAC Tentative Rules for Nomenclature of Organic Chemistry. Section E. Fundamental Stereochemistry, *J. Org. Chem.* **35**, 2849 (1970).
7. E. Jedlicki, G. Jacob, F. Faini, and O. Cori, *Arch. Biochem. Biophys.* **152**, 590 (1972).
8. D. V. Banthorpe, G. N. J. Le Patourel, and M. J. O. Francis, *Biochem. J.* **130**, 1045 (1972).
9. L. Chayet, R. Pont-Lezica, C. George-Nascimento, and O. Cori, *Phytochemistry* **12**, 95 (1973).
10. W. E. Shine and W. D. Loomis, *Phytochemistry* **13**, 2095 (1974).
11. G. Popják, P. W. Holloway, R. P. M. Bond, and M. Roberts, *Biochem. J.* **111**, 333 (1969).
12. G. Popják, P. W. Holloway, and J. M. Baron, *Biochem. J.* **111**, 325 (1969).
13. G. Popják, J. L. Rabinowitz, and J. M. Baron, *Biochem. J.* **113**, 861 (1969).
14. K. Ogura, T. Koyama, T. Shibuya, T. Nishino, and S. Seto, *J. Biochem. (Tokyo)* **66**, 117 (1969).
15. T. Nishino, K. Ogura, and S. Seto, *Biochim. Biophys. Acta* **235**, 322 (1971).
16. T. Nishino, K. Ogura, and S. Seto, *J. Am. Chem. Soc.* **93**, 794 (1971).
17. K. Ogura, T. Nishino, T. Koyama, and S. Seto, *J. Am. Chem. Soc.* **92**, 6036 (1970).
18. T. Nishino, K. Ogura, and S. Seto, *J. Am. Chem. Soc.* **94**, 6849 (1972).
19. K. Ogura, T. Koyama, and S. Seto, *J. Chem. Soc., Chem. Commun.* p. 881 (1972).
20. T. Koyama, K. Ogura, and S. Seto, *Chem. Lett.* p. 401 (1973).
21. T. Nishino, K. Ogura, and S. Seto, *Biochim. Biophys. Acta* **302**, 33 (1973).
22. K. Ogura, A. Saito, and S. Seto, *J. Am. Chem. Soc.* **96**, 4037 (1974).
23. G. Popják, *in* "The Enzymes" (P.O. Boyer, ed.), 3rd ed., Vol. 2, pp. 115–215. Academic Press, New York, 1970.
24. A. Saito, K. Ogura, and S. Seto, *Chem. Lett.* p. 1013 (1975).
25. T. Shinka, K. Ogura, and S. Seto, *Chem. Lett.* p. 111 (1975).
26. T. Shinka, K. Ogura, and S. Seto, *J. Biochem. (Tokyo)* **78**, 1177 (1975).
27. A. A. Kandutsch, H. Paulus, E. Levin, and K. Bloch, *J. Biol. Chem.* **239**, 2507 (1964).
28. D. L. Nandi and J. W. Porter, *Arch. Biochem. Biophys.* **105**, 7 (1964).
29. C. R. Benedict, J. Kett, and J. W. Porter, *Arch. Biochem. Biophys.* **110**, 611 (1965).
30. J. K. Dorsey, J. A. Dorsey, and J. W. Porter, *J. Biol. Chem.* **241**, 5353 (1966).
31. K. Ogura, T. Nishino, and S. Seto, *J. Biochem. (Tokyo)* **64**, 197 (1968).
32. K. Ogura, T. Shinka, and S. Seto, *J. Biochem. (Tokyo)* **72**, 1101 (1972).
33. T. R. Green and C. A. West, *Biochemistry* **13**, 4720 (1974).
34. B. C. Reed and H. C. Rilling, *Biochemistry* **14**, 50 (1975).
35. N. L. Eberhardt and H. C. Rilling, *J. Biol. Chem.* **250**, 863 (1975).
36. H. Gutfreund, *Isr. J. Chem.* **12**, 319 (1974).
37. F. W. Hemming, *Biochem. Soc. Symp.* **29**, 105–117 (1970).
38. T. Suga, T. Shishibori, S. Kosela, Y. Tanaka, and M. Itoh, *Chem. Lett.* p. 771 (1975).
39. C. M. Allen, W. Alworth, A. Macrae, and K. Bloch, *J. Biol. Chem.* **242**, 1895 (1967).
40. J. G. Christenson, S. K. Cross, and P. W. Robbins, *J. Biol. Chem.* **244**, 5436 (1969).
41. T. Kurokawa, K. Ogura, and S. Seto, *Biochem. Biophys. Res. Commun.* **45**, 251 (1971).
42. I. F. Durr and M. Z. Habbal, *Biochem. J.* **127**, 345 (1972).
43. M. V. Keenan and C. M. Allen, Jr., *Arch. Biochem. Biophys.* **161**, 375 (1974).

44. K. J. Thorne and D. C. Barker, *Biochem. J.* **122**, 45P (1971).
45. M. V. Keenan and C. M. Allen, Jr., *Biochem. Biophys. Res. Commun.* **61**, 338 (1974).
46. D. P. Gough, A. L. Kirby, J. B. Richards, and F. W. Hemming, *Biochem. J.* **118**, 167 (1970).
47. J. B. Richards and F. W. Hemming, *Biochem. J.* **128**, 1345 (1972).
48. J. A. Hamilton and G. B. Cox, *Biochem. J.* **123**, 435 (1971).
49. IUPAC-IUB Commission on Biochemical Nomenclature, *Eur. J. Biochem.* **53**, 15 (1975).
50. K. Momose and H. Rudney, *J. Biol. Chem.* **247**, 3930 (1972).
51. P. Hagel and H. Kindl, *FEBS Lett.* **59**, 120 (1975).
52. H. Rudney, *Biochem. J.* **128**, 13P (1972).
53. I. G. Young, R. A. Leppik, J. A. Hamilton, and F. Gibson, *J. Bacteriol.* **110**, 18 (1972).
54. G. Thomas and D. R. Threlfall, *Biochem. J.* **134**, 811 (1973).
55. G. Thomas and D. R. Threlfall, *Phytochemistry* **13**, 1825 (1974).
56. F. Gibson, *Biochem. Soc. Trans.* **1**, 317 (1973).
57. P. Dansette and R. Azerad, *Biochem. Biophys. Res. Commun.* **40**, 1090 (1970).
58. Z. E. Hachimi, O. Samuel, and R. Azerad, *Biochimie* **56**, 1239 (1974).
59. S. S. Alam, A. M. D. Nambudiri, and H. Rudney, *Arch. Biochem. Biophys.* **171**, 183 (1975).
60. N. Schechter, T. Nishino, and H. Rudney, *Arch. Biochem. Biophys.* **158**, 282 (1973).
61. I. Shechter, *Biochim. Biophys. Acta* **362**, 233 (1974).
62. G. H. Dialameh, K. G. Yekundi, and R. E. Olson, *Biochim. Biophys. Acta* **223**, 332 (1970).
63. F. C. Lee, R. M. Houser, and R. E. Olson, *Fed. Proc., Fed. Am. Soc. Exp. Biol.* **34**, 655 (1975).
64. A. B. Woodruff, *Symp. Soc. Gen. Microbiol.* **16**, 22 (1966).
65. J. W. Hendrix, *Annu. Rev. Phytopathol.* **8**, 111 (1970).
66. W. J. Lennarz and M. G. Scher, *Biochim. Biophys. Acta* **265**, 417 (1972).
67. P. F. Heinstein, S. L. Lee, and H. G. Floss, *Biochem. Biophys. Res. Commun.* **44**, 1244 (1971).
68. S. L. Lee, *Diss. Abstr. Int. B* **35**, 2617 (1974).
69. C. M. Allen, Jr., *Biochemistry* **11**, 2154 (1972).
70. C. M. Allen, Jr., *J. Am. Chem. Soc.* **95**, 2386 (1973).
71. C. L. Deyrup and C. M. Allen, Jr., *Phytochemistry* **14**, 971 (1975).
72. P. S. Steyn, *Tetrahedron* **29**, 107 (1973).
73. R. M. McGrath, P. S. Steyn, and N. P. Ferreira, *J. Chem. Soc. Chem. Commun.* p. 812 (1973).
74. R. M. McGrath, P. S. Steyn, N. P. Ferreira, and D. C. Neethling, *Bioorg. Chem.* **5**, 1 (1976).
75. R. M. McGrath, P. N. Nourse, N. P. Ferreira, and D. C. Neethling, *Bioorg. Chem.* (in press).
76. B. E. Ellis and S. A. Brown, *Can. J. Biochem.* **52**, 734 (1974).
77. T. J. Lin and P. Heinstein, *Phytochemistry* **13**, 1817 (1974).
78. N. Ojima, K. Ogura, and S. Seto, *J. Chem. Soc. Chem. Commun.*, p. 717 (1975).
79. J. K. Bartz and D. Söll, *Biochimie* **54**, 31 (1972).
80. N. Rosenbaum and M. Z. Gefter, *J. Biol. Chem.* **247**, 5675 (1972).
81. J. Holtz and D. Klämbt, *Hoppe-Seyler's Z. Physiol. Chem.* **356**, 1459 (1975).
82. R. H. Hall, *Annu. Rev. Plant Physiol.* **24**, 415 (1973).

83. J. H. Lister, *in* "Fused Pyrimidines" (D. J. Brown, ed.), Part 2, pp. 1 and 309. Wiley (Interscience), New York, 1971.
84. N. J. Leonard and J. A. Deyrup, *J. Am. Chem. Soc.* **84**, 2148 (1962).
85. D. M. G. Martin and C. B. Reese, *J. Chem. Soc. C* p. 1731 (1968).
86. N. J. Robins, R. H. Hall, and R. Thedford, *Biochemistry* **6**, 1837 (1967).
87. R. H. Davies, *Annu. Rev. Genet.* **9**, 39 (1975).

11

Enzymatic Catalysis of Decarboxylation*

Marion H. O'Leary

INTRODUCTION

A variety of decarboxylation reactions are important in metabolism. To a first approximation, the mechanisms of action of the decarboxylases that catalyze these reactions are well understood. Various enzymes utilize a metal ion, pyridoxal 5'-phosphate, thiamine pyrophosphate, or an abnormally reactive lysine residue to aid in the decarboxylation. Organic analogs for all of these processes are known, and reasonable mechanistic outlines can be given.

However, these schemes do not fully account for the catalytic power of enzymes. Even after these aids are considered, the model reactions fall several powers of 10 short of the rates of enzymatic reactions. At the present state of knowledge, these models can serve as a starting point for probing in greater detail the intricacies of enzymatic catalysis. Specifically, we would like to know how enzymes increase the efficiency of these basic mechanisms. Are the rate-determining steps the same in enzymatic reactions as in the corresponding model reactions? What is the nature of the decarboxylation step, and to what extent is it accelerated? What ultimately limits the rates of enzymatic reactions?

Kinetic isotope effects are particularly useful tools for elucidating the details of enzymatic reaction mechanisms in cases for which reasonable model reactions are available [1,2]. In the case of decarboxylations, carbon isotope

* Dedicated to Professor F. H. Westheimer.

effects can provide information about the decarboxylation step and the rates of other steps relative to the rate of that step. When compared with data from appropriate model reactions, this information provides an estimate of the power of the enzyme to catalyze various individual steps in the reaction mechanism.

KINETIC ISOTOPE EFFECTS AND RATE-DETERMINING STEPS

Substitution of carbon-13 for carbon-12 in the carboxyl carbon of a compound undergoing decarboxylation often decreases the decarboxylation rate by a few percent [3,4]. In nonenzymatic decarboxylations in which the decarboxylation step is rate determining this rate difference usually amounts to 3–7% near room temperature; that is, $k^{12}/k^{13} = 1.03$–1.07. When decarboxylation is not rate determining no isotope effect is observed; $k^{12}/k^{13} = 1.00$. Intermediate values of the isotope effect indicate that the decarboxylation step is partially, although not entirely, rate determining.

Heavy-atom isotope effects in many reactions are difficult to understand and even more difficult to predict [3,4]. Carbon isotope effects on decarboxylations, on the other hand, are well behaved and predictable because the isotope effect arises primarily from a single factor—the stretching of the bond between the carboxyl carbon and the adjacent carbon. Variations in the magnitude of the carbon isotope effect reflect variations in the degree of breaking of the carbon–carbon bond at the transition state.

Because they are measured by competitive methods, heavy-atom isotope effects on enzymatic reactions always represent isotope effects on V_{max}/K_m rather than isotope effects on V_{max}. These isotope effects thus reflect the rates of all steps up through the first irreversible step—usually the decarboxylation step in the case of decarboxylases. Strictly speaking, it is not correct to interpret heavy-atom isotope effects in terms of the rate-determining step for the overall reaction unless other information is available indicating that the step which is rate determining in the overall reaction occurs prior to the first irreversible step. It is usually found in practice that this condition holds, but exceptions exist [5].

The kinetic isotope effect observed in an enzymatic reaction may be similar in magnitude to or smaller than that observed in the corresponding organic reaction, depending on the extent to which the step in which the change in bonding to the isotopic atom occurs is rate determining. For example, in the case of enzymatic decarboxylations, a carbon isotope effect, $k^{12}/k^{13} = 1.03$–1.07, is expected if the decarboxylation step is the rate-determining step.

For a more general case, such as Eq. (1), the observed isotope effect is given

$$E + S \underset{k_2}{\overset{k_1}{\rightleftharpoons}} ES \overset{k_3}{\longrightarrow} E + CO_2 + P \tag{1}$$

by Eq. (2), under the assumption that only the decarboxylation step is subject

$$\frac{k^{12}}{k^{13}} \text{ (observed)} = \frac{k_3^{12}/k_3^{13} + k_3/k_2}{1 + k_3/k_2} \tag{2}$$

to a significant isotope effect. More complex kinetic schemes give equations of the same form in which the term k_3/k_2 in Eq. (2) is replaced by a more complex function of rate constants [5].

Thus, according to Eq. (2) our interpretation of a carbon isotope effect in an enzymatic reaction depends on our ability to estimate the carbon isotope effect on the decarboxylation step (k_3^{12}/k_3^{13} in this case). It is logical to assume that this isotope effect is similar to that observed in a corresponding model reaction; that is, $k_3^{12}/k_3^{13} = 1.03-1.07$. This is equivalent to assuming that the structure of the transition state (or, more specifically, the force constants around the site of isotopic substitution) is the same in the enzymatic reaction as in the model reaction. The role of the enzyme then must be to aid in the attainment of that transition state. Unfortunately, it has not been possible to subject this assumption to rigorous test. In fact, one can envision circumstances under which this assumption might be invalid. In the case of a decarboxylation, it is possible that the enzyme might somehow stabilize the transition state such that it occurs earlier along the reaction coordinate; that is, the degree of carbon–carbon bond breaking at the transition state might be smaller in the enzymatic reaction than in the corresponding model reaction. Such a stabilization would lower the isotope effect on the decarboxylation step.

ENZYMATIC DECARBOXYLATIONS

To a first approximation the enzymatic catalysis of decarboxylation can be understood in terms of the functioning of an "electron sink." Decarboxylation involves the departure of carbon dioxide from the substrate, leaving behind the pair of electrons that previously constituted the bond to the carboxyl group. The catalytic mechanism provides for the absorption of this pair of electrons. The substrates in enzymatic decarboxylations ordinarily have a second functional group, commonly an amino, carbonyl, or incipient carbonyl group, contiguous to the carboxyl group. This second group serves either as the electron sink or as a handle that binds the substrate to the electron sink.

TABLE 1

Types of Enzymatic Decarboxylations

Type of electron sink	Type of Substrate	Example	Reference
Metal ion	β-Keto acids	Oxaloacetate decarboxylase, isocitrate dehydrogenase	6
Lysine amino group	β-Keto acids	Acetoacetate decarboxylase	7
Pyridoxal 5′-phosphate	α-Amino acids	Glutamate decarboxylase, arginine decarboxylase	8
Pyruvate	α-Amino acids	Histidine decarboxylase	8
Thiamine pyrophosphate	α-Keto acids	Pyruvate decarboxylase	9,10

The classification of decarboxylases according to the type of electron sink present is shown in Table 1 [6–10]. Information about kinetics and mechanism for model reactions corresponding to all these types is available. Kinetic data and carbon isotope effects are available for enzymes of all types except the pyruvate-dependent decarboxylases. In the following paragraphs we will discuss these data and compare them with appropriate model reactions in order to assess the extent to which the decarboxylation step is rate determining in both enzyme and model. This information will then be used to understand how enzymes catalyze decarboxylation.

Metal Catalysis

β-Keto acids decarboxylate slowly at room temperature, particularly when the pH is sufficiently low that an appreciable amount of free acid is present. The decarboxylation is catalyzed by a variety of metal ions. Metals appear to function by chelating to the ketone oxygen and stabilizing the enolate which is formed as a result of decarboxylation. The uncatalyzed decarboxylation of oxaloacetic acid [Eq. (3)] shows a carbon isotope effect $k^{12}/k^{13} = 1.045$ at

$$^-O_2{}^*CCH_2COCO_2{}^- \longrightarrow {}^*CO_2 + CH_3COCO_2{}^- \qquad (3)$$

25°C in aqueous solution [11]. The Dy^+-catalyzed decarboxylation of the same acid shows a carbon isotope effect $k^{12}/k^{13} = 1.034$ under the same conditions [11]. The Mn^{2+}-catalyzed decarboxylation shows a carbon isotope effect [10] $k^{12}/k^{13} = 1.06$ in water at 10°C. In all these cases, the decarboxylation step is wholly rate limiting, and the variation in the observed isotope effects reflects variations in transition-state structure.

The oxaloacetate decarboxylase from *Micrococcus lysodeikticus* requires a divalent metal ion for activity but has no other cofactors. There is no evidence for the involvement of a Schiff base in the catalytic mechanism. The metal ion appears to serve as an electron sink in the enzymatic reaction just as it does in the model reaction. The carbon isotope effect on the enzymatic decarboxylation [12] is $k^{12}/k^{13} = 1.002$ at 10°C in the presence of Mn^{2+}. An isotope effect in the range 1.03–1.07 would be expected if decarboxylation were rate determining. Thus, even though the decarboxylation of the enzyme–oxaloacetate complex is at least 50,000 times faster than decarboxylation of the Mn^{2+}–oxaloacetate complex [12], the decarboxylation step itself has been accelerated by considerably more than this figure, and decarboxylation is not even partially rate limiting.

It is not clear what the rate-determining step in the enzymatic decarboxylation is. There is no chemical step preceding decarboxylation which might be rate determining. It is possible that dissociation of the substrate from the enzyme is very slow and that the enzyme–substrate complex, once formed, virtually always undergoes decarboxylation. Under those circumstances k_3/k_2 in Eq. (2) is very large, and no isotope effect is observed. The Michaelis constant for oxaloacetate [12] is approximately 2 mM, but under those circumstances the Michaelis constant would not be a true dissociation constant.

The alternate possibility is that the decarboxylation step is reversible and that the rate-determining step is the release of carbon dioxide. The overall reaction is known to be reversible and is inhibited slightly by carbon dioxide [13]. Although one might not ordinarily expect carboxylation to compete effectively with carbon dioxide release, carboxylation of a metal-chelated enolate by a properly oriented carbon dioxide molecule might be quite fast.

This particular ambiguity with respect to the rate of carboxylation of an enzyme–product complex is not well understood, and the argument given above might explain the small isotope effects observed with a variety of decarboxylases. In the case of glutamate decarboxylase, decarboxylation of unlabeled glutamic acid in the presence of $^{14}CO_2$ is reported to produce [^{14}C] glutamic acid [14], conceivably by means of such a carboxylation.

Isocitrate dehydrogenase also catalyzes a metal-ion-dependent decarboxylation of a β-keto acid, but in this case the decarboxylation is preceded by the $NADP^+$-dependent oxidation of the β-hydroxy acid substrate to a β-keto acid [6] [Eq. (4)]. Under optimum conditions (pH 7.5, Mg^{2+} present) the carbon isotope effect [15,16] is $k^{12}/k^{13} = 0.999$. Thus, decarboxylation is not rate determining. Under the same conditions the hydrogen isotope effect for the hydrogen being transferred in the oxidation is approximately unity [16,17], indicating that hydride transfer is not rate determining. The Michaelis constants for substrates and products are in the submicromolar range [18],

$$\underset{\overset{|}{*CO_2^-}}{\overset{\overset{OH}{|}}{^-O_2CCH_2CHCHCO_2^-}} + NADP^+ \;\rightleftharpoons\; \underset{\overset{|}{*CO_2^-}}{\overset{\overset{O}{\|}}{^-O_2CCH_2CHCCO_2^-}} + NADPH$$

$$\downarrow \qquad\qquad\qquad\qquad\qquad\qquad\qquad\qquad\qquad (4)$$

$$\overset{\overset{O}{\|}}{^-O_2CCH_2CH_2CCO_2^-} + {}^*CO_2$$

and other kinetic evidence suggests that the slowest steps in the overall sequence are the product dissociation steps [18].

This enzyme is active with a variety of metal ions. If Ni^{2+} is used in place of Mg^{2+} the activity of the enzyme is reduced by approximately a factor of 10 [19], the carbon isotope effect is $k^{12}/k^{13} = 1.005$, and the hydrogen isotope effects are still approximately unity [19]. Thus, the decrease in rate of the enzymatic reaction is not primarily the result of a less efficient oxidation or a less efficient decarboxylation. The metal ion effect on the rate must result from metal effects on enzyme conformation changes or product release rates.

However, hidden from view in these data is the fact that the metal also decreases the rate of the decarboxylation step. This conclusion arises in the following way. Substitution of Ni^{2+} for Mg^{2+} decreases the activity of the enzyme by a factor of 10 and increases the carbon isotope effect from essentially unity to 1.005. Thus, Ni^{2+} must have slowed the decarboxylation step even more than it slows the overall reaction in order that decarboxylation might contribute in a small way to the observed rate and a small carbon isotope effect might be observed. This metal-dependent deceleration can be estimated to be at least a factor of 50 and might be much larger. Thus, it is possible that the metal may play the same role in this enzyme that it does in the nonenzymatic metal-catalyzed decarboxylations.

Other evidence is consistent with this proposed role of the metal ion. Isocitrate binds more tightly to the enzyme in the presence of Mg^{2+} than in its absence, and Mg^{2+} binds more tightly to the enzyme in the presence of isocitrate than in its absence [20]. A corresponding synergism between metal and nucleotide is not observed. Nuclear magnetic resonance and electron spin resonance studies indicate that one functional group of isocitrate, probably the hydroxyl group, is in the inner coordination sphere of the metal in the ternary enzyme–Mn^{2+}–isocitrate complex [20].

Amine Catalysis

There are two common mechanisms for the catalysis of decarboxylation of β-keto acids, the metal-ion-dependent mechanism discussed above and the amine-catalyzed mechanism. The decarboxylation of acetoacetic acid catalyzed by primary amines [Eq. (5)] has been studied extensively [7]. The reaction proceeds by way of a Schiff base intermediate, and decarboxylation

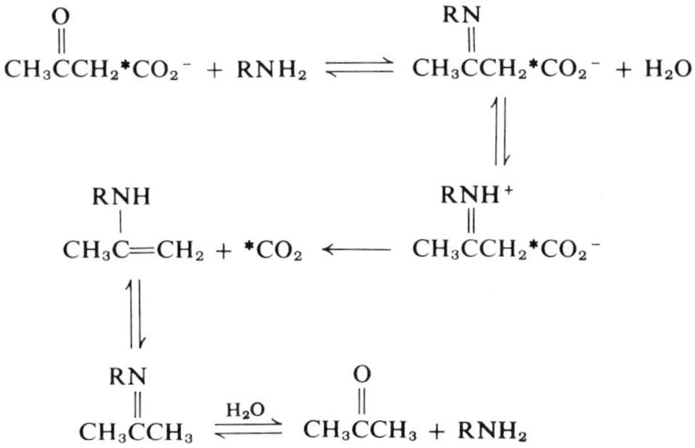

$$
\underset{\substack{\|\\ O}}{CH_3CCH_2*CO_2^-} + RNH_2 \rightleftharpoons \underset{\substack{\|\\ RN}}{CH_3CCH_2*CO_2^-} + H_2O
$$

$$
\underset{\substack{|\\ RNH}}{CH_3C=CH_2} + *CO_2 \longleftarrow \underset{\substack{\|\\ RNH^+}}{CH_3CCH_2*CO_2^-} \tag{5}
$$

$$
\underset{\substack{\|\\ RN}}{CH_3CCH_3} \xrightarrow{H_2O} \underset{\substack{\|\\ O}}{CH_3CCH_3} + RNH_2
$$

occurs from the protonated Schiff base. The carbon isotope effect [21] on the decarboxylation catalyzed by cyanomethylamine (R = CH_2CN) in aqueous solution at 30°C is k^{12}/k^{13} = 1.031 at pH 3.6, 1.032 at pH 4.1, and 1.036 at pH 5.0. At low pH, nucleophilic attack on the carbonyl carbon atom of the substrate is rate determining, and a small isotope effect is observed. At higher pH, decarboxylation becomes rate determining, and a larger isotope effect is observed. Results of other kinetic studies are consistent with this interpretation [22].

Acetoacetate decarboxylase from *Clostridium acetobutylicum* contains no cofactors or metal ions [7]. A single lysine residue of the enzyme [23] has a pK_a near 6.0 and functions in a manner analogous to that shown above for the reaction catalyzed by primary amines. The carbon isotope effect [21] on the enzymatic decarboxylation is k^{12}/k^{13} = 1.018 and is pH independent over the range pH 5.3–7.2. The magnitude of this isotope effect indicates that decarboxylation is at least partially rate determining. Oxygen exchange studies are consistent with this conclusion [24].

Explanation of the catalytic efficiency of acetoacetate decarboxylase is somewhat easier than for most other enzymes. The difference in rate between the enzymatic reaction and the nonenzymatic reaction is approximately 10^9, but all except about 10^3 is explained by the primary amine catalysis [25]. Possible explanations of this remaining factor are considered in a later section.

Pyridoxal Catalysis

Most amino acid decarboxylases contain pyridoxal 5′-phosphate but lack metal ions or other cofactors [8]. The coenzyme functions by means of a Schiff base mechanism (Fig. 1).

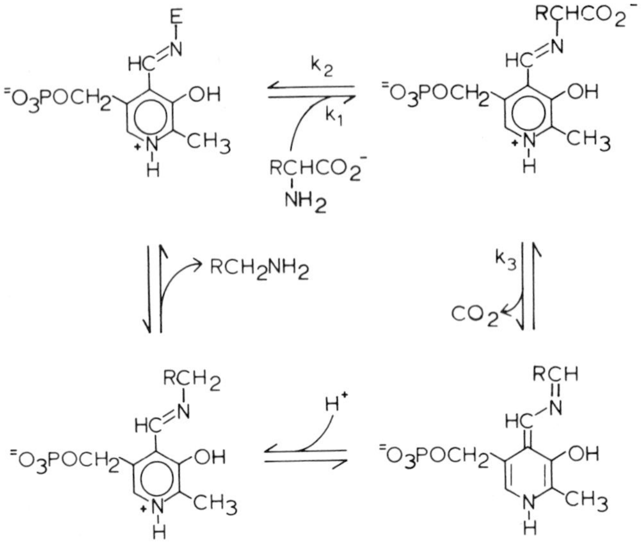

Fig. 1 Mechanism of action of pyridoxal 5'-phosphate-dependent decarboxylases. In the absence of substrate the coenzyme is bound to the protein by means of a Schiff base linkage. Formation of the initial substrate–pyridoxal 5'-phosphate Schiff base takes place in two steps which are not shown separately here.

Model reactions for pyridoxal-dependent enzymatic reactions have been studied extensively [26]. Schiff bases between pyridoxal and most amino acids are formed rapidly at room temperature in aqueous solution. Decarboxylation does not occur under these conditions. At slightly higher temperatures transamination of the Schiff base is observed, but appreciable decarboxylation is not observed even at much higher temperatures. If transamination is prevented by use of an α-methyl amino acid, decarboxylation occurs [27], but only at temperatures near 100°C. Clearly, decarboxylation must be considered to be the rate-determining step in the model reaction.

Notwithstanding the slowness of the model reaction, enzymatic decarboxylations of amino acids occur at reasonable rates. The decarboxylation of glutamic acid by an enzyme from *Escherichia coli* has been studied particularly extensively. The enzyme is active [28,29] in the range pH 3.5–5.5. The activity drops off rapidly above pH 5.5. Between pH 3.5 and 4.5 the carbon isotope effect [29] at 37°C is $k^{12}/k^{13} = 1.015$ and is independent of pH. Above this pH, the isotope effect increases, reaching a value of 1.022 at pH 5.5.

Both the magnitude of the carbon isotope effect and its pH dependence indicate that decarboxylation is not entirely rate limiting in the enzymatic reaction. Presumably the carbon isotope effect on the decarboxylation step itself is pH independent; that is, there is only one transition state for decarboxylation, even though the rate of passage through that transition state may vary with pH. The pH dependence arises because of pH-dependent changes in the partitioning of the Schiff base intermediate [k_3/k_2 in Eq. (2)]. Formation of the Schiff base can occur independent of whether the pyridine nitrogen of the coenzyme is protonated, whereas decarboxylation requires that the nitrogen be protonated. Thus, at low pH all reactions involve protonated pyridine, the observed isotope effect is 1.015, and k_3/k_2 is 2–4. At higher pH the rate of decarboxylation effectively decreases, and k_3/k_2 decreases. Unfortunately, it is not possible to calculate the pK_a of the pyridine nitrogen from these data.

When the decarboxylation of glutamic acid is conducted in 100% D_2O the product γ-aminobutyric acid contains 1.0 nonlabile deuterium atom. When the decarboxylation is conducted in 50% H_2O–50% D_2O the product contains 0.48 nonlabile deuterium atom [30]. Thus, there is no hydrogen isotope discrimination in the protonation of the quinoid intermediate, in spite of the fact that such proton transfers to carbon are ordinarily subject to large hydrogen isotope effects. This result requires that the proton that is added to the carbon atom not be in equilibrium with the solvent. That is, this proton must be donated by a specific group of the enzyme, and this group must be shielded from hydrogen exchange with the solvent in the quinoid intermediate. This group must be a monoprotic acid, probably an imidazole group or a carboxyl group.

Whereas no isotope effect is observed on the protonation of the quinoid intermediate, a large solvent isotope effect is observed in the overall reaction [30]; $k_{H_2O}/k_{D_2O} = 5$ at pH 5 in the presence of saturating levels of substrate. This indicates that the protonation step is not rate determining in the overall reaction. The rate-determining step must be the same as the step that determines the carbon isotope effect.

Arginine decarboxylase from *E. coli* also contains pyridoxal 5'-phosphate and operates by the mechanism shown in Fig. 1. At pH 5.25 the carboxyl carbon isotope effects on the decarboxylation of arginine [31] are $k^{12}/k^{13} = 1.027$ at 5°C, 1.014 at 25°C, 1.012 at 37°C, and 1.012 at 50°C. The magnitude of the isotope effect indicates that decarboxylation is partially rate limiting but not entirely so. The large temperature dependence is probably indicative of a large temperature dependence in the ratio k_3/k_2.

The kinetic and isotope effect data above indicate that pyridoxal-dependent enzymes accelerate the decarboxylation step by a much greater factor than they accelerate the Schiff base formation step.

Thiamine Catalysis

Enzymatic decarboxylation of α-keto acids is effected by means of thiamine pyrophosphate, which provides an electron sink for decarboxylation (Fig. 2) [9,10]. A carbanion is formed at the 2 position of the thiazolium ring, and

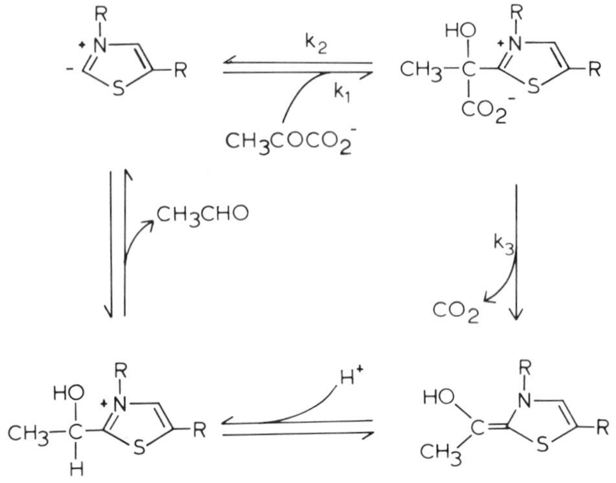

Fig. 2 Mechanism of action of pyruvate decarboxylase.

this anion condenses with the carbonyl group of the substrate. Decarboxyla-tion of the condensation product is aided by the positive charge in the thia-zolium ring. Analogous decarboxylations in models have been studied [32].

Yeast pyruvate decarboxylase requires thiamine pyrophosphate and a divalent metal ion for activity. The enzyme is active with a variety of α-keto acids [9,10]. A number of coenzyme analogs have been shown to bind to the enzyme, and a few of these have appreciable catalytic activity [10]. Based on studies of product inhibition, it has been suggested that product release is rate determining in the overall reaction [33].

The carboxyl carbon isotope effect on the enzymatic decarboxylation of pyruvate [34] is $k^{12}/k^{13} = 1.0083$ at pH 6.8, 25°C. The possibility that product release may be rate determining in the overall reaction has no bearing on the observed isotope effect because the isotope effect is governed by the steps up to and including carbon dioxide release. The small size of the isotope effect indicates that decarboxylation is not a slow step; it is presumably much faster than formation of the enzyme-bound thiamine pyrophosphate–pyruvate adduct.

The enzymatic decarboxylation of pyruvate catalyzed by this enzyme is faster by a factor of 10^5–10^6 than the corresponding nonenzymatic decarboxy-lation [32]. However, decarboxylation is the rate-determining step in the nonenzymatic decarboxylation, whereas decarboxylation must be faster than the rate-determining step by at least an order of magnitude (and conceivably much more) in the enzymatic decarboxylation. Thus, pyruvate decarboxylase accelerates the decarboxylation step by at least 10^6–10^7.

Conclusion

A variety of enzymatic decarboxylations have been studied by means of heavy-atom isotope effects, and it is striking that not in a single case has the decarboxylation step been observed to be entirely rate limiting. In some cases (for example, glutamate decarboxylase and acetoacetate decarboxylase) the decarboxylation step is not very much faster than the overall rate, but in other cases (pyruvate decarboxylase, oxaloacetate decarboxylase, and isocitrate dehydrogenase) the decarboxylation is so much faster than the overall rate that almost no isotope effect is observed.

This result is in striking contrast to results of corresponding model studies, in most of which the decarboxylation step is rate determining. Thus, enzymatic acceleration of the decarboxylation step is generally greater than enzymatic acceleration of the overall reaction.

OTHER ASPECTS OF DECARBOXYLASES

Before moving on to a discussion of the means by which enzymes may catalyze decarboxylations, we must consider two other aspects of decarboxylases.

Carboxyl Group Binding

The carboxyl group that is to undergo decarboxylation during the enzymatic reaction is usually an important binding point for attaching substrate to enzyme. Cases that have been studied include glutamate decarboxylase [28] and acetoacetate decarboxylase [35,36]. In no case has the identity of the carboxyl binding group been established.

Stereochemistry

Decarboxylation consists of the replacement of a carboxyl group by a single hydrogen. In every case in which the stereochemistry of this replacement has been determined, retention of configuration has been observed. Cases studied include tyrosine decarboxylase [37], lysine decarboxylase [38], isocitrate dehydrogenase [39], malic enzyme [40], and the oxaloacetate decarboxylase activity of codfish muscle pyruvate kinase [41].

The protonation that occurs following decarboxylation is probably mediated by a catalytic group of the enzyme. In the case of glutamate decarboxylase, evidence for the existence of such a catalytic group was given above. Since decarboxylation occurs with retention of configuration, this catalytic group must occupy roughly the same site occupied by the carboxyl binding group mentioned above. In fact, both functions might be served by the same group. This group probably also serves to orient the substrate for decarboxylation (see below).

CATALYSIS OF DECARBOXYLATION

What factors are responsible for the ability of enzymes to catalyze decarboxylations at very high rates? In this section we will attempt to enumerate these factors and assess their importance.

The Electron Sink

The first level of understanding of decarboxylations in enzymes and in models is clear: Enzymes and their congruent models function as electron sinks, obviating the necessity to form highly unstable carbanions. The reactions that serve as models for the enzyme cases have been extensively studied within the past 25 years, and our knowledge of these reactions is now sufficiently complete that they can serve as starting points for a discussion of enzymatic catalysis.

These electron sinks provide an important source of catalytic power in enzymatic reactions. However, the rates observed in appropriate model reactions are still several orders of magnitude short of those observed in the corresponding enzymatic reactions. In the case of acetoacetate decarboxylase, the enzymatic reaction is about 10^3 times faster than the reaction catalyzed by primary amines [25]. Catalysis of decarboxylation by oxaloacetate decarboxylase [12] is more efficient than that by metal ions by a factor of approximately 10^4. For pyruvate decarboxylase [32] the figure is near 10^6. The acceleration is large but unknown for decarboxylations dependent on pyridoxal 5'-phosphate.

Polarity Effects

One very important consequence of the electron sink concept is the potential for large medium effects in decarboxylations. The electron sink mechanism works by charge neutralization. In every case the intermediate preceding the decarboxylation step has a negatively charged carboxyl group and a positively charged group separated by a fairly large distance. In the decarboxylation transition state the neutralization of positive and negative charge is in process, and following the decarboxylation it is complete. The intermediate after decarboxylation is very much less polar than the preceding intermediate.

Each type of decarboxylation mechanism has its own characteristic positively charged group, but in each case the group is there, and in each case the role of the group is the same.

Charge neutralization reactions of this type may be subject to very large medium effects. Although medium effects have been studied in a great variety of ionic reactions, few studies have been devoted specifically to charge separations of the type under discussion. The decarboxylation of the thiamine analog shown in Eq. (6) has been studied extensively [32]. Decarboxylation

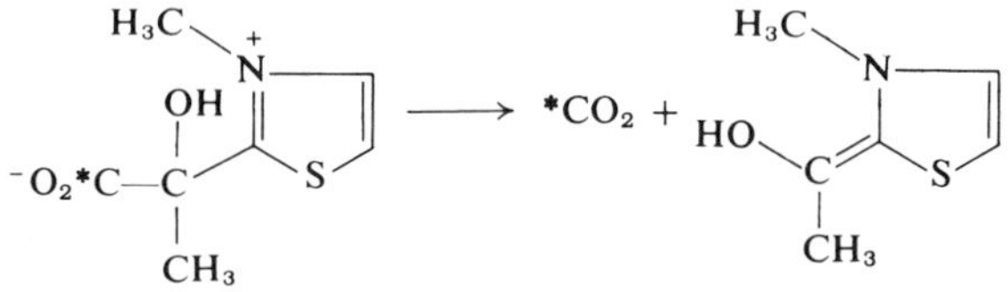

$$
\text{(6)}
$$

of the zwitterionic form occurs readily slightly above room temperature. The reaction occurs 9000 times more rapidly in ethanol than in water at 26°C in the presence of 1 M LiCl. Less polar media would presumably give rise to even larger rate accelerations.

Decarboxylation of the zwitterion of 4-pyridylacetic acid [Eq. (7)] also

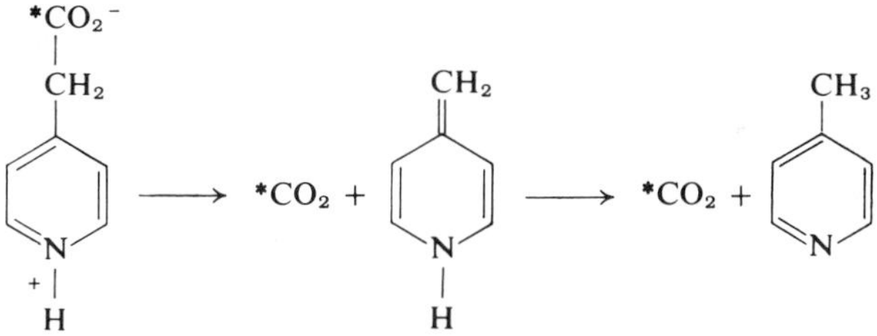

$$
\text{(7)}
$$

shows large medium effects. The rate of decarboxylation of this compound and the proportion of zwitterion present at equilibrium were determined for various isopropyl alcohol–water mixtures. From these data rate constants for decarboxylation of the zwitterion (the only form that undergoes decarboxylation) could be determined [42]. As the solvent changed from 20% (by volume) isopropyl alcohol–80% water to 80% isopropyl alcohol–20% water, the rate of decarboxylation of the zwitterion increased by a factor of 1200.

Thus, enzymes may accelerate decarboxylations significantly by providing nonpolar environments. The rate accelerations available by this means are quite large and are limited primarily by the ability of the enzyme to maintain a nonpolar active site. Further studies of the potential magnitude of this important effect are needed.

Conformational Control

In all of the reactions cited above, the step in which the actual decarboxylation takes place can be considered to be an elimination reaction in which carbon dioxide serves as a leaving group for the formation of a carbon–carbon double bond between the atoms α and β to the carboxyl group. Because of the formation of the π bond, a very particular orientation of the carboxyl group relative to the adjacent atoms must occur in order for decarboxylation to occur. The carboxyl group must be above the plane of the forming double bond, as shown in Fig. 3.

In the case of decarboxylations dependent on pyridoxal 5'-phosphate, the steric consequences of this geometric requirement have been considered in detail by Dunathan [43]. It appears that enzymatic control of transamination vs. decarboxylation is a result of each enzyme's ability to control the stereochemistry at the reactive site. Decarboxylation can be prevented in pyridoxal-dependent model reactions by the presence of metal ions [26]. The metal chelates the carboxyl group, the Schiff base nitrogen, and the phenolic hydroxyl. The carboxyl group in this chelate is in the wrong conformation for decarboxylation.

In the case of acetoacetate decarboxylase, evidence for steric control can be seen in studies of enzyme-catalyzed hydrogen exchange. The enzyme catalyzes hydrogen exchange between acetone and the solvent at a significant rate [44]. The mechanism is presumably the reverse of the last two steps in Eq. (5), and the hydrogen exchange step itself is probably mediated by the same enzyme catalytic group that donates a hydrogen for imine formation following decarboxylation in the normal catalytic mechanism. The enzyme fails to catalyze the corresponding hydrogen exchange in the ketophosphonate $CH_3COCH_2PO_3^{2-}$, even though this compound is a potent competitive inhibitor of the enzyme [36]. The stereochemistry of the ketophosphonate at the active site probably mimics the stereochemistry of bound acetoacetate, with the phosphonate occupying the site normally occupied by the carboxylate. Hydrogen exchange is prevented because this stereochemistry prevents the formation of a planar enamine after loss of a proton.

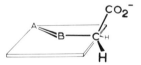

Fig. 3   The proper conformation for decarboxylation. Before decarboxylation, there is a double bond between A and B. Following decarboxylation, there is a double bond between B and C. In order for decarboxylation to occur the carboxyl group must be above the A–B–C plane.

The possibility that enzymes accelerate reactions by freezing their substrates in the proper conformations has been discussed many times [45,46] and has been the subject of a number of controversies. The loss in rotational entropy on freezing a single bond into a particular conformation generally amounts to the equivalent of about a factor of 10 in rate acceleration. However, this estimate is correct only in cases where the freely rotating substrate has no preferred conformation. In cases where a preferred conformation exists and this conformation is not the proper one for reaction, the entropic contribution may be much greater than a factor of 10. A number of such examples are known [45]. The conformational acceleration of enzymatic decarboxylations may often fit into this last category because hydrogen-bonding and chelation effects often appear to favor unreactive conformations.

In the case of acetoacetate decarboxylase, the protonated Schiff base that is formed between acetoacetate and the amino group of the catalyst prefers in solution to exist in the planar conformation shown in Fig. 4b, in which the protonated imine is adjacent to the carboxyl group and has ionic and perhaps hydrogen-bonding interactions with it. However, this conformation is not proper for decarboxylation; instead, conformation (a), Fig. 4, is required. Undoubtedly, the conformation of acetoacetate bound to acetoacetate decarboxylase corresponds to (a), and the enzyme probably prevents the assumption of any other conformation. In a case such as this the locking of this key bond in the proper conformation may give rise to much more than a factor of 10 increase in rate because conformation (b), Fig. 4, is favored for the free Schiff base.

A similar argument can be made regarding metal-catalyzed decarboxylations of β-keto acids. In solution, chelation of both the carbonyl group and the carboxyl group can occur, leading to a conformation analogous to (b) in Fig. 4, from which decarboxylation cannot occur. Enzymatic control of conformation results in production of a conformation analogous to (a), in which the carboxyl group is no longer coordinated to the metal. These reactions, too, may show large conformational rate accelerations.

Thus, it appears that the binding of the carboxyl group to a specific site on the enzyme may serve not only a specificity function but a catalytic function

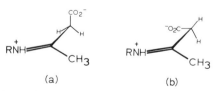

(a)   (b)

Fig. 4 Conformations of the zwitterionic acetoacetate–primary amine Schiff base. Nonplanar conformation (a) is required for decarboxylation, but the planar conformation (b) is favored in solution because of polar interactions and hydrogen bonding.

as well. Few data exist which might provide an estimate of the magnitude of the conformational acceleration factor in enzymatic decarboxylations, but it is clear that this factor may be sizable.

CONCLUSION

Like other enzymes, decarboxylases are efficient catalysts. The decarboxylation step itself is accelerated by the enzyme to a point where it is not wholly rate determining, although in some cases this step is partially rate determining.

The catalytic efficiency of decarboxylases appears to be understandable in terms of three effects:

1. *The electron sink principle.* This factor provides a very significant rate acceleration which can be understood in terms of the chemistry of appropriate model systems.

2. *The medium effect.* Decarboxylation steps appear to involve neutralization of the charge of the carboxylate ion by a distal positive charge. Medium effects on such reactions are large. An active site of proper polarity might provide a large driving force for decarboxylation.

3. *The orientation effect.* Enzymatic control of the conformation of the substrate at the active site may provide a large rate acceleration.

Although the above factors clearly provide the majority of the driving force for enzymatic decarboxylation, it is impossible at present to estimate the accelerations achieved by medium and orientation with any certainty. It is also possible that other factors not included in this discussion may contribute to the observed rate acceleration.

ACKNOWLEDGMENT

Work in the author's laboratory was supported by the National Science Foundation and the National Institutes of Health.

REFERENCES

1. H. Simon and D. Palm, *Angew. Chem., Int. Ed. Engl.* **5**, 920 (1966).
2. J. H. Richards, *in* "The Enzymes" (P. D. Boyer, ed.), 3rd ed., Vol. 2, p. 321. Academic Press, New York, 1970.
3. A. Fry, *in* "Isotope Effects in Chemical Reactions" (C. J. Collins and N. S. Bowman, eds.), p. 364. Van Nostrand-Reinhold, Princeton, New Jersey, 1970.
4. A. MacColl, *Annu. Rep. Chem. Soc.* **71B**, 77 (1974).
5. M. H. O'Leary, *in* "Transition States of Biochemical Processes" (R. L. Schowen and R. Gandour, eds.). Plenum, New York, 1977.
6. G. W. E. Plaut, *in* "The Enzymes" (P. D. Boyer, H. Lardy, and K. Myrbäck, eds.), 2nd ed., Vol. 7, p. 105. Academic Press, New York, 1963.

7. I. Fridovich, in "The Enzymes" (P. D. Boyer, ed.), 3rd ed., Vol. 6, p. 255. Academic Press, New York, 1972.
8. E. A. Boeker and E. E. Snell, in "The Enzymes" (P. D. Boyer, ed.), 3rd ed., Vol. 6, p. 217. Academic Press, New York, 1972.
9. J. Ullrich, Y. M. Ostrovsky, J. Eyzaguirre, and H. Holzer, *Vitam. Horm.* (*N. Y.*) **28**, 365 (1970).
10. A. Schellenberger, *Angew. Chem., Int. Ed. Engl.* **6**, 1024 (1967).
11. A. Wood, *Trans. Faraday Soc.* **60**, 1263 (1964).
12. S. Seltzer, G. A. Hamilton, and F. H. Westheimer, *J. Am. Chem. Soc.* **81**, 4018 (1959).
13. L. O. Krampitz, H. C. Wood, and C. H. Werkman, *J. Biol. Chem.* **147**, 243 (1943).
14. R. Koppelman, S. Mandeles, and M. E. Hanke, *J. Biol. Chem.* **230**, 73 (1958).
15. M. H. O'Leary, *Biochim. Biophys. Acta* **235**, 14 (1971).
16. M. H. O'Leary and J. A. Limburg, *Biochemistry* **16**, 1129 (1977).
17. N. Ramachandran, M. Durbano, and R. F. Colman, *FEBS Lett.* **49**, 129 (1974).
18. M. L. Uhr, V. W. Thompson, and W. W. Cleland, *J. Biol. Chem.* **249**, 2920 (1974).
19. D. B. Northrop and W. W. Cleland, *Fed. Proc., Fed. Am. Soc. Exp. Biol.* **19**, 408 (1970).
20. J. J. Villafranca and R. F. Colman, *J. Biol. Chem.* **247**, 209 (1972).
21. M. H. O'Leary and R. L. Baughn, *J. Am. Chem. Soc.* **94**, 626 (1972).
22. J. P. Guthrie and F. Jordan, *J. Am. Chem. Soc.* **94**, 9136 (1972).
23. D. E. Schmidt, Jr. and F. H. Westheimer, *Biochemistry* **10**, 1249 (1971).
24. G. A. Hamilton, Ph.D. Dissertation, Harvard University, Cambridge, Massachusetts (1959).
25. J. P. Guthrie and F. H. Westheimer, *Fed. Proc., Fed. Am. Soc. Exp. Biol.* **26**, 562 (1967).
26. T. C. Bruice and S. J. Benkovic, "Bioorganic Mechanisms," Vol. 2, Chapter 8. Benjamin, New York, 1966.
27. G. D. Kalyankar and E. E. Snell, *Biochemistry* **1**, 594 (1962).
28. M. L. Fonda, *Biochemistry* **11**, 1304 (1972).
29. M. H. O'Leary, D. T. Richards, and D. W. Hendrickson, *J. Am. Chem. Soc.* **92**, 4435 (1970).
30. H. Yamada and M. H. O'Leary, *J. Am. Chem. Soc.* **99**, 1660 (1977).
31. M. H. O'Leary and G. Piazza, unpublished results.
32. J. Crosby, R. Stone, and G. E. Lienhard, *J. Am. Chem. Soc.* **92**, 2891 (1970).
33. G. Hübner and A. Schellenberger, *Hoppe-Seyler's Z. Physiol. Chem.* **351**, 1435 (1970).
34. M. H. O'Leary, *Biochem. Biophys. Res. Comm.* **73**, 614 (1976).
35. I. Fridovich, *J. Biol. Chem.* **243**, 1043 (1968).
36. R. Kluger and K. Nakaoka, *Biochemistry* **13**, 910 (1974).
37. B. Belleau and J. Burba, *J. Am. Chem. Soc.* **82**, 5751 (1960).
38. E. Leistner and I. D. Spenser, *J. Chem. Soc., Chem. Commun.* p. 378 (1975).
39. G. E. Lienhard and I. A. Rose, *Biochemistry* **3**, 185 (1964).
40. I. A. Rose, *J. Biol. Chem.* **245**, 6052 (1970).
41. D. J. Creighton and I. A. Rose, *J. Biol. Chem.* **251**, 61 (1976).
42. R. G. Button and P. J. Taylor, *J. Chem. Soc., Perkin Trans. 2* p. 557 (1973).
43. H. C. Dunathan, *Adv. Enzymol.* **35**, 79 (1971).
44. W. Tagaki and F. H. Westheimer, *Biochemistry* **7**, 901 (1968).
45. W. P. Jencks, *Adv. Enzymol.* **43**, 219 (1975).
46. T. C. Bruice, in "The Enzymes" (P. D. Boyer, ed.), 3rd ed., Vol. 2, p. 217. Academic Press, New York, 1970.

CHAPTER

12

The Catalytic Mechanism of
Thymidylate Synthetase

Alfonso L. Pogolotti, Jr., and Daniel V. Santi

INTRODUCTION

Thymidylate synthetase catalyzes the reductive methylation of 2'-deoxy-uridylate (dUMP) to 2'-deoxythymidylate (dTMP) with the concomitant conversion of 5,10-methylenetetrahydrofolic acid (CH_2—H_4folate) to 7,8-dihydrofolic acid (H_2folate). This enzyme is unique in that CH_2—H_4folate serves the dual function of both one-carbon carrier and reductant, and the regeneration of H_4folate is required for continued dTMP synthesis.

$$dUMP + CH_2—H_4folate \longrightarrow H_2folate + dTMP$$

It was recognized some time ago that a minimal mechanism for this enzyme must involve at least two steps. On the basis of experiments which demonstrated that H_4folate served as both carbon carrier and reductant [1,2], Friedkin [3] proposed that condensation of CH_2—H_4folate with dUMP results in a 5-thymidylyl—H_4folate intermediate, which subsequently undergoes disproportionation via a 1,3-hydride shift to give the products dTMP and H_2folate (Fig. 1). The first step was viewed as an electrophilic substitution reaction in which the methylene carbon of CH_2—H_4folate replaces the hydrogen at the 5 position of dUMP without a change in oxidation level. The second step of this mechanism was proposed to be a nucleophilic attack at the incipient methyl group of dTMP by hydride originating from the 6

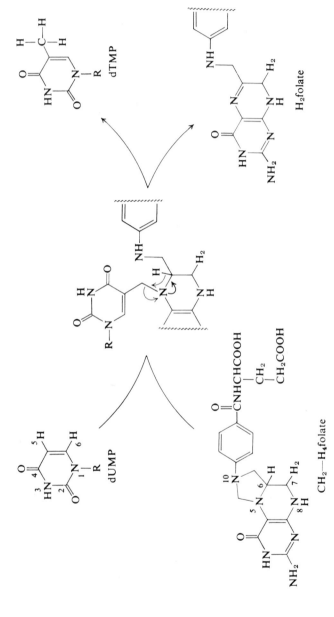

Fig. 1 Reaction catalyzed by thymidylate synthetase as proposed by Friedkin [3]; R = 5-phospho-2′-deoxyribosyl.

position of the cofactor to give dTMP and H_2folate. The important fact that the cofactor serves as the reductant has been verified in a number of laboratories by experiments demonstrating that tritium of CH_2—[6-^{3}H]H_4folate is transferred to the methyl group of dTMP [4–7]. A significant finding with regard to the mechanism of thymidylate synthetase was reported by Lomax and Greenberg [8], who observed that in the presence of CH_2—H_4folate the enzyme also catalyzes exchange of the hydrogen at the 5 position of dUMP for protons of solvent. This reaction represents an electrophilic substitution and, in all likelihood, is mechanistically related to the first step of the overall reaction catalyzed by thymidylate synthetase.

The mechanism proposed by Friedkin provided the main outlines of the thymidylate synthetase reaction and has stimulated much effort directed toward understanding the detailed chemical mechanism of catalysis. From chemical considerations each of the two reactions depicted in Fig. 1 involves conversions that are difficult to envision as single-step processes. For example, displacement of the 5-H of dUMP, by cofactor or protons of water, is difficult to reconcile with the extremely low acidity of vinylic carbon acids [9]. The second step of the proposed mechanism is in essence a S_N2 displacement of a tertiary amine by hydride, but tertiary amines are not generally susceptible to nucleophilic attack. Based on this type of reasoning, it was clear that substrates and intermediates of the reaction possessed unique and interesting chemical reactivities, or the mechanism originally proposed would require expansion and modification. As a result of investigations by a number of workers it has become apparent that both of the above are correct; nevertheless, the salient features of the reaction are contained in the mechanism first proposed some 18 years ago.

An in-depth review of the literature and status of thymidylate synthetase prior to 1973 has recently appeared [10]. The purpose of this chapter is to review the evidence that led to the current proposed mechanism of thymidylate synthetase and to present recent data supporting this mechanism. The major focus of this account is on approaches used in this laboratory, and no attempt has been made to provide a complete review of all aspects of this enzyme. A number of laboratories are engaged in studies of many facets of this enzyme, and we regret that space does not permit discussion of the many interesting reports that have appeared since Friedkin's review [10].

MODEL STUDIES

Prerequisite to understanding the mechanism of any enzyme is a fundamental knowledge of the physical organic chemistry of related chemical counterparts. A decade ago, model systems of the thymidylate synthetase reaction had not been established, and the required chemical precedent for the proposed mechanism was lacking. In view of the complexity of the enzymatic reaction *in toto*, it appeared unlikely that a single model system would be useful in describing more than a fraction of the mechanism; thus, separate systems would be required to obtain useful information regarding separate parts of the enzymatic reaction. As an initial simplifying assumption, the overall reaction was segregated into the two steps suggested by Friedkin [3], which defined it in terms of well-known reaction types: (1) electrophilic substitution of the 5-H of dUMP by CH_2—H_4folate or protons of water and (2) oxidation–reduction via nucleophilic (hydride) attack by the 6-H of H_4folate at the incipient methyl group of dTMP. Candidate chemical counterparts of each of these reactions were chosen and investigated in detail. The results of such model studies performed in this and other laboratories which are pertinent to the mechanism of thymidylate synthetase are summarized below. From conclusions drawn, a pathway depicting the catalytic mechanism of thymidylate synthetase can be proposed which is in complete accord with all chemical and biochemical data thus far reported.

Electrophilic Substitution at the 5 Position of dUMP

Appropriate chemical counterparts for the "first step" of the thymidylate synthetase reaction include the reaction of 1-substituted uracils with formaldehyde and exchange of 5-hydrogen for protons of water. In fact, before their possible relevance as enzyme models was recognized, both of these reactions were known to occur. (1) Acid- and base-catalyzed 5-hydroxymethylation of uracils [11–13], as well as Mannich reactions of uracil [14], were described as synthetic procedures some time ago. (2) In a study of the isotopic distribution of preparations of tritiated pyrimidines, Fink [15] observed that vigorous treatment of tritiated uridine with strong acid or base resulted in the loss of tritium from the 5 but not the 6 position. Although the high temperatures and pH extremes required for these reactions might raise questions regarding their appropriateness as enzyme models, the fact that they were directly analogous to the "first step" of the thymidylate synthetase reaction prompted investigations of their mechanisms.

In 1968, our laboratory [16] and another [17] independently observed that the rate of base-catalyzed exchange of uracil nucleosides was dependent on the nature of the sugar moiety. A catalytic role for sugar hydroxyl groups was apparent from the observation that 5-H exchange of uracil nucleosides in MeONa–MeOD may proceed up to 10^3 times faster than 1MeUra or derivatives that do not possess one or more of the free hydroxyl groups (Table 1) [16,18]. The rate enhancement of 5-H exchange of nucleosides requires the

TABLE 1

**5-H Exchange Rates for 1-Substituted Uracils at 60°C
in 0.47 N CH$_3$ONa–CH$_3$OD**

Compound	Abbreviation	$k \times 10^3$ (hr^{-1})	k_{rel}[a]
1-Methyluracil	1MeUra	0.466	1.0
2′,3′-O-Isopropylideneuridine	IpUrd	243	522
5′-Deoxy-2′,3′-O-isopropylideneuridine	d⁵IpUrd	[b]	[b]
Uridine	Urd	3.62	7.77
2′-Deoxyuridine	d²Urd	4.03	8.65
5′-Deoxyuridine	d⁵Urd	[b]	[b]
Arabinosyluracil	Ara-Ura	290	623
5′-Deoxyarabinosyluracil	d⁵Ara-Ura	547	1174
1-(3-Hydroxypropyl)uracil	1(HOPr)Ura	21.9	47
1-(4-Hydroxybutyl)uracil	1(HOBu)Ura	0.456	0.98
1-(5-Hydroxypentyl)uracil	1(HOPe)Ura	[b]	[b]
1-(3-Phenoxypropyl)uracil	1(PhOPr)Ura	[b]	[b]

[a] Rate of 5-H exchange relative to 1MeUra.
[b] No exchange observed after as long as 2 weeks.

presence of a 5′-hydroxyl or the 2′-"up"-hydroxyl of arabinofuranosides. The 67-fold rate enhancement observed for IpUrd compared to Urd is probably a result of the rigidity imposed on the furanoside ring by the acetonide group [19,20]. These structural effects are reminiscent of reactions of nucleosides that involve covalent bond formation between groups of the sugar moiety with C-6 of the heterocycle. For example, at neutral and acidic pH, 20% of 5′-thiouridine (**1a**) exists as its 6,5′-cyclic episulfide (**2a**) [21], whereas the corresponding 2′,3′-O-acetonide (**1b**) is completely cyclized to (**2b**) [22]. The base-catalyzed cleavage of 5-halouridines proceeds poorly,

(**1a**), R = H
(**1b**), R = R = C(CH$_3$)$_2$

(**2a**), R = H
(**2b**), R = R = C(CH$_3$)$_2$

but conversion to its 2',3'-*O*-acetonide results in facile reaction via 6,5'-cyclic nucleosides [23]; similarly, corresponding arabinofuranosides undergo rapid hydrolysis via 6,2'-cyclic nucleosides [24,25].

From the above considerations, a general mechanism for 5-H exchange was proposed which involved intramolecular catalysis by a nucleophile attached to the 1 substituent (Fig. 2). Here, the oxyanion **ii** attacks the 6 position of the heterocycle to form the carbanionic intermediate **iii**; nonstereospecific protonation at C-5 by solvent followed by reversal of these steps accounts for 5-H exchange. Support for this mechanism was obtained from comparison of base-catalyzed exchange rates of 1-(ω-hydroxyalkyl)uracils (Table 1). Of the compounds examined, only 1(HOPr)Ura showed a significant enhancement in the rate of 5-H exchange as compared to 1MeUra. Formation of a cyclic carbanionic intermediate **iii** (Fig. 2) from 1(HOPr)Ura is in accord with the ease of formation and stability of six-membered rings. 1-(4-Hydroxy-

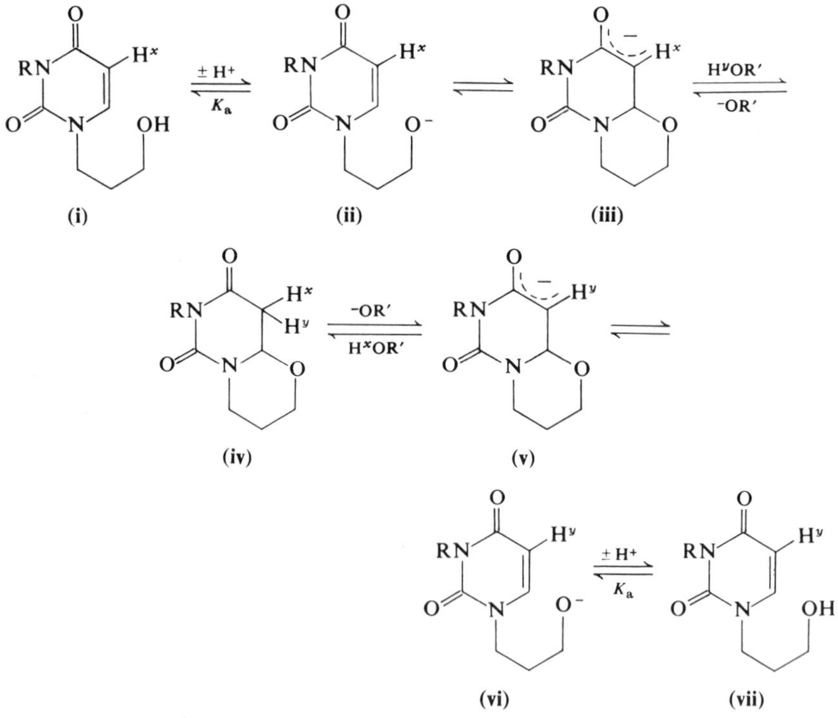

Fig. 2 General scheme for mechanism of 5-H exchange of 1-substituted uracils involving neighboring group participation by oxyanion, where R = H or ⁻, R' = H or CH₃, H^x refers to the hydrogen isotope initially present at C-5, and H^y is the isotope present in the solvent.

butyl)uracil and 1(HOPe)Ura, which would have to form unfavorable seven- and eight-membered ring intermediates, show no rate enhancement.

5-Hydroxymethylation of 1-substituted uracils represents a direct chemical counterpart for the initial condensation of dUMP with CH_2—H_4folate. Rates of the base-catalyzed hydroxymethylation are dependent on the presence of a hydroxyl group on the 1-substituent (Fig. 3) and parallel those for the

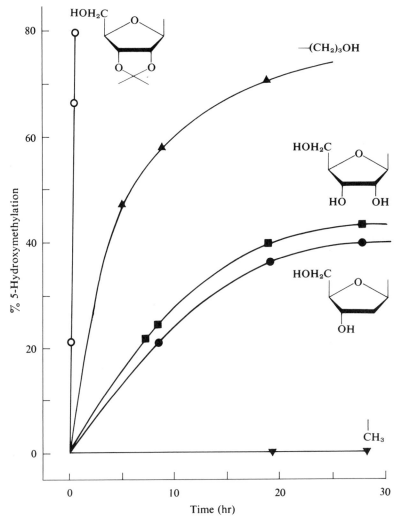

Fig. 3 Extent of 5-hydroxymethylation of 1-substituted uracils. All reactions were performed at 60°C and were 0.2 M in substrate, 3.4 M formaldehyde, and 0.5 N in NaOH.

exchange reactions given in Table 1. Thus, as with 5-H exchange, 5-hydroxy-methylation apparently requires addition of a nucleophile to the 6 position of the heterocycle and is susceptible to intramolecular catalysis [16,18]. The mechanism proposed is similar to that for 5-H exchange in Fig. 2 except that the carbanionic intermediate **iii** reacts with formaldehyde rather than water. It is noted that this reaction is analogous *in toto* to that catalyzed by dUMP-hydroxymethylase [26], which also catalyzes labilization of the 5-H of dUMP in the presence of H$_4$folate [27].

The kinetics of hydroxide-catalyzed 5-H exchange of uracil-5-*d* and its *N*-methylated derivatives [28] indicated that when the N-1 position of the heterocycle is unsubstituted (e.g., Ura or 3MeUra), the reaction proceeds via anchimeric assistance of the N-1 anion (**3b**) as depicted below. In contrast,

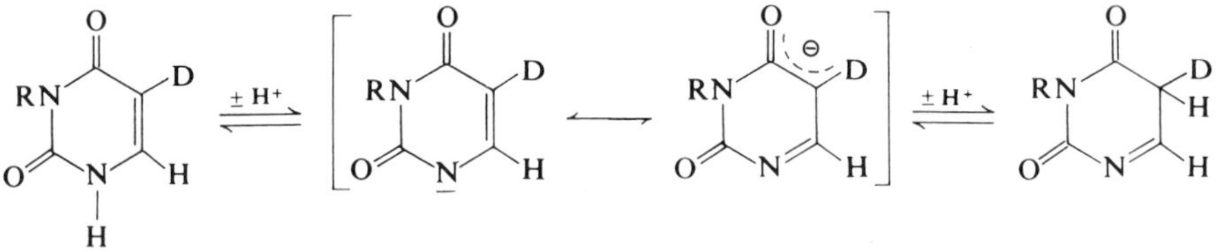

(3a)   (3b)   (3c)

5-H exchange of 1MeUra and 1,3Me$_2$Ura (**4a**, R = H, CH$_3$) proceeds via bimolecular attack of hydroxide at the 6 position to produce the enolate (**4b**); exchange occurs by nonstereospecific protonation of the latter to give the dihydrouracil intermediate (**4c**), followed by reversal. It has been observed

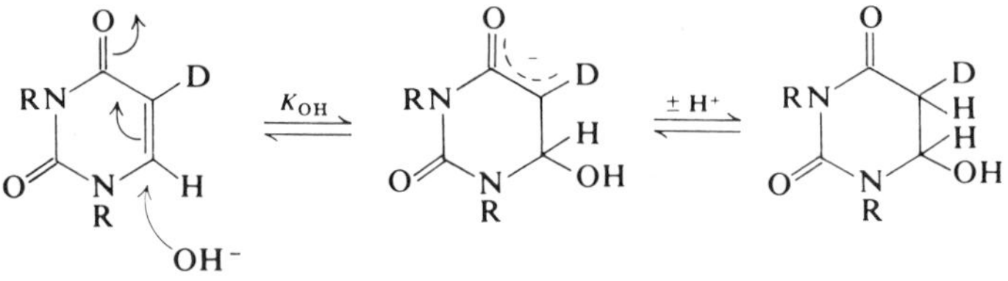

(4a)   (4b)   (4c)

that thiols such as 2-mercaptoethylamine [29,30], glutathione [31], and cysteine [30] are much more effective catalysts of 5-H exchange of 1-substituted uracils than hydroxide, cysteine being the most effective. As with hydroxide-catalyzed exchange, kinetic studies are in accord with a mechanism involving reversible addition–elimination of SH across the 5,6 double bond of the heterocycle [30,31]. These reactions are directly related to the afore-mentioned conversion of 5'-thiouridines to their 6,5'-cyclic episulfides [21,22], which would be of interest to examine in detail as an intramolecular model of thiol-catalyzed 5-H exchange of uracil derivatives.

If either thiols [30,31] or hydroxide [28] are used as intermolecular nucleophilic catalysts, kinetic evidence indicates that the reactive species of 1-substituted uracils is the neutral form and not the N-3 anion. Since the pK_a values of nucleoside hydroxyl groups are some five units higher than that of the 3-NH, the species possessing both oxyanion and neutral heterocycle is a minor component at any pH. For this reason, a rigorous kinetic analysis of 5-H exchange of [5-^{3}H]IpUrd in aqueous media at 25°C was undertaken [18]. Since the pK_a of the 3-NH was known ($pK_a = 9.0$) and the pK_a of the 5'-hydroxyl was estimated to be 13.9 by a kinetic method [32], the pH–log k_{obs} profile for lyate-catalyzed 5-H exchange could be analyzed and values for rate constants involved in oxyanion-participated exchange of the neutral (5a) and anionic species (5b) of the heterocycle were calculated; the values ob-

(5a) k_1 k_2 (5b)

[5-^{1}H]IpUrd

tained were $k_1 = 5.0 \times 10^2\ hr^{-1}$ and $k_2 = 7.5 \times 10^{-3}\ hr^{-1}$, a 67,000-fold difference. Thus, although the apparent exchange rates of these models are slow, the reactive ionic species undergoes a facile intramolecular-catalyzed 5-H exchange. In fact, it is considerably faster than the enzyme-catalyzed 5-H exchange ($k_{cat} = 0.72\ hr^{-1}$) described later. The 5-H exchange of IpUrd has also been shown to be susceptible to general base catalysis [18]; this has been interpreted as evidence that the rate-determining step of this reaction is not cyclization, but rather proton abstraction from a saturated dihydro-pyrimidine intermediate (iv, Fig. 2).

The reaction of bisulfite with uracils yields stable 5,6-dihydro-6-sulfonates (6) [33,34] which are analogous to the transient 5,6-dihydrouracil intermediates proposed in the 5-H exchange of 1-substituted uracils and provides strong evidence for their existence. Moreover, the adduct (6) undergoes 5-H exchange in the presence of amine buffers [35], which may be explained by a

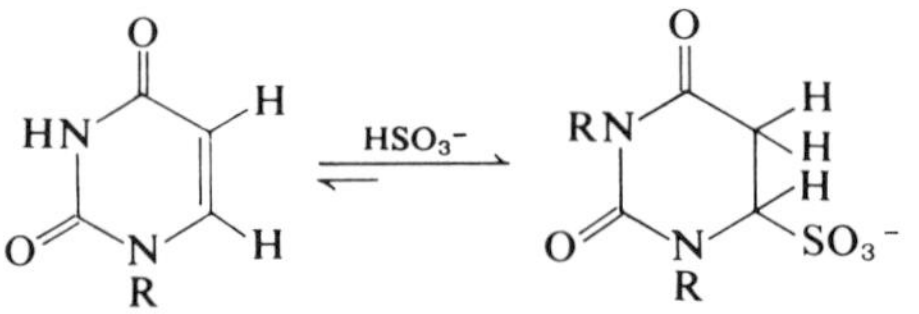

(6)

mechanism similar to that depicted for reversible interconversion of (4c) $\rightleftharpoons$ (4b). Bisulfite also catalyzes the dehalogenation of 5-bromo-, 5-iodo-, and 5-chlorouracil derivatives (7) [36–38]. The initial step of this reaction has been established to be nucleophilic attack of bisulfite at the 6 position of the heterocycle to give a 5-halo-5,6-dihydrouracil 6-sulfonate intermediate (8). The subsequent steps are less well understood but involve either S_N2 displacement of the halogen from C-5 by sulfite or direct attack at the halogen to produce a halosulfonic acid intermediate. Since the reaction of bisulfite with

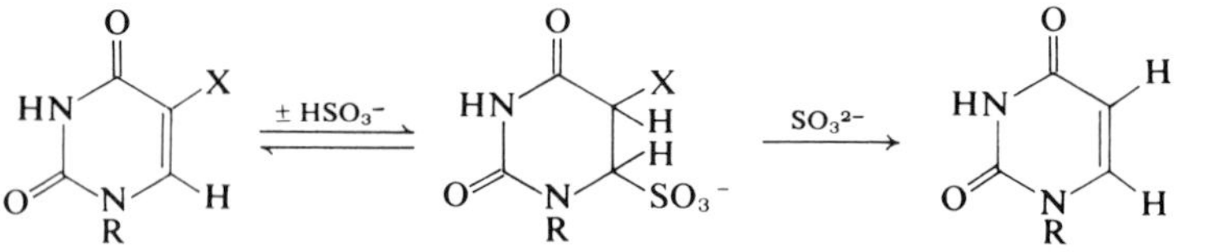

(7)   (8)

X = I, Br, Cl

+ HX, SO_4^{2-}, HSO_3^-

uracil derivatives has been the subject of a recent review [39], further mechanistic details are not considered here. Interestingly, thiols such as cysteine also catalyze the dehalogenation of 5-iodo- and 5-bromouracil nucleosides by a mechanism that appears to be directly analogous to that of bisulfite [40,41]. A priori, these reactions may seem to have little relevance to the thymidylate synthetase reaction; however, as discussed in a subsequent section, 5-bromo-2′-deoxyuridylate (BrdUMP) and 5-iodo-2′-deoxyuridylate (IdUMP) are rapidly dehalogenated by this enzyme by a mechanism that appears to be similar to the chemical counterparts.

From the data thus far discussed, minimal mechanistic features can be suggested for the first step of the thymidylate synthetase reaction (Fig. 4). It is proposed that the reaction is initiated by attack by a nucleophilic group of the enzyme (:X) at the 6 position of dUMP to produce an enolate intermediate with high electron density at the 5-carbon of the heterocycle (9) or the corresponding enol. The activated intermediate could react with CH_2—H_4folate (or an equivalent species of formaldehyde) to give (10), which would undergo further chemical changes within the central complex to

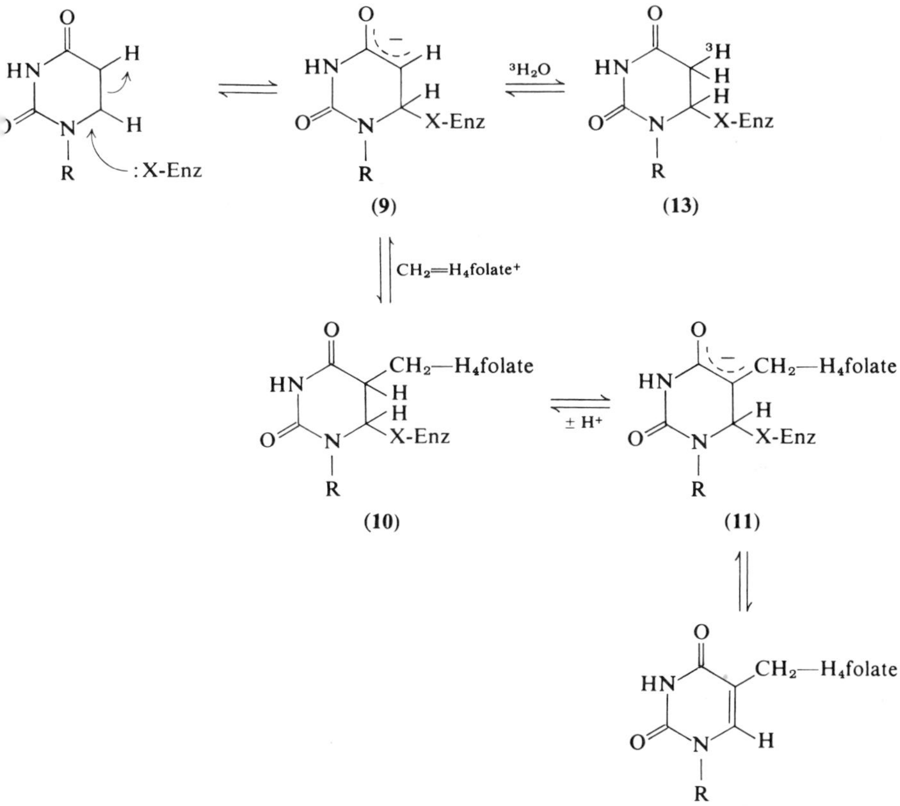

Fig. 4 A proposed minimal mechanism for the first step of the thymidylate synthetase reaction showing a possible pathway for formation of 5-thymidylyl—H₄folate; also shown is a dihydro-dUMP intermediate (13) which may be involved in the 5-H exchange of dUMP.

provide the observed reaction products. It is noted that intermediate (10) differs from that proposed by Friedkin [3] in that it is covalently bound to the enzyme, and the 5,6 double bond of the heterocycle is saturated. The formation of thymidylyl—H₄folate (12) would require proton abstraction from (10) to give (11) followed by β elimination of the nucleophilic catalyst.

In this mechanism, an additional step must be proposed to occur prior to formation of intermediate (10) in which the cofactor is converted to a reactive electrophilic species. Although there is little doubt that the thermodynamically stable adduct formed between formaldehyde and H₄folate is CH₂—H₄folate, it can be stated with a fair degree of certainty that CH₂—H₄folate is *not* the immediate donor of the one-carbon unit. The sp^3 hybridized adducts of aldehydes are unreactive toward direct nucleophilic displacements; rather, such reactions occur by elimination–addition pathways in which the aldehyde carbon undergoes sp^2 rehybridization before reaction with nucleophiles [42]. Kinetic evidence has been obtained which demonstrates that the formation and, through microscopic reversibility, the hydrolysis of CH_2—H_4folate [43] and related models [44] proceed via the sp^2 hybridized iminium cation (5-CH_2= H_4folate$^+$). As depicted in Fig. 5, three general possibilities exist for the electrophilic species of formaldehyde, which undergoes reaction with the nucleophilic enolate (9): (1) It may be associated with H_4folate as in 5- or 10-CH_2=H_4folate$^+$; (2) free formaldehyde may be released from the cofactor before reaction with the substrate; and (3) the one-carbon unit could be transferred from CH_2—H_4folate to a nucleophilic group of the enzyme prior to transfer to the final acceptor. Although the correct reactive electrophilic species cannot be ascertained from model studies, evidence is presented in the last section which implicates 5-CH_2=H_4folate$^+$.

The enzyme-catalyzed 5-H exchange of dUMP reported to occur in the presence of CH_2—H_4folate [8] can be envisioned to occur by two distinct mechanisms. The first involves nonstereospecific reaction of enolate (9) with solvent protons to provide (13), an intermediate not on the pathway leading to dTMP; reversal of this sequence would account for the exchange. Alternatively, incorporation of solvent protons in dUMP could occur by loss of the

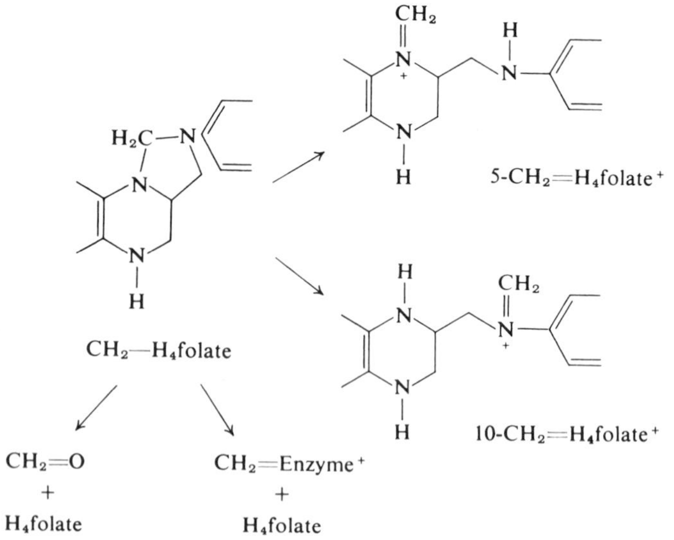

Fig. 5 Possibilities for the reactive electrophilic species of CH_2—H_4folate.

5-proton of (10), reprotonation by solvent, and reversal; this mechanism utilizes intermediates on the normal reaction pathway and requires that equilibration of solvent protons with that lost from (10) be competitive in rate with the overall reaction. The mechanism of enzyme-catalyzed 5-H exchange is discussed in a later section.

Chemical Counterparts of Thymidylyl—H₄folate

The previous section dealt with chemical counterparts of the condensation of dUMP and CH_2—H_4folate. These studies led to the conclusion that the reaction is initiated by nucleophilic attack at the 6 position of dUMP, resulting in activation of the 5 position toward a reactive species of the cofactor. The second step of the mechanism proposed by Friedkin involves rearrangement of 5-thymidylyl—H_4folate to dTMP and H_2folate (Fig. 1); this has been depicted as hydride attack at the incipient methyl group of dTMP.

In an attempt to ascertain whether the proposed intermediate did possess the unusual chemical properties ascribed, Gupta and Huennekens [45] prepared 5- and 10-thyminyl—H_4folate; however, under a variety of conditions these models did not undergo the intramolecular rearrangement to yield thymine. Only recently has definitive chemical precedent been established for the proposed rearrangement of 5-thymidylyl—H_4folate. Wilson and Mertes [46,47] cleverly designed a series of N-thyminyl derivatives of 1,2-dihydro- and 1,2,3,4-tetrahydroquinoline which contained all the salient chemical features of the intermediate proposed by Friedkin. When 1,2-dihydro-N-thyminylquinoline (14a) was heated neat, in diglyme, or in water, thymine and quinoline were obtained. In elegant experiments utilizing reactants deuterated at specific positions and mass spectral analysis, these workers demonstrated that hydrogen from the 2 position of the N-thyminyl-dihydroquinoline (14a) was transferred to the methyl group of thymine, both

(14a), R = H
(14b), R = CH₃

inter- and intramolecularly. Similarly, heating 1,2-dihydro-(1-methylthy-minyl)quinoline (**14b**) provided 1-methylthymine and quinoline, but only in 5% yield. In the tetrahydroquinoline models, the N-thyminyl analog provided a low yield of thymine and substitution of the N-1 position of the thymine moiety prevented the reaction from occurring.

The Wilson and Mertes model established that rearrangement of 5-thymi-dylyl—H_4folate to dTMP and H_2folate was indeed a feasible reaction. Nevertheless, in view of the poor leaving group potential of amines, it remained difficult to envision the mechanistic features underlying the susceptibility of the N-methylene group of the intermediate or its chemical counterparts toward nucleophilic attack by hydride. It appeared reasonable to approach this problem by simply investigating the mechanism of nucleo-philic substitution reactions of thyminyl derivatives of type (**15**). There was reason to believe that such compounds were unusually susceptible toward nucleophilic displacement reactions at the 5-methylene group. For example, 5-hydroxymethyluracil (HmUra) readily alkylates poorly nucleophilic aromatic amines in aqueous base [48]. Ethers of HmUra are formed under mild conditions and are hydrolyzed easily at neutral pH [11]. Esters and ethers of HmUra are unusually reactive toward nucleophilic displacement, the former undergoing O-alkyl rather than O-acyl bond cleavage upon treatment with methoxide [49,50]; in addition, when treated with hydride reagents, such compounds rapidly gave rise to corresponding thymine

(15)

derivatives. On the basis of these observations, we investigated the mechanism of nucleophilic displacement reactions of derivatives of 5-p-nitrophenoxy-methyluracils (NPmUra) (**15**, X $= p$-$NO_2C_6H_4O^-$) as chemical counterparts of the second step of the thymidylate synthetase reaction [51]. Although the leaving group in the model system greatly differs from H_4folate in structure and reactivity, we believed the salient features of the reaction resided in the uracil heterocycle and would be retained regardless of the nature of the leaving group.

Hydrolysis of NPmUra derivatives that do not possess a substituent at the 1-nitrogen to the corresponding HmUra derivatives (**18**) was found to proceed by a mechanism in which the N-1 anion (**16**) participates in the facile liberation of p-nitrophenoxide ($k = 66.0$ min^{-1}) to form an intermediate (**17**) having a highly reactive exocyclic methylene group. When reactants specifically labeled with deuterium at the methylene group were used, a kinetic secondary isotope effect (k_H/k_D = 1.28) indicative of sp^3 to sp^2 rehybridization was observed and provided strong support for the proposed intermediate (17).

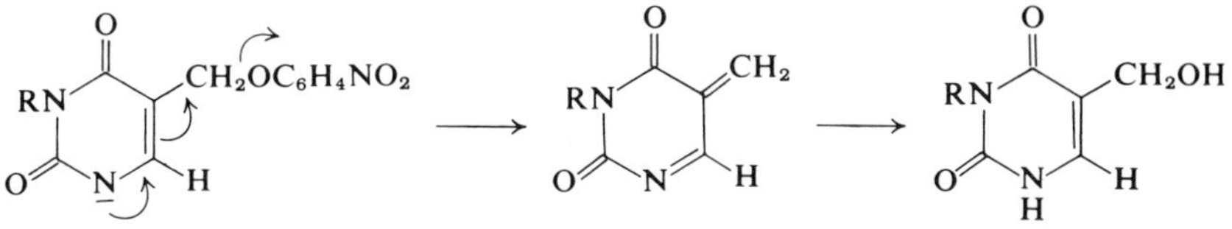

    (16)   (17)   (18)

When the N-1 position of NPmUra derivatives has a methyl substituent (19), the apparent rates of base-catalyzed hydrolysis are much slower than observed for unsubstituted derivatives. A similar reactive exocyclic methylene (21) intermediate is formed, but the driving force appears to be nucleophilic attack at the 6 position by hydroxide ion. As with 5-H exchange of 1-substituted uracils, the reaction is initiated by nucleophilic attack at the 6 position to form an enolate anion (20); this intermediate eliminates p-nitrophenoxide to give the exocyclic methylene intermediate (21), which rapidly reacts with water to give the 5-hydroxymethyluracil (22).

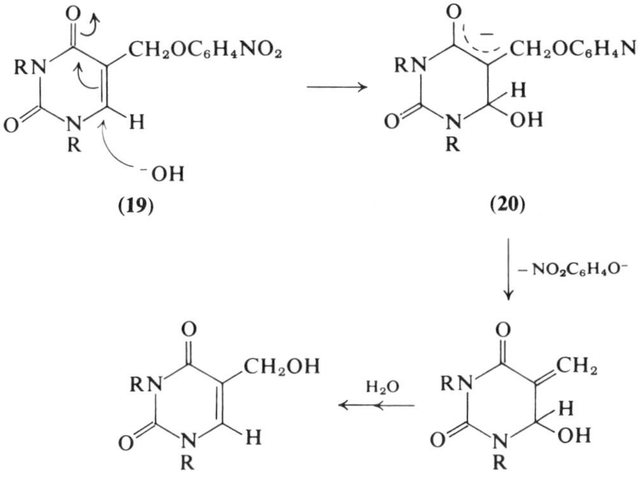

    (19)   (20)

                            $- NO_2C_6H_4O^-$

    (22)   (21)

This mechanism was supported by a secondary deuterium isotope effect with the 1-methyl derivative labeled at the methylene position (k_H/k_D = 1.43) and an inverse secondary isotope effect with $1,3Me_2NPmUra$ ($k_H/k_D = 0.89$) labeled with deuterium at the 6 position; the latter is indicative of sp^2 to sp^3 rehybridization and provides convincing evidence for nucleophilic attack at the 6 position of this heterocycle. In addition, as in related intramolecular models of 5-H exchange and 5-hydroxymethylation of 1-substituted uracils, placement of a nucleophile on the 1-substituent results in intramolecular catalysis of hydrolysis of the p-nitrophenyl ether. The amino group of 1-(3-aminopropyl)-3-methyl-5-p-nitrophenoxymethyluracil (23) can form a favorable six-membered ring (24) by attack at the 6-carbon of the heterocycle which rapidly eliminates p-nitrophenoxide to give the highly reactive (25); as a result, (23) hydrolyzes some 13,000 times faster than $1,3Me_2NPmUra$.

The conclusions drawn from these studies which are pertinent to the enzymatic reaction are as follows. (1) Nucleophilic displacement reactions of models of thymidylyl—H_4folate proceed exclusively via S_N1-type reactions involving intermediates in which the methylene group is sp^2 hybridized; in no situation is there any indication of direct (S_N2) displacement reactions. (2) The driving force for formation of such intermediates may be provided by nucleophilic attack at the 6 position to form a reactive enolate anion which readily undergoes elimination to give the highly reactive exocyclic methylene derivatives.

Although it may be unwarranted to relate the above mechanisms to the pyrolytic disproportionation of the Wilson–Mertes model of thymidylyl—H_4-folate, it is likely that the disproportionation of 1,2-dihydro-N-thyminyl-quinoline (14a) in water or diglyme proceeds in an analogous manner to nucleophilic displacements of thyminyl derivatives (15) not possessing an N-1 substituent. In this fashion (14a) would lose a proton from N-1 to generate an exocyclic methylene intermediate analogous to (17) and 1,2-dihydroquinoline. Hydride or its equivalent could be transferred from 1,2-dihydroquinoline to this intermediate, accounting for the apparent rearrangement. It would be worthwhile to examine this model system in more detail.

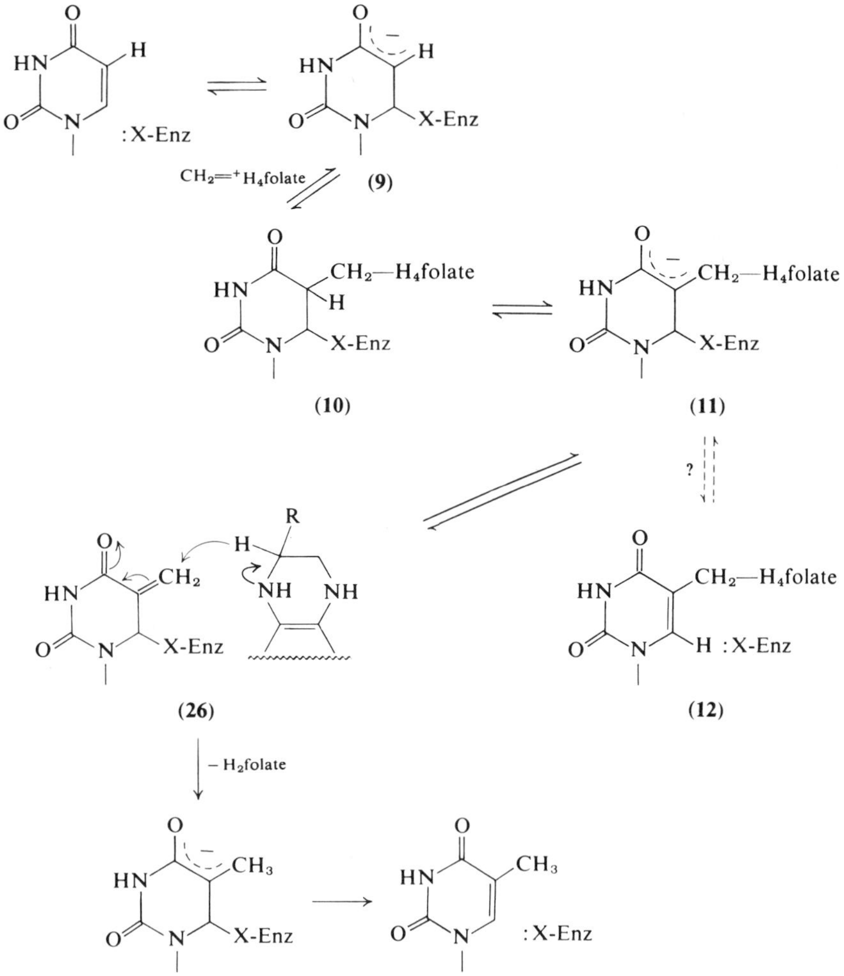

Fig. 6 Suggested sequence for the thymidylate synthetase reaction based on model studies. Intermediate (12) is that proposed by Friedkin [3], which, according to the scheme shown, is not an obligatory intermediate. All pyrimidine structures have a 1-(5-phospho-2'-deoxyribosyl) substituent and R = CH₂—NHC₆H₄COGlu.

If direct evidence for the intermediacy of (17) could be obtained with this closely related chemical counterpart of the Friedkin intermediate, much credence would be given to the validity of nucleophilic substitution reactions of NPmUra as models of the second step of the thymidylate synthetase reaction. If (14a) does give rise to (17) in aqueous media, the latter should be easily trapped by nucleophiles in the reaction medium. It would be of further interest to design analogs of (14) possessing suitably positioned nucleophilic groups on the 1-substituent of the thyminyl moiety which might attack the 6 position and provide an effective intramolecular model of the disproportionation of thymidylyl—H_4folate.

The combined studies described thus far permit construction of a chemically reasonable mechanism for thymidylate synthetase (Fig. 6). As previously described, model systems of electrophilic substitution at the 5 position of dUMP suggest that the reaction is initiated by attack of a nucleophile to the 6 position. The initial condensation product between dUMP and CH_2—H_4- folate is covalently bound to the enzyme and saturated across the 5,6 double bond of dUMP (10). The formation of thymidylyl—H_4folate (12) would require proton abstraction from (10) to give (11) followed by β elimination of the nucleophilic catalyst. From the studies described in this section, the conversion of (12) to products would proceed by addition of a nucleophile to the 6 position of the uracil heterocycle to form the enolate intermediate (11). As with the chemical models, (11) should readily undergo a β elimination to produce the highly reactive exocyclic methylene intermediate (26) and H_4folate, bound to the enzyme in close proximity. Intermolecular hydride transfer from H_4folate to (26) followed by β elimination would yield dTMP, H_2folate, and the native enzyme. Since the enolate intermediate (11) is common to both the formation of a thymidylyl—H_4folate intermediate and its conversion to reaction products, it follows that thymidylyl—H_4folate (12) is not an obligatory intermediate of the reaction in this formulation.

ENZYME STUDIES

Investigations of model systems have permitted the proposal of a chemically reasonable mechanism for the thymidylate synthetase reaction. Of course, the validity of conclusions derived from chemical studies depends on the demonstration that intermediates and mechanisms proposed do in fact occur in the enzymatic reaction. The finding that methotrexate- and dichloromethotrexate-resistant strains of *Lactobacillus casei* are rich sources of a stable thymidylate synthetase [52,53] which can be readily purified [53–55] has greatly facilitated studies of this enzyme. In the ensuing section, selected experiments using *L. casei* thymidylate synthetase are described which provide convincing evidence for a number of the salient features of the catalytic mechanism proposed from model systems.

Thymidylate Synthetase-Catalyzed Exchange of the 5-H of dUMP for Protons of Water

As mentioned in a previous section, Lomax and Greenberg [8] have reported that thymidylate synthetase from *Escherichia coli* catalyzes an exchange between the 5-H of dUMP and protons of water. The exchange reaction was shown to require CH_2—H_4folate and, under optimal conditions, proceeded at 5–10% of the rate of dTMP formation. In fact, this was the observation that prompted studies of 5-H exchange of dUMP as a model system for the thymidylate synthetase reaction. As discussed earlier, two pathways can be envisioned for this exchange (Fig. 4). The first involves condensation of dUMP and the cofactor along the normal reaction pathway to provide an intermediate in which the 5-H of dUMP has been displaced, such as (11), followed by reversal of these steps. The second mechanism by which 5-H exchange might occur involves formation and reversal of a dihydro-dUMP intermediate (13), which is not on the pathway to dTMP. If the former pathway is correct, then the slow step of the reaction must occur after formation of (11), and the proton released from the 5 position must exchange with solvent protons at a rate competitive with reversal of (11) to free reactants. However, the more interesting of the two possibilities would be if the dihydro-dUMP intermediate (13) is formed. This would provide strong evidence for nucleophilic catalysis and the existence of the carbanionic intermediate (9) in the reaction pathway.

If 5-H exchange proceeds via a dihydro-dUMP intermediate (13), the requirement for CH_2—H_4folate may be one of providing or inducing an environment necessary for catalysis. In principle, an approach toward distinguishing the two aforementioned mechanisms for 5-H exchange would be to attempt to observe tritium release from [5-³H]dUMP in the absence of CH_2—H_4folate or in the presence of a cofactor analog that is incapable of undergoing condensation with dUMP. However, the assumption that tritium release from [5-³H]dUMP is synonymous with exchange can be misleading, since such assays do not distinguish between replacement by protons and other electrophiles. This is especially pertinent when H_4folate analogs are present, since they yield unpredictable by-products upon degradation [56] which can conceivably cause displacement of tritium from [5-³H]dUMP. This was first encountered by Lomax and Greenberg [8], who found it necessary to resort to the tedious technique of monitoring tritium incorporation from ³H_2O to demonstrate that the *E. coli* enzyme catalyzed exchange of

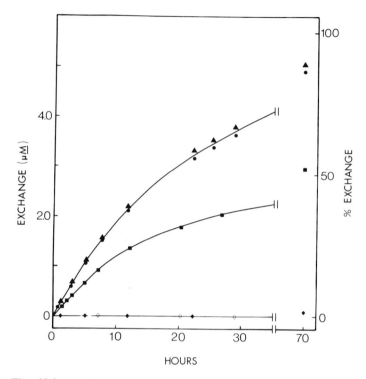

Fig. 7 Thymidylate synthetase-catalyzed exchange of the 5-H of dUMP for protons of water. All reactions were performed at 22°C and contained 0.4 μM enzyme and 6 μM [2-^{14}C,5-^{3}H]dUMP; ▲, 1 mM folic acid and 10 mM formaldehyde; ●, 1 mM folic acid; ■, no additions; ◆, 1 mM folic acid and 50 μM FdUMP; ◇, minus enzyme.

the 5-H of dUMP for protons of water in the presence of H$_4$folate. However, since their preparation of H$_4$folate contained traces of formaldehyde, these experiments did not authenticate an uncoupling of the 5-H exchange from the normal enzymatic reaction. Using the same enzyme source, Friedkin did not observe tritium labilization in the presence of the nonsubstrate H$_4$homo-folate [57]. In early experiments performed in this laboratory we observed that *L. casei* thymidylate synthetase catalyzed tritium release from [2-^{14}C,5-^{3}H]dUMP in the presence of excess 5-CH$_3$—H$_4$folate, which had been prepared and purified by reported methods. Unexpectedly, product analysis demonstrated that the tritium release was a consequence of dTMP formation which might have resulted from contamination of our 5-CH$_3$—H$_4$-folate preparation by less than 1% CH$_2$—H$_4$folate. It has been reported [58] that in the presence of 10-CH$_3$—H$_4$folate, thymidylate synthetase catalyzes release of tritium from [5-^{3}H]dUMP. This result represented a verification of the Lomax and Greenberg experiment which indicated that 5-H release from dUMP might be uncoupled from dTMP formation. Unfortunately, product analysis was not performed to insure that tritium release resulted from exchange with protons of water rather than displacement by some other electrophile.

We have found [59] that *L. casei* thymidylate synthetase catalyzes an exchange of the 5-H of dUMP for protons of water in the *absence* of H_4folate or its analogs (Fig. 7). In these experiments, tritium release from [2-^{14}C,5-^{3}H]dUMP was monitored as a decrease in the ^{3}H/^{14}C ratio of the nonvolatile nucleotide. In addition, dUMP isolated from reaction mixtures by a chromatographic method that readily separates 5-substituted 2'-deoxyuridylates was demonstrated to have undergone 5-H exchange in an amount identical to that obtained using the tritium release assay. Although the reaction is slow, it does not occur when the enzyme is omitted or inhibited with specific nucleotide inhibitors. In the presence of folic acid, a stimulation of the rate of tritium release from [2-^{14}C,5-^{3}H]dUMP was observed (Fig. 7). As before, product analysis verified that dUMP was the sole nucleotide present and that the tritium release assay reflected exchange for protons of water. As shown in Table 2, V_{max} of the exchange reaction is not affected by folate, but K_m is decreased by about 10-fold. Thus, it appears that folate increases the affinity of the enzyme for dUMP but has no effect on catalytic events leading to exchange. This is in accord with other studies indicating a synergism in binding of ligands to this enzyme. The K_m for dUMP in the exchange reaction (16 μM) is quite similar to the dissociation constant for the binary enzyme–dUMP complex as determined by equilibrium dialysis [60], while K_m for the exchange reaction in the presence of folate is similar to the K_m of dUMP for dTMP formation (3 μM) and dissociation constants of dUMP from complexes formed in the presence of other folate analogs [60].

Stereochemical considerations of the enzyme-catalyzed 5-H exchange of dUMP are also of interest since, *a priori*, the reaction appears to require non-stereospecific protonation of the enolate (9). This follows from the fact that if

TABLE 2

Kinetic Parameters for Thymidylate Synthetase-Catalyzed
5-H Exchange of dUMP and for dTMP Formation

Addition	K_m (μM)	V_{max} (nmoles/min mg)
dUMP	16	0.15
dUMP + folic acid	1.2	0.17
dUMP + CH_2—H_4folate	3.0	7040

(9) were protonated stereospecifically on one face of the pyrimidine, the same proton would be removed upon reversal to dUMP and 5-H exchange would not be observed. If both faces of (9) were accessible to solvent, albeit to different degrees, the observed rate of exchange would reflect but a fraction of the rate of interconversion of dUMP and the dihydro-dUMP intermediate (13). This might account for the large differences in V_{max} observed for 5-H exchange and dTMP formation. Further, since V_{max} for exchange is independent of the presence of folate, it can be concluded that the accessibility of (9) to solvent protons is unaffected by binding of this analog.

The mechanism of 5-H exchange of dUMP can also be envisioned as a direct counterpart to that proposed for the normal enzymatic reaction. Figure 8 depicts a general mechanism for substitution of the 5-H of dUMP by an electrophile (R^+). A stereochemical consequence of this mechanism is that addition of the nucleophile and R^+ across the 5,6 double bond (shown here as *trans* addition) proceeds in an opposite fashion to the subsequent elimination of proton and nucleophile (depicted here as *cis* elimination); by microscopic reversibility, the reverse reaction would proceed by *cis* addition followed by *trans* elimination. In the case of 5-H exchange of dUMP, (27) could receive a proton ($R^+ = H^+$) from a single face of the pyrimidine to give the dihydro-dUMP intermediate (28). The hydrogen that was originally present as the 5-tritium of dUMP would be abstracted to provide the carbanion (29), and subsequent β elimination would provide the unaltered

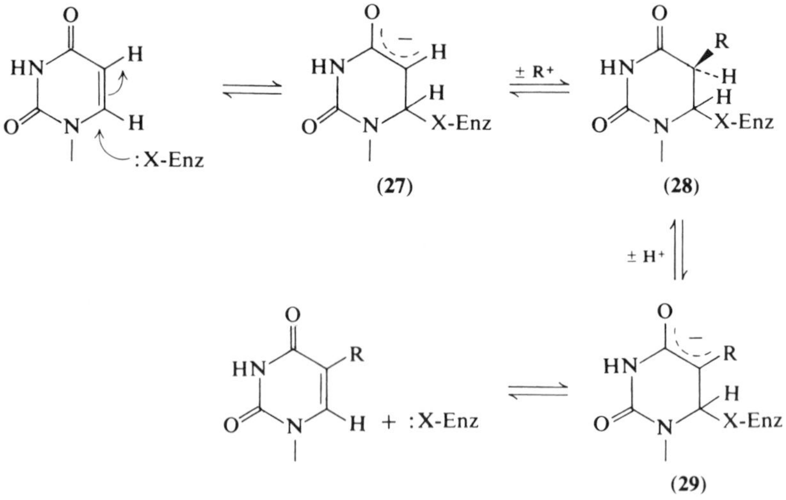

Fig. 8   General mechanism for the thymidylate synthetase-catalyzed 5-H exchange of dUMP by an electrophile (R^+).

enzyme and 5-protio-dUMP. It is to be noted that, in displacement reactions where R^+ is not a proton, (27) and (29) are not equivalent and the nature of the electrophile would dictate the partitioning of intermediates. When R^+ is a proton, intermediates (27) and (29) are identical and their protonation to give (28), albeit stereospecific in any one direction, would occur from both faces of the pyrimidine with equal facility; the "irreversibility" of the tritium exchange experiments described simply results from dilution of tritium from [5-^{3}H]dUMP into water. Although it is more difficult to rationalize the large difference in rates of tritium exchange and dTMP formation by this "stereospecific" mechanism, it is attractive from the standpoint that it utilizes catalytic features of the enzyme which are analogous to the conversion of dUMP to dTMP.

In summary, the above-mentioned experiments clearly demonstrate that 5-H exchange of dUMP may be "uncoupled" from dTMP formation. Together with the chemical studies of the mechanism of 5-H exchange, they provide convincing evidence for the existence of the dihydro-dUMP intermediate (13) and the precursor enolate (9) (or its corresponding enol). The enolate (9) is also the nucleophilic form of dUMP, which is believed to react with the cofactor in the pathway leading to dTMP, and its existence provides direct support for the notion that nucleophilic attack at the 6 position of dUMP is a requisite catalytic feature of reactions catalyzed by thymidylate synthetase.

Thymidylate Synthetase-Catalyzed Dehalogenation of
5-Bromo- and 5-Iodo-2'-deoxyuridylate

Recently, we have observed [61] that treatment of BrdUMP and IdUMP with catalytic amounts of thymidylate synthetase results in their conversion to dUMP. Kinetic parameters for these reactions, as monitored by the decrease in absorbance at 285–290 nm, are given in Table 3. The reaction requires the presence of thiols, such as dithiothreitol or 2-mercaptoethanol,

TABLE 3

**Kinetic Parameters for Thymidylate Synthetase-Catalyzed
Dehalogenation of BrdUMP and IdUMP[a]**

Substrate	K_m (μM)	V_{max} (μmoles/min mg)
IdUMP	1.38	0.030
BrdUMP	5.7	0.084

[a] Reaction conditions were as described in Wataya and Santi [61].

and it has been demonstrated that oxidation to disulfide is stoichiometric with dehalogenation of BrdUMP. Thus, the overall reaction can be written as

$$\text{BrdUMP} + 2\text{RSH} \longrightarrow \text{dUMP} + (\text{RS})_2 + \text{HBr}$$

Unlike the normal enzymatic reaction, dehalogenation of IdUMP or BrdUMP does not require the presence of cofactor and questions may arise as to whether the same sites of the enzyme are involved in binding the halogenated substrates and dUMP. Since the dehalogenation reactions proceed slower than the normal enzymatic reaction, we could demonstrate that IdUMP ($K_i = 1.6\,\mu M$) and BrdUMP ($K_i = 1.4\,\mu M$) are competitive inhibitors with respect to dUMP; similarly, dUMP is a competitive inhibitor ($K_i = 2.3\,\mu M$) of the dehalogenation reaction. This strongly supports the view that the same site of the enzyme is involved in binding of dUMP, BrdUMP, and IdUMP.

As described in the section on model systems, thiols and bisulfite catalyze the dehalogenation of 5-bromo- and 5-iodouracil derivatives. These reactions appear to require nucleophilic attack at the 6 position of the heterocycle to form 5-halo-5,6-dihydrouracil-6-thiol(sulfonate) intermediates, which subsequently undergo thiol reduction to give halide, oxidized thiol (or sulfonic acid), and the uracil derivative. Recently, it has been reported that treatment of 5-bromo-5,6-dihydro-6-methoxyuracils with bisulfite results in facile dehalogenation [62]. Similarly, reaction of 5-bromo-5,6-dihydro-6-methoxythymine with cysteine results in the formation of bromide ion, cystine, and thymine [63]. From all data reported, it appears that the salient requirements for dehalogenation of 5-halouracil derivatives is saturation of the 5,6 double bond of the heterocycle and the presence of nucleophilic reductant, such as thiol or bisulfite.

Thus, we proposed that thymidylate synthetase-catalyzed dehalogenation of BrdUMP and IdUMP is initiated by attack of a nucleophile at the 6 position of the heterocycle to produce the 5-halo-5,6-dihydro-dUMP intermediate (30). As proposed for the chemical models, two general mechanisms, with variations, may account for the subsequent steps. The first (E2 Hal) involves abstraction of bromonium (Br^+) or iodonium (I^+) ion by thiol, or an intermediary, to provide intermediate (31) and a sulfenyl halide. The latter would rapidly disproportionate to the halide ion and oxidized thiol, and (31) would undergo a β elimination to yield dUMP and the unmodified enzyme. The second mechanism (S_N2) involves nucleophilic displacement of Br^- from (30) by thiolate to give the intermediate (32). Further reaction with RS^- would yield the oxidized thiol (R—SS—R) and intermediate (31), which is common with the E2 Hal mechanism, and would yield dUMP upon β elimination of the enzyme.

Direct evidence for nucleophilic attack at the 6 position of BrdUMP has been obtained by demonstrating an inverse kinetic secondary isotope effect during its conversion to dUMP [64]. In this experiment, $[2\text{-}^{14}C,6\text{-}^{3}H]$BrdUMP was treated with thymidylate synthetase; during the course of the reaction, reactant was separated from product and the isotope content of each was determined. As shown in Fig. 9, the ^{3}H content of the product dUMP is highly enriched in the initial stages of the reaction and decreases until the specific activity approaches that initially present in the reactant. Depletion of tritium from BrdUMP is not initially apparent because of the small amount reacted, but the decrease becomes apparent at later stages of the reaction. From these data, we calculate $k_H/k_T = 0.83$, which corresponds to $k_H/k_D = 0.88$. Since an inverse secondary isotope of ca. 10% is accepted as evidence for sp^2 to sp^3 rehybridization at a pre-rate-determining or rate-determining step, these data provide strong evidence for nucleophilic attack at the 6 position of BrdUMP during its enzyme-catalyzed dehalogenation.

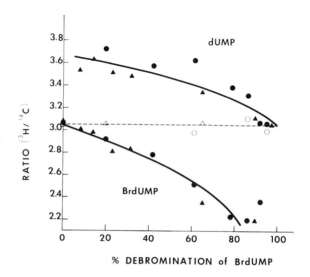

Fig. 9 Plot of ^{3}H/^{14}C content of [2-^{14}C,6-^{3}H]dUMP and [2-^{14}C,6-^{3}H]BrdUMP as thymidylate synthetase-catalyzed dehalogenation proceeds. Points (●, ▲) represent data from two separate experiments. The control (○, △) omitted enzyme and represents the ^{3}H/^{14}C in BrdUMP analyzed throughout the reaction. The solid line represents the theoretical curve calculated for $k_H/k_T = 0.83$.

Although the dehalogenation of BrdUMP and IdUMP has not yet been extensively investigated, it is clear that the simplicity of this reaction should permit investigations of aspects of the mechanism and interactions of thymidylate synthetase not previously possible.

Interaction of Thymidylate Synthetase with FdUMP

Direct support for major aspects of the proposed mechanism of thymidylate synthetase has been derived from studies of its interaction with 5-fluoro-2′-deoxyuridylate (FdUMP). It has been known for some time [65,66] that this nucleotide is an extremely potent inhibitor of thymidylate synthetase, but the nature of inhibition was a topic of considerable controversy. Since the 6 position of 1-substituted 5–fluorouracils was known to be quite susceptible to nucleophilic attack [21,25,67], it was suspected that FdUMP might exert its inhibitory effect by reaction with the proposed nucleophilic catalyst of thymidylate synthetase. In 1972 we demonstrated [68] that the dimeric enzyme from *L. casei* interacts with 2 equivalents of FdUMP, presumably 1 per sub-unit [53], in a reaction that was dependent on the presence of CH_2—H_4folate. The complex formed was stable toward protein denaturants and showed a complete loss of absorbance at 269 nm, the absorption maximum of the pyrimidine chromophore of FdUMP. Based on these findings and the model studies previously discussed, it was proposed that a covalent bond was formed between the enzyme and the 6 position of FdUMP. This finding was independently verified by Langenbach et al. [69], who further suggested that the 10-nitrogen of the cofactor was attached to the 5 position of FdUMP via a methylene bridge.

Shortly thereafter, several lines of evidence were reported which conclusively demonstrated that a reversible covalent bond is formed between FdUMP and thymidylate synthetase within the complex. (1) From studies of the relative rates of association and dissociation of FdUMP with the enzyme–CH_2—H_4folate binary complex, the dissociation constant (K_d) of the complex was calculated to be ca. 10^{-13} M [70]. The kinetically determined K_d is

$$CH_2\text{—}H_4\text{folate–Enz} + [^3H]FdUMP \rightleftharpoons [^3H]FdUMP\text{–}CH_2\text{—}H_4\text{folate–Enz}$$

approximately 10^8-fold lower than that for the binary FdUMP–enzyme complex; in effect, the presence of cofactor increases the affinity of the enzyme for FdUMP by over 10 kcal/mole in binding energy. Clearly, a most pertinent feature of the interaction of FdUMP and thymidylate synthetase involves changes that occur within the bound ternary complex. (2) The enzyme–FdUMP–CH_2—H_4folate complex can be treated with a number of protein denaturants (urea, guanidine hydrochloride, etc.) without apparent dissociation of protein-bound ligands [58,68,70]. With few exceptions, such treatment is sufficient to disrupt noncovalent interactions between low molecular weight ligands and their protein receptors. (3) Although denaturation of the enzyme yields stable protein-bound ligands, ligands bound to the native complex slowly dissociate in an unchanged form [7,70]. Thus, covalent bonds formed between ligands and proteins are reversible and their dissociation requires the integrity of catalytic groups of the enzyme. (4) Upon formation of the ternary complex, there is a decrease of absorbance at 269 nm which corresponds to stoichiometric loss of the pyrimidine chromophore of FdUMP [58,70]. This result strongly suggests that the 5,6 double bond of the pyrimidine is saturated in the bound complex. (5) Dissociation of [6-^{3}H]FdUMP from the complex shows a secondary tritium isotope effect (k_H/k_T) of 1.23 [70]. This would correspond to $k_H/k_D = 1.15$ and clearly demonstrates that the 6-carbon of the heterocycle undergoes sp^3 to sp^2 rehybridization during the process as required if the 5,6 double bond of FdUMP is saturated in the complex. (6) Proteolytic digestion of the complex yields a peptide that is covalently bound to FdUMP and CH_2—H_4folate [70,71].

From these lines of evidence, together with information gathered from model chemical counterparts, the structure of the enzyme–FdUMP–CH_2—H_4folate complex is currently believed to be that depicted in structure (**33**).

Here, a nucleophile of the enzyme has added to the 6 position of FdUMP, and the 5 position of the pyrimidine is coupled to the 5 position of H_4folate via the methylene group of the cofactor. The remainder of this section deals with the structure of the enzyme–FdUMP–CH_2—H_4folate complex and its relevance to the catalytic reaction. Other aspects of the interaction of thymidylate synthetase with FdUMP and CH_2—H_4folate have been reported since Friedkin's review [10] which are not discussed here [7,58,70–77].

(33)

The most definitive evidence for the structure of the FdUMP–CH_2—H_4-folate–thymidylate synthetase complex rests on an analysis of the peptide covalently bound to both FdUMP and cofactor which is obtained upon extensive proteolysis [71,78]. As with the native ternary complex, there is no evidence of ultraviolet absorption of the FdUMP chromophore, indicating that the 5,6 double bond is saturated when bound to the peptide. However, upon vigorous acid hydrolysis of the peptide, the 5,6 double bond is regenerated and 5-fluorouracil is recovered in near quantitative yield. The 94 MHz ^{19}F nuclear magnetic resonance spectrum consists of a quintet with intensity ratio 1:2:2:2:1 located 87.2 ppm upfield of a trifluoroacetic acid external reference. Our current interpretation of this spectrum is as follows. The apparent quintet is caused by an overlapping doublet of triplets, implying that the fluorine-bonded carbon is flanked by CH and CH_2 groups (i.e., $CHCFCH_2$). The CH is assigned to the 6 position of the nucleotide, which is attached to a nucleophile of the peptide; the 6-proton results in splitting of the ^{19}F resonance into a doublet with a coupling constant of 34.0 Hz. Each component of the doublet is split further into a triplet (intensity ratio 1:2:1) caused by coupling of the fluorine with adjacent CH_2 protons, the magnitude of the coupling constant being 19.5 Hz. The only logical assignment for the CH_2 is the bridging group between the nucleotide and the cofactor as depicted for the native complex in structure (33). Finally, one of the FdUMP–CH$_2$—H$_4$-folate peptides prepared in this laboratory has been shown to be stable after storage for over 6 months without precautions to avoid oxidation [78]. Since H$_4$folates with a free 5-NH are extremely susceptible to oxidation, whereas 5-substituted derivatives are relatively stable [79], it may be concluded that the covalent linkage to H$_4$folate is at the 5- and not the 10-nitrogen as suggested by other workers [69].

An aspect of thymidylate synthetase that has received much attention is the nature of the nucleophile that attacks dUMP in the normal enzymatic reaction and FdUMP in the formation of ternary complexes. As with numerous other sulfhydryl-requiring enzymes, classic experiments [80] involving the relationship of free sulfhydryl groups to enzyme activity, protection against sulfhydryl reagents by substrates, etc. [53,58,81,82], have prompted a number of workers to take the position that a sulfhydryl group of Cys is the nucleophile of interest [31,58,82]. Considering the high nucleophilicity of thiols and the fact that sulfhydryl groups are necessary for activity [10], it was obvious that Cys should be considered as a prime candidate for this role. Nevertheless, experimental evidence had not been reported which warranted preemptive assignment of thiol as this nucleophile.

The availability of FdUMP–CH$_2$—H$_4$folate peptides provides an approach to unequivocal assignment of the nucleophile that is covalently attached to FdUMP and by inference the nucleophilic catalyst of the thymidylate synthetase reaction. The FdUMP–CH$_2$—H$_4$folate peptide obtained upon Pronase digestion of the ternary complex has been shown to have the sequence Ala-Leu-Pro-Pro(Cys,His)Thr [78]. For reasons unclear to us, analysis of hydrolyzed samples failed to definitively reveal Cys as half-cystine [71,78], but prior performic oxidation clearly indicated the presence of one residue of cysteic acid [78]. Since His and Cys were known to be present from amino acid analysis, they were assigned to positions corresponding to the two open Edman cycles where no Dns amino acids could be identified, but their relative positions were unknown. From these data the three candidates for the nucleophile are His, Cys, and Thr. The most direct approach toward identifying the nucleophile would be to obtain a single amino acid residue attached to FdUMP. Unfortunately, the bulky ligands attached near the C-terminus and the Leu-Pro-Pro sequence at the N-terminus block further digestion by the least specific exopeptidases which were available to us, and chemical hydrolysis cleaves the linkage to FdUMP under conditions required for peptide bond hydrolysis [71]. However, we were able to demonstrate that this FdUMP–CH$_2$—H$_4$folate peptide quantitatively reacts with the Pauly reagent and isolated a peptide possessing both monoazoimidazole and FdUMP. Since diazotized sulfanilic acid forms chromophoric derivatives with imidazoles [83–85] but not with N-substituted imidazoles [86], the covalent linkage to

FdUMP cannot involve the His residue, leaving Cys and Thr as remaining possibilities.

An independent and much more extensive analysis of the peptide sequence associated with the FdUMP binding site of *L. casei* thymidylate synthetase has been reported by Bellisario *et al.* [87]. These workers carboxymethylated free thiols of the FdUMP–CH_2—H_4folate–thymidylate synthetase complex and after treatment with cyanogen bromide obtained a fragment of ca. 10,000 daltons which was covalently bound to FdUMP. A similar fragment was also obtained from *S*-carboxymethylated enzyme which had not been exposed to FdUMP. The sequence of the first 13 amino acids of both these peptides was determined to be

Ala-Leu-Pro-Pro-Cys-His-Thr-Leu-Tyr-Gln-Phe-Tyr-Val-

The first seven amino acids in this sequence correspond to the peptide obtained upon Pronase digestion of the FdUMP–CH_2—H_4folate–enzyme complex [78] and establishes the relative positions of Cys and His. These workers were also unable to detect Cys upon direct acid hydrolysis of the FdUMP-bound peptide but did detect it as *S*-carboxymethylcysteine using the peptide obtained from CNBr cleavage of the carboxymethylated thymidylate synthetase which had not been previously bound to FdUMP. Most important, these workers obtained *direct* evidence that the sulfhydryl of cysteine is the most likely candidate for the nucleophile. Thymidylate synthetase and the FdUMP–CH_2—H_4folate–enzyme complex were both carboxymethylated under denaturing conditions. Amino acid analysis showed that the FdUMP-bound enzyme possessed one-half the number of carboxymethylcysteines as did the unbound enzyme. These results contradicted similar experiments in which we found the same number of *N*-ethylmaleimide (NEM) titratable groups [88] in denatured thymidylate synthetase and the denatured FdUMP complex. However, from a combination of NEM titration and subsequent amino acid analysis, Bellisario *et al.* [87] found that amino acids other than the cysteines of thymidylate synthetase react with NEM. Upon amino acid analysis, approximately one-half of the residues of *N*-succinylcysteine were found with the FdUMP-treated enzyme as compared to the untreated enzyme. Thus, some 5 years after sulfhydryl was first proposed to be the nucleophilic catalyst in the thymidylate synthetase reaction [31], experimental evidence has been obtained which directly supports this notion. The isolation of a single amino acid attached to FdUMP remains a worthwhile pursuit and will provide irrefutable evidence for the identity of the amino acid residue of thymidylate synthetase which appears to play a key role in catalysis.

The above-mentioned characteristics of FdUMP–CH_2—H_4folate peptides

are in complete accord with structure (33), which was independently proposed from properties of the ternary complex and model studies. At this point, the structural assignment is well supported by experimental evidence and there appears to be little reason to question its authenticity. From this structure, pertinent aspects of the mechanism of thymidylate synthetase can be inferred.

Referring to structure (33), it is noted that the assigned structure for the enzyme–FdUMP–CH_2—H_4folate complex is analogous to one of the proposed steady-state intermediates of the normal enzymatic reaction [namely, (10), Fig. 6]. They differ in that (10) possesses a proton at the 5 position of the nucleotide which is abstracted in a subsequent step, whereas the enzyme–FdUMP–CH_2—H_4folate complex (33) possesses a stable fluorine at the corresponding position. Thus, it appears that FdUMP behaves as a "quasi substrate" for this reaction. That is, it enters into the catalytic reaction as depicted for the substrate dUMP in Fig. 6 up to the point where an intermediate is formed which can proceed no further; in effect, a complex is trapped which resembles a steady-state intermediate [namely, (10), Fig. 6] of the normal catalytic reaction. Furthermore, as discussed in a previous section, the reactive electrophilic species of CH_2—H_4folate may be the 5- or 10-iminium ion of the cofactor, free formaldehyde, or a methylated form of the enzyme. One of these undoubtedly exists on the reaction pathway of the normal enzymatic reaction and is a direct precursor to the FdUMP–CH_2—H_4-folate–enzyme complex. Since the 5-nitrogen of the cofactor is linked to the 5 position of the nucleotide in the peptide, it can be concluded that the 5-iminium ion of CH_2—H_4folate is the reactive electrophilic species; the close analogy of this reaction to the normal enzymatic reaction suggests that this intermediate is probably also involved in dTMP formation. Thus, after long standing as candidate methyl donors in this reaction, methylated enzyme [89] and methylated derivatives of H_2folate [90] can now be excluded from consideration.

PERSPECTIVES

Although fundamental features of the mechanism of the thymidylate synthetase reaction are understood, many directions for future investigation remain; in this section we consider some of those that we feel would be most enlightening.

With the exception of the requirement for a nucleophilic catalyst, we know little concerning the catalytic role of the enzyme. It may now be worthwhile to reinvestigate model systems to obtain ideas as to other possible catalytic groups and attempt to verify their existence and function in the enzymatic reaction. Furthermore, using rapid reaction technology and quasi substrates such as FdUMP, BrdUMP, and IdUMP, it should be feasible to unravel intricacies of individual catalytic events of the mechanism. A focal point for future studies of this enzyme should be the nature of its interactions with ligands. It is clear that the dimeric *L. casei* enzyme shows unusual environmental perturbations or that conformation changes occur upon formation of complexes [7,70,74]; in addition, heterotropic and negative-homotropic effects are manifested upon binding of ligands [7,60]. Clearly, these interactions may play pertinent roles in catalysis and/or regulation and should be delineated. Furthermore, reports of structural and physical investigations of this enzyme are sparse at this time but undoubtedly represent a fruitful area for future study.

A most important reason for obtaining a detailed understanding of the mechanism of thymidylate synthetase is the possibility (likelihood?) that it may serve as a paradigm for a large number of enzymes. We believe that nucleophilic attack at the 6 position of the uracil or cytosine heterocycle may be a common mechanistic feature utilized by many enzymes to enhance the reactivity at various sites of the heterocycle. From what is known thus far, this is the most reasonable mechanism for a variety of enzyme-catalyzed electrophilic substitution reactions occurring at the 5 position of the uracil heterocycle. These include the dUMP- and dCMP-hydroxymethylases, the pyrimidine methylases of RNA and DNA (the latter including certain restriction enzymes), pseudo uridylate synthetase, and the large number of yet uncharacterized enzymes that alkylate the 5 position of minor bases found in tRNA. In addition, although not extensively studied, it appears that compared to the unaltered base, 5,6-dihydrouracil and cytosine derivatives are chemically more reactive toward nucleophilic substitution at the 4 position of the heterocycle and to glycosidic bond cleavage. It is not unreasonable to suggest that at least some of the enzymes that catalyze such processes might also operate via nucleophilic attack at the 6 position of the heterocycle to achieve saturation of the 5,6 double bond. From teleological considerations, this mechanism is extremely attractive since a primary event in the catalysis of a large and diverse group of enzymes may involve a single relatively simple manipulation of the heterocycle.

ACKNOWLEDGMENTS

This work was supported by USPHS Grant CA-14394 from the National Cancer Institute. D.V.S. is the recipient of an NIH Career Development Award. The authors are grateful to the members of our research group for their comments and to K. Ramos for her patience and diligence in preparation of this account.

REFERENCES

1. G. K. Humphreys and D. M. Greenberg, *Arch. Biochem. Biophys.* **78**, 275 (1958).
2. M. Friedkin, *Fed. Proc., Fed. Am. Soc. Exp. Biol.* **18**, 230 (1959).
3. M. Friedkin, *in* "The Kinetics of Cellular Proliferation" (F. Stohlman, Jr., ed), p. 97. Grune & Stratton, New York, 1959.
4. E. Pastore and M. Friedkin, *J. Biol. Chem.* **237**, 3802 (1962).
5. R. L. Blakley, B. V. Ramasastri, and B. M. McDougall, *J. Biol. Chem.* **238**, 3075 (1963).
6. M. G. Lorenson, G. F. Maley, and F. Maley, *J. Biol. Chem.* **242**, 3332 (1967).
7. S. S. Lam, V. A. Pena, and D. V. Santi, *Biochim. Biophys. Acta* **438**, 324 (1976).
8. M. I. S. Lomax and G. R. Greenberg, *J. Biol. Chem.* **242**, 1302 (1967).
9. D. J. Cram, "Fundamentals of Carbanion Chemistry," Chapter 1, pp. 1–20. Academic Press, New York, 1965.
10. M. Friedkin, *Adv. Enzymol.* **38**, 235 (1973).
11. R. E. Cline, R. M. Fink, and K. Fink, *J. Am. Chem. Soc.* **81**, 2521 (1959).
12. B. R. Baker, T. J. Schwan, and D. V. Santi, *J. Med. Chem.* **9**, 66 (1966).
13. K. H. Scheidt, *Tetrahedron Lett.* p. 1031 (1965).
14. R. J. Burckhalter, R. J. Seiwald, and H. C. Scarborough, *J. Am. Chem. Soc.* **82**, 991 (1960).
15. R. M. Fink, *Arch. Biochem. Biophys.* **107**, 493 (1964).
16. D. V. Santi and C. F. Brewer, *J. Am. Chem. Soc.* **90**, 6236 (1968).
17. R. J. Cushley, S. R. Lipsky, and J. J. Fox, *Tetrahedron Lett.* **52**, 5393 (1968).
18. D. V. Santi and C. F. Brewer, *Biochemistry* **12**, 2416 (1973).
19. L. D. Hall, *Adv. Carbohydr. Chem.* **19**, 51 (1964).
20. R. J. Abraham, L. D. Hall, L. Hough, and K. A. McLauchlan, *J. Chem. Soc.* p. 3699 (1962).
21. E. J. Reist, A. Benitez, and L. Goodman, *J. Org. Chem.* **29**, 554 (1964).
22. B. Bannister and F. Kagan, *J. Am. Chem. Soc.* **82**, 3363 (1960).
23. B. A. Otter, E. A. Falco, and J. J. Fox, *Tetrahedron Lett.* p. 2967 (1968).
24. B. A. Otter and J. J. Fox, *J. Am. Chem. Soc.* **89**, 3663 (1967).
25. B. A. Otter, E. A. Falco, and J. J. Fox, *J. Org. Chem.* **34**, 1390 (1969).
26. J. G. Flaks and S. S. Cohen, *J. Biol. Chem.* **234**, 2981 (1959).
27. R. B. Dunlap, N. G. L. Harding, and F. M. Huennekens, *Ann. N. Y. Acad. Sci.* **186**, 153 (1971).
28. D. V. Santi, C. F. Brewer, and D. Farber, *J. Heterocycl. Chem.* **7**, 903 (1970).
29. S. R. Heller, *Biochem. Biophys. Res. Commun.* **32**, 998 (1968).
30. Y. Wataya, H. Hayatsu, and Y. Kawazoe, *J. Biochem. (Tokyo)* **73**, 871 (1973).
31. T. I. Kalman, *Biochemistry* **10**, 2567 (1971).
32. T. C. Bruice, T. H. Fife, J. J. Bruno, and N. E. Brandon, *Biochemistry* **1**, 7 (1962).
33. R. Shapiro, R. E. Servis, and M. Welcher, *J. Am. Chem. Soc.* **92**, 422 (1970).
34. H. Hayatsu, Y. Wataya, K. Kai, and S. Iida, *Biochemistry* **9**, 2858 (1970).
35. Y. Wataya and H. Hayatsu, *Biochemistry* **11**, 3583 (1972).
36. E. G. Sander and C. A. Deyrup, *Arch. Biochem. Biophys.* **150**, 600 (1972).
37. G. S. Rork and I. H. Pitman, *J. Am. Chem. Soc.* **97**, 5559 (1975).
38. F. A. Sedor, D. G. Jacobson, and E. G. Sander, *Bioorg. Chem.* **3**, 221 (1974).
39. H. Hayatsu, *Prog. Nucleic Acid Res. Mol. Biol.* **16**, 75 (1976).
40. Y. Wataya, K. Negishi, and H. Hayatsu, *Biochemistry* **12**, 3992 (1973).
41. F. A. Sedor, D. G. Jacobson, and E. G. Sander, *Bioorg. Chem.* **3**, 154 (1974).

310   ALFONSO L. POGOLOTTI, JR., AND DANIEL V. SANTI

42. W. P. Jencks, *Prog. Phys. Org. Chem.* **2**, 63 (1964).
43. R. G. Kallen and W. P. Jencks, *J. Biol. Chem.* **241**, 5821 (1966).
44. S. J. Benkovic and W. P. Bullard, *Prog. Bioorg. Chem.* **2**, 134 (1973).
45. B. S. Gupta and F. M. Huennekens, *Biochemistry* **6**, 2168 (1967).
46. R. S. Wilson and M. P. Mertes, *J. Am. Chem. Soc.* **94**, 7182 (1972).
47. R. S. Wilson and M. P. Mertes, *Biochemistry* **12**, 2879 (1973).
48. D. V. Santi, *J. Heterocycl. Chem.* **4**, 475 (1967).
49. D. V. Santi and A. L. Pogolotti, *Tetrahedron Lett.* p. 6159 (1968).
50. D. V. Santi and A. L. Pogolotti, *J. Heterocycl. Chem.* **8**, 265 (1971).
51. A. L. Pogolotti and D. V. Santi, *Biochemistry* **13**, 456 (1974).
52. T. C. Crusberg, R. Leary, and R. L. Kisliuk, *J. Biol. Chem.* **245**, 5292 (1970).
53. R. B. Dunlap, N. G. L. Harding, and F. M. Huennekens, *Biochemistry* **10**, 88 (1971).
54. R. P. Leary and R. L. Kisliuk, *Prep. Biochem.* **1**, 47 (1971).
55. J. H. Galivan, G. F. Maley, and F. Maley, *Biochemistry* **14**, 3338 (1975).
56. Y. Yeh and G. R. Greenberg, *J. Biol. Chem.* **242**, 1307 (1967).
57. M. Friedkin, *Adv. Enzymol.* **38**, 248 (1973).
58. P. V. Danenberg, R. J. Langenbach, and C. Heidelberger, *Biochemistry* **13**, 926 (1974).
59. A. L. Pogolotti, C. Weill, and D. V. Santi, *Fed. Proc., Fed. Am. Soc. Exp. Biol.* (abstr.) **35**, 1705 (1976).
60. J. H. Galivan, G. F. Maley, and F. Maley, *Biochemistry* **15**, 357 (1976).
61. Y. Wataya and D. V. Santi, *Biochem. Biophys. Res. Commun.* **67**, 818 (1975).
62. G. S. Rork and I. H. Pitman, *J. Am. Chem. Soc.* **97**, 5566 (1975).
63. F. A. Sedor and E. G. Sander, *J. Am. Chem. Soc.* **98**, 2314 (1976).
64. C. Garrett, Y. Wataya, and D. V. Santi, *Abstr.*, 126 (*Biol. Chem.*) *172nd Nat. Meet. Am. Chem. Soc.* (1976).
65. C. Heidelberger, G. Kaldor, K. L. Mukherjee, and P. B. Danenberg, *Cancer Res.* **20**, 903 (1960).
66. R. L. Blakley, "The Biochemistry of Folic Acid and Related Pteridines," p. 246. Am. Elsevier, New York, 1969.
67. J. J. Fox, N. C. Miller, and R. J. Cushley, *Tetrahedron Lett.* p. 4927 (1966).
68. D. V. Santi and C. S. McHenry, *Proc. Natl. Acad. Sci. U.S.A.* **69**, 1855 (1972).
69. R. J. Langenbach, P. V. Danenberg, and C. Heidelberger, *Biochem. Biophys. Res. Commun.* **48**, 1565 (1972).
70. D. V. Santi, C. S. McHenry, and H. Sommer, *Biochemistry* **13**, 471 (1974).
71. H. Sommer and D. V. Santi, *Biochem. Biophys. Res. Commun.* **57**, 689 (1974).
72. J. L. Aull, J. A. Lyon, and R. B. Dunlap, *Arch. Biochem. Biophys.* **165**, 805 (1974).
73. R. K. Sharma and R. L. Kisliuk, *Biochem. Biophys. Res. Commun.* **64**, 648 (1975).
74. H. Donato, J. L. Aull, J. A. Lyon, J. W. Reinsch, and R. B. Dunlap, *J. Biol. Chem.* **251**, 1303 (1976).
75. R. K. Sharma and R. L. Kisliuk, *Fed. Proc., Fed. Am. Soc. Exp. Biol.* **31**, 591 (1973).
76. R. K. Sharma, R. L. Kisliuk, S. P. Verma, and D. F. H. Wallach, *Biochim. Biophys. Acta* **391**, 19 (1975).
77. R. K. Sharma and R. L. Kisliuk, *Fed. Proc., Fed. Am. Soc. Exp. Biol.* **33**, 1546 (1974).
78. A. L. Pogolotti, K. I. Ivanetich, H. S. Sommer, and D. V. Santi, *Biochem. Biophys. Res. Commun.* **70**, 972 (1976).
79. R. L. Blakley, "The Biochemistry of Folic Acid and Related Pteridines," pp. 78–91. Am. Elsevier, New York, 1969.
80. P. D. Boyer, *in* "The Enzymes" (P. D. Boyer, H. Lardy, and K. Myrbäck, eds.), 2nd ed., Vol. 1, pp. 511–586. Academic Press, New York, 1959.

81. R. P. Leary, N. Beaudette, and R. Kisliuk, *J. Biol. Chem.* **250**, 4864 (1975).
82. T. I. Kalman, *Biochem. Biophys. Res. Commun* **49**, 1007 (1971).
83. H. Z. Pauly, *Hoppe-Seyler's Z. Physiol. Chem.* **42**, 508 (1904).
84. J. M. Stewart and J. D. Young, "Solid Phase Peptide Synthesis," p. 56. Freeman, San Francisco, California, 1969.
85. J. F. Riordan and B. L. Vallee, *in* "Methods in Enzymology" (C. H. W. Hirs and N. Timasheff, eds.), Vol. 25, Part B, p. 521. Academic Press, New York, 1972.
86. R. W. Cowgill, *Anal. Chem.* **27**, 1519 (1955).
87. R. L. Bellisario, G. F. Maley, J. H. Galivan, and F. Maley, *Proc. Natl. Acad. Sci. U.S.A.* **73**, 1848 (1976).
88. C. S. McHenry and D. V. Santi, *Biochem. Biophys. Res. Commun.* **57**, 204 (1974).
89. A. J. Wahba and M. Friedkin, *J. Biol. Chem.* **237**, 3794 (1962).
90. W. Wilmanns, B. Rucker, and L. Jaenicke, *Hoppe-Seyler's Z. Physiol. Chem.* **322**, 283 (1960).

13

Studies of Amine Catalysis via Iminium Ion Formation

Thomas A. Spencer

INTRODUCTION

One of the most prevalent types of catalytic mechanism in biochemical processes involves condensation of a primary amine in an enzyme, usually that of a lysine residue, with a carbonyl group of the substrate to form an imine, or Schiff base. A principal purpose of such condensations is to facilitate removal, without its bonding pair of electrons, of a substituent α to the carbonyl from the protonated form of the Schiff base, the iminium ion. This pathway is shown in a generalized form in Scheme 1, as is the "simpler" alternate pathway involving removal of the α-substituent Q directly from the carbonyl compound. Enzymatic examples for which there is good evidence of the amine–carbonyl condensation pathway include a decarboxylase (Q = —COO$^{\ominus}$) [1], aldolases

$$ Q = -\overset{|}{\underset{|}{C}}-O^{\ominus} $$

[2–4], and dehydratases (Q = H) [5–8].

Such covalent, or nucleophilic, catalysis by amines has, of course, long been recognized in the organic chemistry laboratory as well. The pioneering work of Westheimer includes early detection of this phenomenon in simple nonenzymatic reactions [9] as well as delineation of its role in the case of

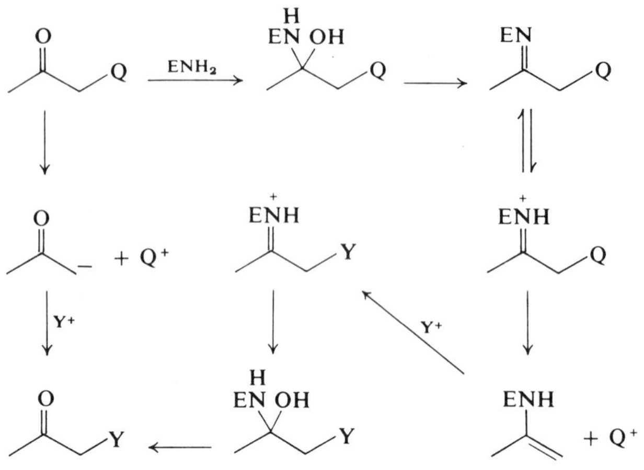

Scheme 1

acetoacetate decarboxylase [1]. An extensive study of amine–carbonyl condensation and its effect on α-substitution has been conducted by Hine [10], and many of his results bear importantly on the work described herein. Numerous others have also studied this type of process, but space limitations preclude an attempt to give due recognition to all of these researchers [11,12].

This chapter describes essentially only our own efforts to gain further understanding of this kind of catalysis. As described below, these efforts have focused on abstraction of protons α to carbonyl groups in β-elimination reactions analogous to the enzymatic dehydrations (Q = H in Scheme 1). Our goal has been to gain insight into why nucleophilic amine catalysis is favorable in biochemical systems. We hope eventually to apply such insight to the construction of molecules that will serve both as adequate enzyme models and as practical catalysts for use in organic synthesis under mild conditions.

Our interest in nucleophilic amine catalysis was first sparked some dozen years ago by the chance observation that pyrrolidine was a better catalyst than piperidine for the conversion of (1) to (2) [13], and this interest was kindled through other studies then in progress of various aspects of (amine-catalyzed) Robinson annelation reactions [14–16]. Considerable effort was expended in demonstrating in a rather imprecise manner (e.g., by following appearance and disappearance of intermediates with thin layer chromatography) that tertiary amines, which cannot condense with carbonyl groups, were less effective than primary or secondary amines in the aldol condensation and dehydration steps of such reactions as (1) → (2) [17]. Subsequently, a similar study with similar conclusions was made of amine-catalyzed epimerization of 2β-methylcholestanone (3) to 2α-methylcholestanone (4) [18],

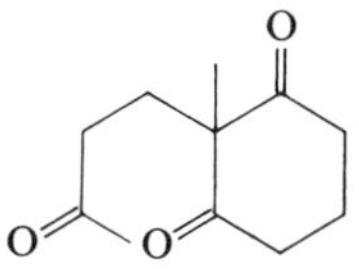

(1) 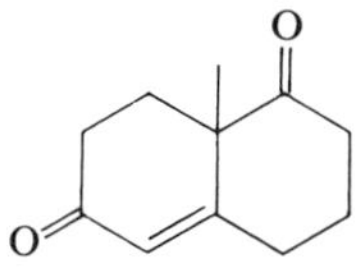   (2)

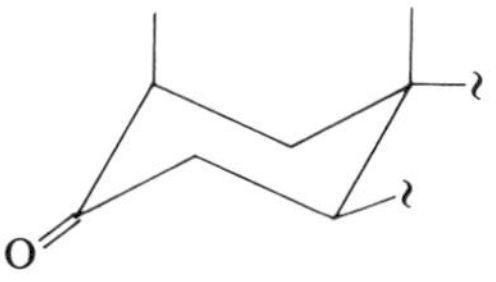

(3) 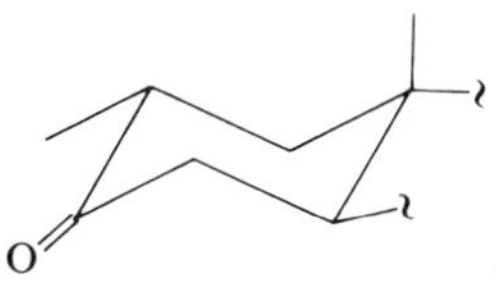   (4)

using optical rotation to follow the reaction. Throughout this early research, pyrrolidine, which is known [19] to condense more readily with carbonyl groups than its homolog piperidine, retained its qualitative status as a superior catalyst.

FINDING AN APPROPRIATE α DEPROTONATION REACTION

What was really needed in order to study the iminium ion pathway effectively, however, was the ability to follow a reaction proceeding via amine–carbonyl condensation quantitatively and conveniently in aqueous solution by ultraviolet (uv) spectroscopy. Water not only has biological relevance, but also permits much easier access to and control over such essential variables as acid and base strengths than do other media. Ultraviolet spectroscopy can permit the use of dilute aqueous solutions and is much more easily adapted for continuous, automatic monitoring of reactions than are other techniques, such as nuclear magnetic resonance (nmr) spectroscopy or polarimetry. In fact, from the present vantage point of an accumulation of several thousand rate constants, it is clear that the studies described below would not have been humanly realizable, at least in the context of our research group, by any technique other than uv spectroscopy.

Happily, it turned out that the type of system we had already been studying allowed us to observe rate-determining α-proton abstraction via iminium ion formation in the desired manner. Specifically, we were able to show, as documented below, that the β-elimination reactions of β-acetoxyketone (5) and β-hydroxyketone (6) to form chromophoric enone (7), ($\varepsilon_{247}^{H_2O}$ 15,500) proceed by just such a mechanism under appropriate catalytic conditions.

Compounds (5) and (6) offer several distinct advantages over apparently simpler model systems. Although (6) and, particularly, (5) are not trivial to prepare [20], they are crystalline and easily purified. More importantly, they react essentially quantitatively to form (7); no equilibrium amount of β-addition product formed from (7) plus hydroxide ion or other nucleophile has ever been observed in these studies. And, unlike some other β-hydroxy-ketones, (6) apparently does not undergo retroaldol reaction to form (8).* Thus, the elimination reactions leading to (7) are very clean and easily monitored. It remained to be demonstrated, however, that these reactions could be catalyzed via iminium ion formation and, if so, whether α-deprotonation would be the slow step.

Catalysis of the conversion of (5) to (7) in aqueous solution in the presence of a primary amine is moderately complex but can be described adequately by the following equation [20]:

$$\frac{d(7)}{dt} = (k_H[H_3O^+] + k_{OH}[OH^-] + k_B[RNH_2]$$
$$+ k_A[RNH_3^+] + k_{AB}[RNH_2][RNH_3^+])[(5)]$$

Determining that this expression could account for the formation of (7) was clearly a major part of this investigation. Doing so required measuring rates of formation of (7) with numerous different catalyst concentrations at numerous different pH values (differing buffer ratios). The results of such experiments with one primary amine buffer concentration at various pH values are shown in Fig. 1 for 0.1 M cyanomethylamine. The characteristic bell-shaped pH–rate profile is a consequence of the k_{AB} term, because $[RNH_2][RNH_3^+]$ is at a maximum when pH = pK_a. The standard "titration" curves for the k_B term proportional to $[RNH_2]$ and the k_A term proportional to $[RNH_3^+]$ are also shown in Fig. 1 [20].

The first question to ask about such observations is whether any of these terms represents catalysis via iminium ion formation. It seemed likely that the

---

* There is no observable indication that an equilibrium between (6) and (8) is established during formation of (7). In the case of trans-fused ketol (22), discussed later, it is certain that retroaldol conversion to (8) does not occur, because (8) cyclizes selectively to (6) (see Spencer et al. [16] and J. A. Marshall and W. I. Fanta, J. Org. Chem. (29), 2501, 1964), and (6) is converted to (7) slightly more slowly than (22).

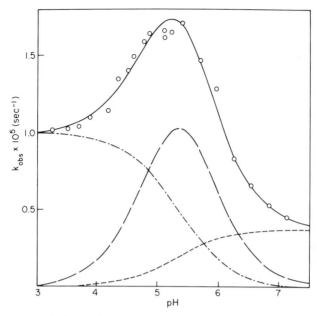

Fig. 1 Plot of pseudo first-order rate constants, k_{obs}, for the conversion of (**5**) to (**7**) in the presence of 0.1 M total cyanomethylamine ($pK_a = 5.34$) buffer vs. pH. The experimental points (○) are shown, as are curves for the $k_B[NCCH_2NH_2]$ term (– – – –), the $k_A[NCCH_2NH_3^+]$ term (— – — – —), the $k_{AB}[NCCH_2NH_2][NCCH_2NH_3^+]$ term (— — —), and their sum (———) calculated using the appropriate rate constants from Hupe *et al.* [20]. Reprinted with permission from *J. Am. Chem. Soc.* **94**, 1254 (1972). Copyright by the American Chemical Society.

k_{AB} term did reflect such nucleophilic catalysis, because concerted general acid–general base catalysis [as in (**9**)] of this magnitude in water is improbable [21]. Support for this conclusion came from studies with tertiary amines as catalysts [20]. Even unhindered [22] tertiary amines showed *only* a k_B term, suggesting that *both* the k_A and k_{AB} terms represented catalysis via iminium ion formation, whereas the k_B term (as well as the k_{OH} term) presumably reflected direct α-proton abstraction.

$$
\begin{array}{cc}
\text{(8)} & \text{(9)}
\end{array}
$$

Scheme 2 delineates the principal steps in the iminium ion pathway for the conversion of (5) to (7). If one assumes that the species B is RNH_2, it can be seen that there are at least the three possibilities shown in (10) → (11), (11) → (12), and (12) → (13) for rate-determining steps which would show the dependence on $[RNH_2][RNH_3{}^+]$ required by the k_{AB} term. Similarly, if B were H_2O, these same three steps would show the dependence on $[RNH_3{}^+]$ required by the k_A term. The observed kinetic behavior is thus readily plausible mechanistically in terms of contributions from nucleophilic catalysis.

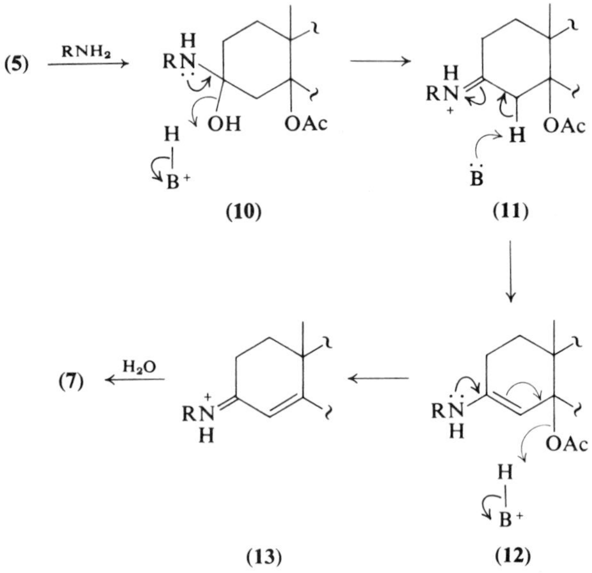

Scheme 2

In order to determine whether, as we hoped, removal of the α-proton from the iminium ion [(11) → (12)] was the rate-determining step, the deuterium-labeled substrate (14), R = Ac, was prepared and its elimination to enone was studied [20]. There was observed a large, invariant primary kinetic isotope effect in the conversion of (14), R = Ac, to deuterated (7), using primary amines under conditions in which the k_{AB} term dominates the catalysis. For example, with cyanomethylamine, the k_{AB} term showed $k_H/k_D = 6.3$ [20]. This demonstrated conclusively that α-proton abstraction was indeed the slow step in the overall process.

Analogous observations were made for the conversion of ketol (6) to enone (7) [23], except that the "k_{OH}" and "k_B" terms were much more complicated, owing to the fact that enolate anion (15) formed from (6) upon reaction with base reprotonates more rapidly than it expels hydroxide ion

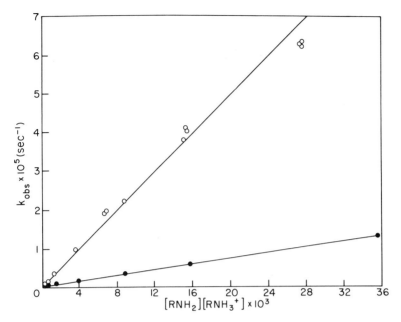

Fig. 2 Plot of pseudo first-order rate constants, k_{obs}, for the conversion of **(6)** (○) and **(14)** (●), R = H, to **(7)** in the presence of different concentrations of cyanomethylamine buffer at pH = pK_a = 5.34 vs. [NCCH$_2$NH$_2$][NCCH$_2$NH$_3^+$] at each buffer concentration. The ratio of the slopes (k_{AB}) is 6.7 = k_H/k_D for **(6)** vs. **(14)**, R = H. Other kinetic terms (e.g., k_A[NCCH$_2$NH$_3^+$]) are essentially negligible under the conditions illustrated [23].

[24],* whereas the enolate anion from **(5)** "always" goes on to **(7)**. Fortuitously, no such complicating reverse reaction was encountered with the k_{AB} or k_A terms for the conversion of **(6)** to **(7)**. All experiments, including a large, invariant primary kinetic isotope effect with **(14)**, R = H, as shown in Fig. 2, indicated that with **(6)**, as with **(5)**, we were able to look directly at rate-determining α-proton abstraction via iminium formation.

It is worth noting also that the rates for **(5)** → **(7)** are only about twice as fast as the rates for **(6)** → **(7)** under conditions in which nucleophilic amine catalysis obtains. This suggests that there is negligible concerted character to the elimination reaction, because a far greater difference in rates would be expected if there were significant involvement of the leaving group in the transition state [25]. The reasonable inference is that the slow step with both

* The complicated kinetic behavior of **(15)** and its conjugate enol was with considerable effort and ingenuity elucidated by D. J. Hupe [24]. This extensive aspect of the behavior of **(6)** is not described in this chapter.

(5) and **(6)** is α-proton removal to form an intermediate enamine, as shown in **(11)** → **(12)** in Scheme 2.

(14)   (15)

Belatedly, direct evidence of the intermediacy of amine–carbonyl condensed species in the conversions of **(5)** and **(6)** to **(7)** emerged. Had we carefully done wavelength scans throughout all our kinetic runs, we would have noticed earlier that the chromophore at 247 nm caused by **(7)** sometimes had a shoulder on the long-wavelength side during the first few percent of reaction. Once this was detected, it was possible to adjust reaction conditions (to high concentrations of catalyst and substrate) so that easily detectable concentrations of species with $\lambda_{max} \sim 270$ nm accumulated at the beginning of runs. These species were isolated and completely characterized as eniminium ions **(13)** [e.g., **(13)**, R = $CH_2CH_2OCH_2CH_3$, has $\varepsilon_{270}^{H_2O}$ 16,000] [26,27]. The eniminium ions were shown to behave in their rates of appearance and disappearance as intermediate species in the consecutive reaction series A = **(6)** → B = **(13)** → C = **(7)** [26].

The results discussed above established that **(5)** and **(6)** would permit direct investigation of catalysis of α-proton abstraction via iminium ion formation. Having found an appropriate model system, we were in a position to explore this catalytic pathway systematically.

STUDIES OF CATALYSIS VIA
IMINIUM ION FORMATION

To guide our research, we tried to identify factors that presumably contribute to the catalytic effectiveness of enzymes functioning via iminium ion formation. The following areas seemed to be potentially fruitful for investigation:

1. *Binding*: What will enhance amine–carbonyl condensation leading to α-deprotonation?

2. *Approximation*: What is the optimum geometric arrangement of all the atoms involved in the transition state?

3. *Acid–base relationships*: What should be the relationship between the pK_a values of the functional groups involved and the pH of the medium?

These categories are not claimed to be inclusive, independent, or original. However, they have provided a helpful organizational framework. It will be evident in the discussion of each of them that most of our effort and progress to date have been in the third area, acid–base relationships.

Binding

One of the factors contributing to the catalytic efficiency of enzymes is the ability of these proteins to bind the substrate effectively through a variety of covalent and/or other types of interactions such as hydrogen bonding. It is superficially tempting to think in terms of enhancing the equilibrium constant for amine–carbonyl condensation as the analog of improving "binding" in the context of nucleophilic amine catalysis. However, it must be remembered that, for catalysis to occur, there has to be tighter binding (an energetically more favorable interaction) with the activated complex than with the substrate. This point has been effectively illustrated by recent studies of transition-state analogs as enzyme inhibitors [28,29].

It is easy to demonstrate the validity of this principle in the present context of rate-determining α-deprotonation. One way to enhance the equilibrium constant for imine formation is to use an amine of the type displaying the α effect [30], such as methoxylamine. Treatment of ketol (6) with methoxyl-amine did indeed readily lead to amine–carbonyl condensation, but it did not lead to dehydration. A mixture of stable isomeric methoximes, (16) and (17), was obtained instead [31]. These could be induced to dehydrate only under very vigorous conditions [at least 10^5 times slower than (6) itself] [31], and it is clear that methoxylamine, rather than functioning as a catalyst, simply forms a very stable "complex" with the substrate.

(16)   (17)   (18)

What properties should a nucleophilic amine catalyst have in order to stabilize a transition state like (18)? Some of these properties, of course, will be considered when the geometric and acid–base relationships among the functional groups involved in (18) are discussed. But, beyond these considerations, it is not obvious at present how to promote catalysis of α-proton abstraction by better "binding."

Of some relevance to this topic is recent work of Pollack [32], in which an interesting rate enhancement in a related reaction was observed upon decreasing the polarity of the medium. Such an effect might also be found in the present context, because the positive charge in (18) is presumably more spread out than in the starting materials. However, it would be difficult to apply this kind of solvent dependence, if observed, to the improvement of a small-molecule catalyst in aqueous solution. On the other hand, localized medium effects are certainly conceivable as contributors to enzymatic catalysis [32].

Approximation

Perhaps the most important advantage that enzymes have over small-molecule catalysts derives from the fact that the complex three-dimensional structure of proteins allows them in effect to position the appropriate functional groups correctly to participate in the transition state of the reactions they catalyze. Even with polyfunctional small-molecule catalysts it is unlikely that the correct atoms will be in the correct places at the same time. In the case of the α-deprotonation reactions under consideration, one would like to know both what the ideal geometry of a transition state like (18) is and also how advantageous it would be to have such optimum geometry.

With respect to the latter question, there have been several recent approaches to a general analysis of the effect of geometric restriction on rates of reactions, and numerous names, including "approximation" [33], have been used to describe this property. Koshland has analyzed the situation in terms of "proximity" and "orbital steering" and has predicted a maximum total rate acceleration of about 10^5 M (an "effective concentration" in an intramolecular vs. an intermolecular reaction) for a geometrically ideal situation [34]. Jencks and Page have predicted a maximum factor of about 10^8 M on the basis of an analysis of entropy loss upon restriction of translational and rotational motions [35,36].

An indication that there may indeed be a large advantage to be gained from appropriate geometric restriction in the case of a reaction with transition state (18) was obtained in a study of the temperature dependence of the conversion of (5) to (7) [20]. The results in Table 1 show that the k_{AB} term, representing catalysis as in (18), has a considerably larger negative entropy of activation than the k_B term for direct proton abstraction by trimethylamine [20]. This result is not surprising, of course, since (18) requires bringing together two molecules of catalyst with the substrate.

The most obvious geometric restriction appropriate to reducing the large negative $\Delta S^{\ddagger}$ of (18) is placement of the nucleophilic amine and the general

TABLE 1

Values of Activation Parameters Calculated[a] for Three Kinetic
Terms for the Conversion of (5) to (7)

Term	Ea (kcal/mole)	$\Delta H^{\ddagger}$ (kcal/mole)	$\Delta G^{\ddagger}$ (kcal/mole)	$\Delta S^{\ddagger}$ (eu)
$k_{OH}[OH^-]$	18.5	17.9	17.4	2
$k_B[(CH_3)_3N]$	17.4	16.8	19.4	-9
$k_{AB}[HCMA^+][CMA]^b$	10.8	10.2	20.7	-35

[a] Calculated from $\Delta H^{\ddagger} = Ea - RT$; $\Delta G^{\ddagger} = -2.303RT \log(k_2 h/kT)$; $\Delta S^{\ddagger} = (\Delta H^{\ddagger} - \Delta G^{\ddagger})/T$.

[b] CMA, cyanomethylamine.

base in the same molecule. If this could be done without increasing the small $\Delta H^{\ddagger}$ found for the k_{AB} term [20], an excellent catalyst indeed would result. Studies of bifunctional catalysts of this type are clearly desirable, and Hine has already made important findings in this area.

Intramolecular bifunctional catalysis as in (19) escaped detection by investigators looking for it [37] until Hine's study of α-dedeuteration of iso-butyraldehyde-2-d with polyethylenimines [38]. Subsequently, Hine used homologous series of diamines as catalysts for α-dedeuteration of acetone-d_6 [39]. He found that intramolecular bifunctional catalysis was at a maximum when there were three carbon atoms separating the iminium ion nitrogen from the general base, as in (19), $n = 3$ [39]. Among the best catalysts of this type that Hine has yet reported are the 2-(dimethylaminomethyl)cyclopentyl-amines, which effect deuteron exchange from acetone $\sim 10^2$ times faster than would be expected for direct general base-catalyzed exchange at pH 8.76 [39].

Consistent with Hine's results, we have observed catalysis of the conversion of (6) to (7) by histidine methyl ester through transition state (20) [40]. The familiar bell-shaped pH–rate profile was observed, but in this case the kinetic term causing it was proportional to the concentration of mono-protonated histidine methyl ester, not to the square of the buffer concentration [40]. The rate constant characterizing (20) is substantially less than that for Hine's best catalysts, however [40].

Hine reasonably explained the observed preference for the eight-membered transition state of (19), $n = 3$, on the basis of stereoelectronic control in the α-dedeuteration. Only with at least three carbons interposed between the two nitrogens can the general base "reach" an α-deuteron or α-proton oriented so as to permit overlap of its bonding pair of electrons with the π bond of the iminium ion as it is being removed. This geometry is illustrated in (21). When $n > 3$, the entropy requirement of (19) presumably increases.

(19)   (20)   (21)

In cyclohexanones, such as our model systems (5) and (6), stereoelectronic control would mean selective abstraction of the axial α-proton. Evidence has been obtained for such selectivity in proton abstraction directly from carbonyl compounds [41], but not from derived iminium ions. We felt it was highly desirable to try to prove that Hine's inference was correct before embarking on studies designed to assess the approximation factor. Unfortunately, (5) and (6) are *cis*-fused decalins and possess conformational mobility which makes them at best ambiguous candidates for such a study. Therefore, we turned to their *trans*-fused counterparts (22) and (23), in which the α-protons have distinct axial or equatorial geometry, as substrates with which to probe the stereochemistry of proton abstraction.

(22), R = H   (23), R = H
(24), R = D   (25), R = D

In addition to (22) and (23), this study required the selectively α axially monodeuterated (24) and (25). One method used to prepare (22) [42] and (24) is shown for (24) in Scheme 3 as an illustration of the efforts in synthesis that were also required in this research [43]. Acetylation of (22) or (24) was an unreliable source of (23) or (25), just as acetylation of (6) was unsuitable for the preparation of (5) [20], so a more circuitous route, which will not be detailed here, was adopted for synthesis of the *trans*-fused β-acetoxyketones [43].

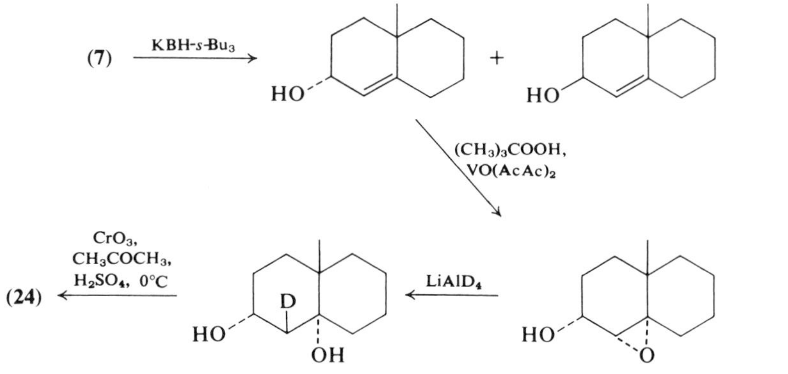

Scheme 3

Conversions of (24) and (25) to (7) under conditions in which the k_{AB} term accounts for nearly all of the product formation showed the familiar large, invariant primary kinetic isotope effects observed in the elimination reactions of cis-fused (14). For example, the reaction of (24) with 0.4 M trifluoroethylamine at pH = pK_a = 5.7, conditions under which 95% of the formation of (7) from (22) occurs via the "k_{AB} pathway," showed a kinetic isotope effect of $k_H/k_D = 7.7$ when compared with (22) [43]. Accordingly, there is little doubt that the α-deprotonation occurs selectively at the axial position, through transition state (26). Efforts are currently in progress to determine an accurate value for the axial/equatorial ratio (stereoelectronic factor) in abstraction of α-protons from iminium ions using mass spectrometric analysis of the enone (7) produced from (24) or (25) under suitable conditions [43].

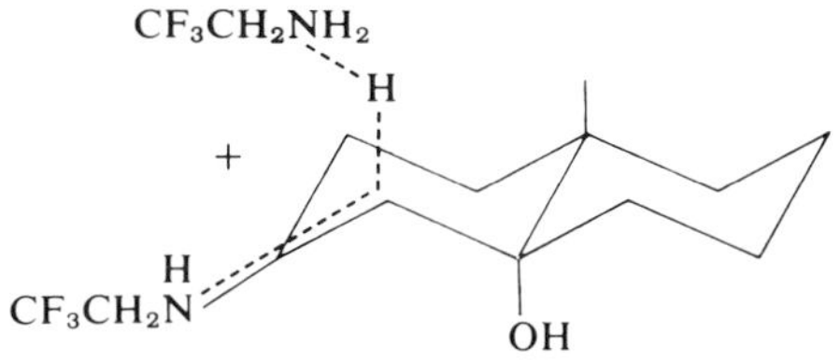

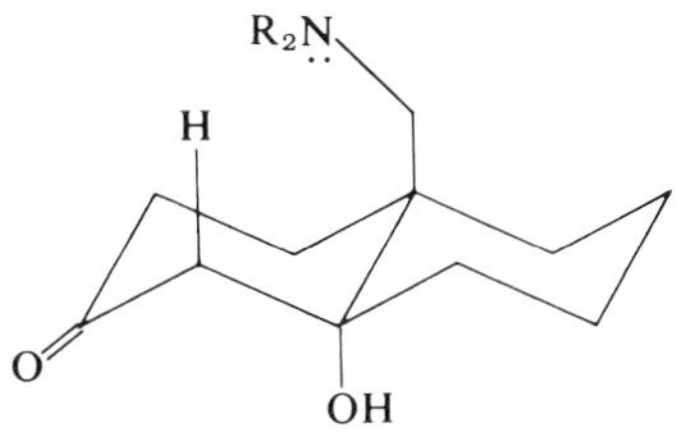

With this confirmation of stereoelectronic control, we know that it is appropriate to concentrate on catalysts designed to abstract the axial α-proton from an iminium ion. At present we are preparing molecules, such as (27), with angular substituents containing a general base which can abstract the axial α-proton intramolecularly. Although the basic atom in (27) is not held rigidly proximate to the α-proton in question, its incorporation into the same molecule in a position where it can interact with that proton may yield most of the advantage that can be gained by geometric restriction [35]. From comparison of such intramolecular α-proton abstraction with appropriate intermolecular reactions we hope to obtain the first experimental information as to how much rate acceleration could possibly be achieved in this type of reaction by correct approximation.

Acid–Base Relationships

The research discussed so far has not dealt with the fundamental question of why the iminium ion pathway is common in enzymatic systems. Some insight into this matter has come from a study of the effectiveness of nucleophilic amine catalysts as a function of their base strengths.

It is well known that stronger bases are more effective than weaker bases for direct abstraction of a proton α to a carbonyl group. Good Brönsted correlations between rate of reaction and pK_a have been obtained in a number of cases, with the value of β usually being ≥ 0.5 [44]. For our β-acetoxyketone (5), the rates for α-proton abstraction by tertiary amines fit a Brönsted plot

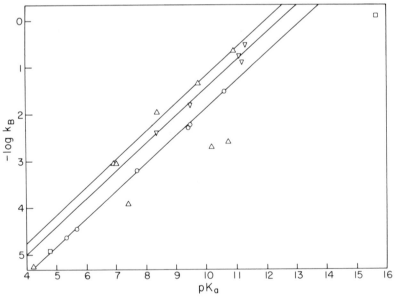

Fig. 3 Brönsted plot of the logarithms of the second-order rate constants, k_B, from Hupe et al. [20] for the reaction of primary (○), secondary (▽), and tertiary (△) amines and hydroxide ion and acetate ion (□) with (5) to give (7) vs. pK_a values of their conjugate acids. The slope of the lines is $\beta = 0.59$. Reprinted with permission from J. Am. Chem. Soc. **94**, 1254 (1972). Copyright by the American Chemical Society.

with $\beta = 0.6$ [20]. Figure 3 shows this Brönsted correlation as well as those for the values of k_B determined with primary and secondary amines. The data for each class of amine are plotted separately on the basis of our observations in the case of general base-catalyzed proton abstraction from 9-fluorenylmethyl chloride (28) to form (29) [22] and on the basis of other precedents [45]. A similar Brönsted β of 0.6 was found for α-proton abstraction from ketol (6) [24].

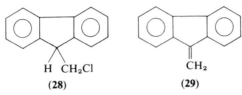

H CH$_2$Cl   CH$_2$
(28)   (29)

At physiological pH only relatively weak bases, such as histidine imidazole residues, would exist in their unprotonated form and be available to abstract protons. The Brönsted correlations discussed above indicate that such weak bases would be relatively ineffective for α-proton abstraction. For example, in the conversion of (5) to (7), quinuclidine ($pK_a = 10.95$) is 250 times more effective than imidazole ($pK_a = 6.95$) [20].

When the k_{AB} terms for the conversions of (5) and (6) to (7) were studied as a function of amine pK_a, however, a strikingly different pattern emerged. In Fig. 4 are shown the pH–rate profiles, dominated by the bell-shaped k_{AB}

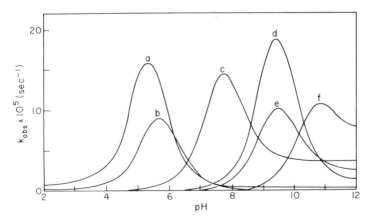

Fig. 4 Six pH–rate profiles calculated for the conversion of (6) to (7) catalyzed by 0.5 M total buffer concentration of primary amines, using data from Hupe *et al.* [23]. The amines are (a) cyanomethylamine, (b) trifluoroethylamine, (c) ethyl glycinate, (d) ethoxyethylamine, (e) allylamine, (f) *n*-butylamine. Reprinted with permission from *J. Am. Chem. Soc.* **95** 2271 (1973). Copyright by the American Chemical Society.

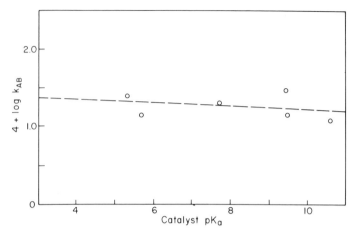

Fig. 5 Brönsted plot of the logarithms of the third-order rate constants, k_{AB} (M^{-2} sec^{-1}), taken from Hupe *et al.* [23], for the conversion of (6) to (7) by the same six amines shown in Fig. 4 vs. the pK_a values of their conjugate acids. The slope of the line is -0.02.

terms, for the conversion of (6)* to (7) using six primary amines with base strengths varying over a range of more than 10^5. There is no diminution of catalytic effectiveness as base strength decreases. Cyanomethylamine, $pK_a = 5.34$, is actually a slightly better catalyst than *n*-butylamine, $pK_a = 10.61$. The values of k_{AB} for these six amines are shown in a Brönsted plot in Fig. 5, and the least-squares line through the points confirms that catalytic effectiveness is essentially independent of amine pK_a. Therefore, the iminium ion pathway has a distinct relative advantage at physiological pH, and it makes sense that evolution would take advantage of this fact by incorporating nucleophilic amine catalysis into enzymatic mechanisms.

How large is this advantage? There have been several earlier attempts to evaluate the relative ease with which a given general base can abstract an α-proton from an iminium ion (30), as opposed to the parent carbonyl compound (31). These evaluations, which necessarily include an estimate of the concentration of the iminium ion, have ranged from a factor of 10^3 [46] to 10^8 [37]. However, these estimates did not take into account our finding that this factor depends on the pK_a of the nucleophilic amine, as is evident from a comparison of the slopes in Figs. 3 and 5. Using the same

---

* Measurement of k_{AB} for amines with $pK_a > 8$ is possible with (6) but not with (5) because the k_B terms are too large with high-pK_a amines to allow measurement of k_{AB} for the β-acetoxyketone. With the ketol, reconversion of enolate (15) to (6) reduces the rate of net observed general base catalysis, allowing measurement of k_{AB} with amines having $pK_a > 8$.

assumptions as others [37,46] to estimate the concentration of iminium ion, we calculated that the cyanomethylamine-catalyzed process shown in (32) is 10^6 times faster than the process shown in (33), whereas with the more strongly basic n-butylamine as catalyst the ratio is only about 10^3 [20].

(30)   (31)   (32), B = H_2NCH_2CN (33)
                            (34), B = H_2O

In order to try to understand why nucleophilic amine catalysts of low pK_a have such a relative advantage, we wished to study separately the two major components of process (30): the proton abstraction and the iminium ion itself. To determine whether proton abstraction from the iminium ion would, as precedent suggests [47], show a dependence on base strength, it is necessary to hold R in the iminium ion constant while varying B. This was accomplished by adding general bases, such as tertiary amines, to cyanomethyl-amine-catalyzed conversions of (5) or (6) to (7) [20,23]. Figure 6 shows the result of one set of such experiments, enhancement of the k_{AB} term. From the magnitude of these enhancements with several general bases, rate constants could be obtained for proton abstraction from the cyanomethyl-iminium ion derived from (5) or (6). These rate constants are shown in a Brönsted plot in Fig. 7. The slope of 0.5 confirms that proton abstraction from an iminium ion, like that directly from a carbonyl compound, has a substantial dependence on base strength.

This finding requires that there be an opposite, compensating effect in the iminium ion portion of (30) to account for the independence of k_{AB} from catalyst pK_a shown in Figs. 4 and 5. Such an effect had, in fact, already been observed by Hine [48] in α-dedeuteration by pyridine of a series of iso-butyraldehyde-2-d iminium ions (RNH^+=CHCDMe$_2$). As the pK_a of the catalyst (RNH_3^+) decreased, the rate of α-dedeuteration increased, with a Brönsted β of -0.4 [48]. Hine's explanation of this trend was that electron-withdrawing R groups have a stronger destabilizing effect on the reactant $R\overset{+}{N}H_3$, which has a full charge on N, than they do on the transition state for dedeuteration, in which the positive charge is dispersed [48].

In the present work, direct detection of this effect could be accomplished using the k_A terms, which depend on [RNH_3^+]. The k_A terms, as noted earlier, can be ascribed to nucleophilic amine catalysis in which water acts as the general base for proton abstraction from an iminium ion, as shown

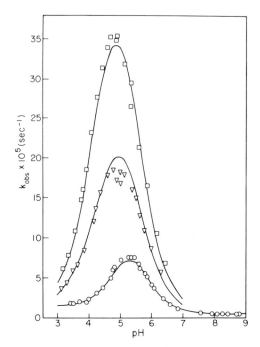

Fig. 6 Three pH–rate profiles for the conversion of (6) to (7) catalyzed by 0.1 M cyanomethylamine (CMA) buffer (○), by 0.1 M CMA plus 0.2 M N,N-dimethylcyano-methylamine (DCMA) (▽), and by 0.1 M CMA plus 0.4 M DCMA (□). The data were taken from Hupe *et al*. [23]. The lines were calculated as the sum of the catalytic terms for CMA and DCMA plus 1.15×10^{-3} [HCMA⁺][DCMA] = k_{AB}[HCMA⁺][DCMA]. Reprinted with permission from *J. Am. Chem. Soc.* **95**, 2271 (1973). Copyright by the American Chemical Society.

in (34). In essence, a study of k_A as a function of catalyst pK_a provides a means to vary R while holding B constant. Because k_A becomes too small to measure reliably with amines having $pK_a > 8$, the evidence is fairly sketchy, but, as shown in Fig. 8, the limited data available do yield a Brön-sted β of -0.5 [23].

Recently, we have used the *trans*-fused substrates (22) and (23) in analogous studies of nucleophilic-amine-catalyzed conversion to (7). The pattern of catalysis is identical to that described for the elimination reactions of (5) and (6), except that with (22) and (23) low-pK_a amines appear to be even slightly better catalysts by the k_{AB} pathway than high-pK_a amines [49].

It is instructive to try to analyze how the pH, the pK_a of the nucleophilic amine (pK_a^N), and the pK_a of the general base (pK_a^B) should be related in order to achieve maximum catalytic effectiveness in a process such as (30).

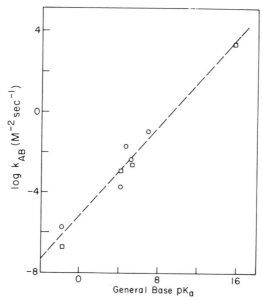

Fig. 7 Brönsted plot of the logarithms of the third-order rate constants, k_{AB}, obtained from $k_{AB}[\text{HCMA}^+][\text{B}]$ terms measured for the conversion of (**5**) (○) and (**6**) (□) to (**7**) vs. the pK_a values of the general bases. The data were taken from Hupe *et al.* [20,23]. The slope of the line is 0.5. Reprinted with permission from *J. Am. Chem. Soc.* **95**, 2271 (1973). Copyright by the American Chemical Society.

Our attempt to do this [23] can be briefly summarized as follows. The rate of process (**30**) is $k_{obs} = k_{AB'}[\text{RNH}_3^+][\text{B}]$, since the transition state requires both protonated nucleophilic amine (RNH_3^+), in rapid prior equilibrium with iminium ion, and free base (B). The product $[\text{RNH}_3^+][\text{B}]$ will be at a maximum at a pH halfway between the pK_a of RNH_3^+ and that of BH^+; i.e., catalysis will be optimized by operating at $pH_{max} = [pK_a^N + pK_a^B]/2$. The ratio of $[\text{RNH}_3^+]$ to [B] at pH_{max} is also important, and this clearly depends on the relationship between pK_a^N and pK_a^B. If $pK_a^N > pK_a^B$ there will be more of the desired combination of RNH_3^+ and B at pH_{max} than if $pK_a^B > pK_a^N$, which would favor RNH_2 and BH^+.

In addition to such considerations about prototropic equilibria, the implications of the Brönsted values for RNH_3^+ and B in process (**30**) must also be taken into account. Greater catalytic effectiveness through a larger value of rate constant $k_{AB'}$ would be obtained by having a high-pK_a general base and a low-pK_a nucleophilic amine, i.e., by having $pK_a^B > pK_a^N$, rather than $pK_a^N > pK_a^B$. When these countervailing influences of the relationship between pK_a^N and pK_a^B at pH_{max} were combined in the cases of (**5**) and (**6**),

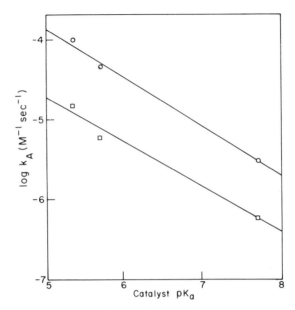

Fig. 8 Brönsted plot of the logarithms of the second-order rate constants, k_A, for the catalysis by protonated primary amines of the conversion of (**5**) (○) and (**6**) (□) to (**7**) vs. their pK_a values. The data were taken from Hupe *et al.* [20,23]. The slope of the line is approximately -0.5. Reprinted with permission from *J. Am. Chem. Soc.* **95**, 2271 (1973). Copyright by the American Chemical Society.

where the Brönsted coefficients describing the dependence of B and RNH_3^+ are 0.5 and -0.5, respectively, the conclusion was reached that optimum catalysis can be achieved at any pH, when pH = pK_a^N = pK_a^B [23].

In cases where the Brönsted coefficients for B and RNH_3^+ are not 0.5 and -0.5, respectively, the situation is more complex, and the relationship pH = pK_a^N = pK_a^B for maximum catalytic effectiveness will not be precise. It is unlikely, however, that the changes in the Brönsted coefficients will be large enough to invalidate this relationship as at least an approximate criterion for the design of a good nucleophilic amine catalyst for α deprotonation.

It is reasonable to assume that enzymes which catalyze reactions via iminium ion formation have evolved essentially in conformity to such a guideline. One would therefore predict that enzymatic nucleophilic amines and general bases should have pK_a values around 6–7. The imidazole moiety of a histidine residue, which has been suggested as a general base in reactions of this type [5], has a pK_a of 6.05 in the parent amino acid [50] and should therefore be well suited to serve as a general base at physiological pH. On the other hand, the usual enzymatic nucleophile, the primary ε-amino group of a lysine residue, has a pK_a of 10.53 in the parent amino acid [50]. Westheimer has found, however, that the imine-forming lysine residue in acetoacetate decarboxylase actually has a pK_a of 6.0 [51], and such lowering of pK_a may be anticipated for other enzymatic nucleophilic amines.

It is also of interest to try to compare the catalytic effectiveness observed in model systems with that of enzymes effecting the same kinds of reaction. Such comparisons always involve a number of assumptions, and the natural tendency is to adopt ones that make the model system look relatively favorable. Nonetheless, simple nucleophilic amine catalysts can be rather impressive as enzyme models. For example, Westheimer [1] has estimated that cyanomethylamine accelerates the decarboxylation of acetoacetate by 10^6 as compared to an acceleration of 10^9 by acetoacetate decarboxylase.

In the context of α-proton abstraction, there have been several comparisons of enzymatic with model systems [23,37,39,52]. Reported rate constants for enzyme-catalyzed α-proton exchange range from about 10^1 sec^{-1} for acetoacetate decarboxylase [52] to about 10^2–10^3 sec^{-1} for aldolases [37,52,53]. Hine [39], in what he acknowledges to be a particularly favorable comparison of one of his best bifunctional catalysts with acetoacetate decarboxylase, estimated that the enzyme is only about 10^2 times more effective for dedeuteration of acetone. For the same process, Westheimer [52] calculated that cyanomethylamine is 10^4–10^5 times less effective than acetoacetate decarboxylase.* We concluded [23] that an effective concentration of general base of 10^4–10^5 M would be required to make cyanomethylamine as effective as the aldolases.

Given the different sources of data and the various assumptions used in formulating these estimates, they are encouragingly consistent. If one compares our value of k_{AB} for cyanomethylamine (4.1×10^{-3} M^{-2} sec^{-1}) with the second-order rate constant that Hine [39] found with his bifunctional catalyst, making appropriate adjustments for an isotope effect and a temperature difference, it can be calculated that Hine's catalyst functions with an effective concentration of intramolecular general base of about 10^2 M. An additional factor of 10^2 [39] or 10^3 would produce enzymatic effectiveness, and the resulting effective concentration of 10^4–10^5 M is in line with our estimate [23]. As discussed earlier in the section on approximation, an effective concentration of this magnitude is not unreasonable, although such a value has yet to be attained in a model system for α-deprotonation of carbonyl compounds.

---

* Westheimer's experiments were run at pH 5.90, which does not show cyanomethylamine quite at its catalytic best, as demonstrated in the present work.

Finally, as an illustration of the power of simple, low-pK_a amines as nucleophilic catalysts, some of our recent results [54] with an amine having a pK_a even lower than cyanomethylamine can be cited. Such amines are of interest because they would permit further testing of Brönsted relationships like those shown in Figs. 5 and 8, and because they might allow us for the first time to detect a rate-limiting step other than α-deprotonation in the overall iminium ion pathway depicted in Scheme 2.

The first new low-pK_a amine to be studied is 3,3,4,4-tetrafluoropyrrolidine (TFP) (35), which was synthesized by reduction of 3,3,4,4-tetrafluoro-succinimide [55] with borane in tetrahydrofuran [54]. 3,3,4,4-Tetrafluoro-pyrrolidine has a pK_a of 4.05 and, like its unfluorinated analog mentioned at the beginning of this chapter, it is exceptionally good at condensing with carbonyl groups. In fact, when TFP is used as a catalyst for the conversion of (23) to (7), enimmonium ion (36) ($\varepsilon_{286}^{H_2O}$ 29,000) is formed in significant equilibrium concentration even at low catalyst concentrations [54]. The kinetics observed are consistent with Scheme 4, in which $k_1 < k_2$ or k_3, and the equilibrium constant $K = [36] [H_2O]/[7] [35 \cdot HCl]$ is about 10^2 [54].

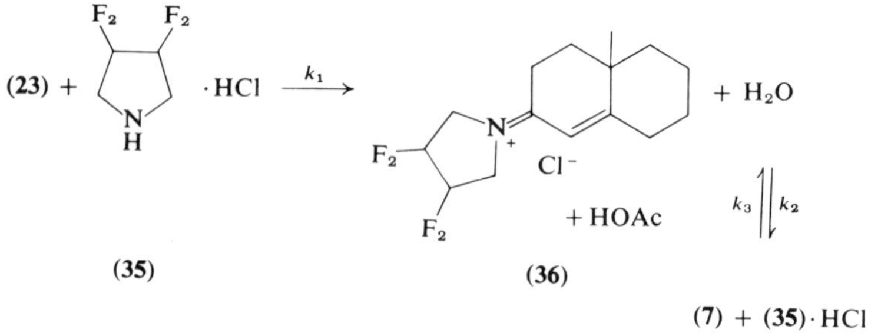

Scheme 4

Formation of (7) from (23) is catalyzed by TFP in exactly the same manner as described above for all the primary amines studied. The familiar k_{AB} and k_A terms were observed, and a large invariant primary kinetic isotope effect was found with (25), establishing that, as usual, α-proton abstraction from the immonium ion derived from (23) is the rate-determining step [54]. The value of k_A for TFP (9.5×10^{-4} M^{-1} sec^{-1}) is significantly larger than any other k_A previously measured, as would be predicted from Fig. 8. The value of k_{AB} for TFP (1.0×10^{-2} M^{-2} sec^{-1}) also is the largest we have yet found.

If these rate constants are translated into reaction times for α-deprotona-tion, the practical potential of low-pK_a nucleophilic amine catalysts becomes more readily apparent. The reaction of (23) with 1 M TFP buffer at pH 4 has a half-time of approximately 4 min. From another perspective, α-proton abstraction under these conditions occurs more than 600 times via immonium

(37), B = TFP or H$_2$O   (38)

ion formation [as in (37)] for every time it occurs by direct general base attack on the carbonyl compound [as in (38)].

ACKNOWLEDGMENTS

The customary tribute to collaborators, indicating that they were indispensable both experimentally and intellectually, is even more than usually true with respect to the research described in this article. The author's background is in organic synthesis, and he has relied very heavily on the able and dedicated associates whose names appear in the cited references, particularly in sorting out some challenging kinetic complexities. Thanks are also due to numerous colleagues at Dartmouth and other institutions for valuable help and advice along the way. This research was supported by PHS Grant AM 11815 and, more recently, by NSF Grants GP-34390X and CHE-7502737. A fellowship from the Alfred P. Sloan Foundation also provided important initial support.

REFERENCES

1. F. H. Westheimer [*Search* 1, 34 (1970)] provides a summary of his investigations of acetoacetate decarboxylase in this publication of his Rivett lecture.
2. C. Y. Lai, O. Tchola, T. Cheng, and B. L. Horecker, *J. Biol. Chem.* 240, 1347 (1965).
3. R. G. Rosso and E. Adams, *J. Biol. Chem.* 242, 5524 (1967).
4. O. M. Rosen, P. Hoffee, and B. L. Horecker, *J. Biol. Chem.* 240, 1517 (1965).
5. D. Portsmouth, A. C. Stoolmiller, and R. H. Abeles, *J. Biol. Chem.* 242, 2751 (1967).
6. D. L. Nandi and D. Shemin, *J. Biol. Chem.* 243, 1236 (1968).
7. R. Jeffcoat, H. Hassall, and S. Dagley, *Biochem. J.* 115, 977 (1969).
8. J. R. Butler, W. L. Alworth, and M. J. Nugent, *J. Am. Chem. Soc.* 96, 1617 (1974).
9. F. H. Westheimer, *Ann. N.Y. Acad. Sci.* 39, 401 (1940).
10. J. Hine and W.-S. Li, *J. Am. Chem. Soc.* 97, 3550 (1975), and previous papers in this series.
11. See M. Brault, R. M. Pollack, and C. L. Bevins, *J. Org. Chem.* 41, 346 (1976), and references cited therein for a particularly pertinent recent series of investigations.

12. D. J. Hupe, Ph.D. Dissertation, Dartmouth College, Hanover, New Hampshire (1972), provides a review of pertinent references to that date.
13. T. A. Spencer and K. K. Schmiegel, *Chem. Ind. (London)* p. 1765 (1963), reporting an observation of M. A. Schwartz.
14. T. A. Spencer, K. K. Schmiegel, and K. L. Williamson, *J. Am. Chem. Soc.* **85**, 3785 (1963).
15. T. A. Spencer, K. K. Schmiegel, and W. W. Schmiegel, *J. Org. Chem.* **30**, 1626 (1965).
16. T. A. Spencer, H. S. Neel, D. C. Ward, and K. L. Williamson, *J. Org. Chem.* **31**, 434 (1966).
17. T. A. Spencer, H. S. Neel, T. W. Flechtner, and R. A. Zayle, *Tetrahedron Lett.* p. 3889 (1965).
18. T. A. Spencer and L. D. Eisenhauer, *J. Org. Chem.* **35**, 2632 (1970).
19. e.g., G. Stork, A. Brizzolara, H. Landesman, J. Szmuszkovicz, and R. Terrell, *J. Am. Chem. Soc.* **85**, 207 (1963).
20. D. J. Hupe, M. C. R. Kendall, and T. A. Spencer, *J. Am. Chem. Soc.* **94**, 1254 (1972).
21. W. P. Jencks, "Catalysis in Chemistry and Enzymology," pp. 201–202. McGraw-Hill, New York, 1969; cf., however, A. F. Hegarty and W. P. Jencks, *J. Am. Chem. Soc.* **97**, 7188 (1975).
22. See T. A. Spencer, M. C. R. Kendall, and I. D. Reingold [*J. Am. Chem. Soc.* **94**, 1250 (1972)] for a study of general base catalyzed proton abstraction which serves as a guide to the behavior of various tertiary amines.
23. D. J. Hupe, M. C. R. Kendall, and T. A. Spencer, *J. Am. Chem. Soc.* **95**, 2271 (1973).
24. D. J. Hupe, M. C. R. Kendall, G. T. Sinner, and T. A. Spencer, *J. Am. Chem. Soc.* **95**, 2260 (1973).
25. R. A. More O'Ferrall and P. J. Warren, *J. Chem. Soc., Chem. Commun.* p. 483 (1975), references therein, and private communication from Dr. More O'Ferrall. For a review which discusses irreversible carbanion (ElcB) vs. E2 elimination mechanisms, see W. H. Saunders, Jr., *Acc. Chem. Res.* **9**, 19 (1976).
26. H. E. Ferran, Jr., D. A. Drake, and T. A. Spencer, *J. Org. Chem.* **40**, 2017 (1975).
27. Analogous intermediates have also recently been identified in amine catalyzed isomerization of β,γ-unsaturated ketones to their α,β-unsaturated isomers: R. H. Kayser and R. M. Pollack, *J. Am. Chem. Soc.* **97**, 952 (1975); W. F. Benisek and A. Jacobson, *Bioorg. Chem.* **4**, 41 (1975).
28. R. Wolfenden, *Acc. Chem. Res.* **5**, 10 (1972).
29. G. E. Lienhard, *Science* **180**, 149 (1973).
30. J. O. Edwards and R. G. Pearson, *J. Am. Chem. Soc.* **84**, 16 (1962).
31. T. A. Spencer and C. W. Leong, *Tetrahedron Lett.* p. 3889 (1975).
32. R. M. Pollack and M. Brault, *J. Am. Chem. Soc.* **98**, 247 (1976).
33. W. P. Jencks, "Catalysis in Chemistry and Enzymology," Chapter 1. McGraw-Hill, New York, 1969.
34. A. Dafforn and D. E. Koshland, Jr., *Proc. Natl. Acad. Sci. U.S.A.* **68**, 2463 (1971).
35. M. I. Page and W. P. Jencks, *Proc. Natl. Acad. Sci. U.S.A.* **68**, 1678 (1971).
36. W. P. Jencks and M. I. Page [*Biochem. Biophys. Res. Commun.* **57**, 887 (1974)] provide a concise and entertaining summary comparison of the "orbital steering" and "entropy loss" approaches to this problem.
37. See, e.g., M. L. Bender and A. Williams, *J. Am. Chem. Soc.* **88**, 2502 (1966).
38. J. Hine, F. E. Rogers, and R. E. Notari, *J. Am. Chem. Soc.* **90**, 3279 (1968).
39. J. Hine, M. S. Cholod, and R. A. King, *J. Am. Chem. Soc.* **96**, 835 (1974).
40. R. C. Crain, D. A. Drake, and T. A. Spencer, unpublished results.

41. E. J. Corey and R. A. Sneen, *J. Am. Chem. Soc.* **78**, 6269 (1956); G. Subrahmanyam, S. K. Malhotra, and H. J. Ringold, *ibid.* **88**, 1332 (1966); G. B. Trimitsis and E. M. Van Dam, *J. Chem. Soc., Chem. Commun.* p. 610 (1974); see, however, F. G. Bordwell and R. G. Scamehorn, *J. Am. Chem. Soc.* **90**, 6749 (1968).
42. Compound **22** was first prepared by another method by H. B. Henbest and J. McEntee, *J. Chem. Soc.* p. 4478 (1961).
43. H. E. Ferran, Jr., R. D. Roberts, and T. A. Spencer, unpublished results.
44. J. Hine, J. G. Houston, J. H. Jensen, and J. Mulders, *J. Am. Chem. Soc.* **87**, 5050 (1965); see, R. P. Bell, "The Proton in Chemistry," 1st ed., p. 172. Cornell Univ. Press, Ithaca, New York, 1959, for a tabular summary of such β's.
45. R. P. Bell and A. F. Trotman-Dickenson, *J. Chem. Soc.* p. 1288 (1949); R. P. Bell and G. L. Wilson, *Trans. Faraday Soc.* **46**, 407 (1950); J. Hine and J. Mulders, *J. Org. Chem.* **32**, 2200 (1967).
46. J. Hine, B. C. Menon, J. H. Jensen, and J. Mulders, *J. Am. Chem. Soc.* **88**, 3367 (1966).
47. J. Hine, J. Mulders, J. G. Houston, and J. P. Idoux, *J. Org. Chem.* **32**, 2205 (1967).
48. J. Hine, B. C. Menon, J. Mulders, and J. P. Idoux, *J. Org. Chem.* **32**, 3850 (1967).
49. R. D. Roberts, H. E. Ferran, Jr., and T. A. Spencer, unpublished results.
50. D. D. Perrin, "Dissociation Constants of Organic Bases in Aqueous Solution." Butterworth, London, 1965.
51. P. A. Frey, F. C. Kokesh, and F. H. Westheimer, *J. Am. Chem. Soc.* **93**, 7266 (1971); F. C. Kokesh and F. H. Westheimer, *ibid.* p. 7270.
52. W. Tagaki and F. H. Westheimer, *Biochemistry* **7**, 901 (1968).
53. M. A. Roseman, Ph.D. Dissertation, Michigan State University, East Lansing (1970).
54. R. D. Roberts and T. A. Spencer, unpublished results.
55. A. L. Henne and W. F. Zimmer, *J. Am. Chem. Soc.* **73**, 1103 (1951).

14

Fatty Acid Synthetase Complexes

James K. Stoops, Michael J. Arslanian,
John H. Chalmers, Jr., V. C. Joshi, and Salih J. Wakil

INTRODUCTION

The *de novo* synthesis of fatty acids consists of two major steps in the conversion of acetate (or acetyl coenzyme A) to a long-chain fatty acid (mainly palmitate). The first step is the conversion of acetyl-CoA to malonyl-CoA [1], a reaction catalyzed by acetyl-CoA carboxylase [reaction (1)]. The second step [reaction (2)] is the conversion of acetyl-CoA and malonyl-CoA in the presence of NADPH to palmitate [2]. The latter reaction is catalyzed by a multienzyme system referred to as fatty acid synthetase (FAS).

$$CH_3COSCoA + CO_2 + ATP \rightleftharpoons \overset{\displaystyle COOH}{\overset{\displaystyle |}{CH_2COSCoA}} + ADP + P_i \qquad (1)$$

$$CH_3COSCoA + 7\overset{\displaystyle COOH}{\overset{\displaystyle |}{CH_2COSCoA}} + 14NADPH + 14H^+ \longrightarrow$$
$$CH_3(CH_2)_{14}COOH + 8CoASH + 14NADP^+ + 7CO_2 + 6H_2O \qquad (2)$$

This review is limited to discussions of the properties of fatty acid synthetase and does not cover the acetyl-CoA carboxylase (for review, see Wakil [3] and Prescott and Vagelos [4]).

The conversion of acetyl-CoA and malonyl-CoA to palmitate involves numerous sequential reactions and acyl intermediates. The nature of these reactions and the intermediates involved became known primarily from studies of fatty acid synthesis in cell-free extracts of *Escherichia coli* [3,4].

The isolation of the individual enzymes that comprise the fatty acid synthesis system in *E. coli* has resulted in the delineation of the biosynthetic pathway. In addition to the isolated enzymes, a pantetheine-containing protein (acyl carrier protein, ACP) was identified. This protein carries the acyl intermediates in fatty acid biosynthesis and acts, therefore, as a coenzyme. The following are the enzymes and reactions resulting in the synthesis of palmitate, the major product:

Acetyl transacylase

$$CH_3COSCoA + ACPSH \rightleftharpoons CH_3COSACP + CoASH \tag{3}$$

Malonyl transacylase

$$\underset{\substack{| \\ CH_2COSCoA}}{COOH} + ACPSH \rightleftharpoons \underset{\substack{| \\ CH_2COSACP}}{COOH} + CoASH \tag{4}$$

Condensing enzyme (β-ketoacyl-ACP synthetase)

$$CH_3COSACP + EnzSH \rightleftharpoons CH_3COSEnz + ACPSH$$

$$\underset{\substack{| \\ CH_3COSEnz + CH_2COSACP}}{COOH} \longrightarrow CH_3COCH_2COSACP + CO_2 + EnzSH \tag{5}$$

β-ketoacyl-ACP reductase

$$CH_3COCH_2COSACP + NADPH + H^+ \rightleftharpoons$$
$$CH_3CHOHCH_2COSACP + NADP^+ \tag{6}$$

β-Hydroxyacyl-ACP dehydratase

$$CH_3CHOHCH_2COSACP \rightleftharpoons CH_3CH{=}CHCOSACP + H_2O \tag{7}$$

Enoyl-ACP reductase

$$CH_3CH{=}CHCOSACP + NADPH + H^+ \rightleftharpoons$$
$$CH_3CH_2CH_2COSACP + NADP^+ \tag{8}$$

The acetyl and malonyl transacylases [reactions (3) and (4)] catalyze the transfer of acetyl and malonyl groups from their respective CoA derivatives to ACP, respectively. The mechanism of this transfer involves the intermediate formation of an acyl enzyme. An active seryl residue on each enzyme is proposed as being involved in the formation of these intermediates. The acetyl or malonyl groups are then transferred from their respective enzymes to ACP to yield acetyl-ACP or malonyl-ACP. In the presence of the condensing enzyme [reaction (5)] the latter substrates are coupled to form acetoacetyl-ACP. This reaction occurs in two steps. First is the formation of acetyl-enzyme, in which an active thiol group is involved. The acetyl group is then coupled to carbon 2 of the malonyl-ACP, yielding

β-ketoacyl-ACP and CO_2. The β-ketoacyl-ACP is reduced by NADPH to form the D-β-hydroxyacyl-ACP [reaction (6)]. The dehydratase removes the elements of water from the β-hydroxyacyl-ACP [reaction (7)] to yield α,β-unsaturated acyl-ACP, which is reduced by NADPH to form butyryl-ACP. The butyryl-ACP produced is elongated through the sequential repetition of reactions (5)–(8) six more times, yielding palmitoyl-ACP. The latter is either hydrolyzed by a specific thioesterase to yield palmitate plus ACP [reaction (9)] or utilized directly in the synthesis of phosphatidic acid by transacylation of glycerol phosphate.

$$CH_3(CH_2)_{14}COSACP + H_2O \longrightarrow CH_3(CH_2)_{14}COOH + ACPSH \qquad (9)$$

Studies of the individual enzymes were carried out in our laboratory and that of Vagelos; they have been extensively reviewed [3,4] and are not considered here. This review is primarily concerned with the fatty acid synthetases of both animal tissues and yeast. We have considered only the more recent developments in this field.

THE MULTIENZYME COMPLEXES OF FATTY ACID SYNTHESIS

The various synthetase systems have been classified according to their physicochemical behavior *in vitro* in two groups (Table 1) [5]. The type I class represents synthetases whose enzymes aggregate and purify together in tight complexes. In the type II group, the enzymes may be loosely associated and can be readily separated by conventional procedures. The fatty acid synthetase system can be reconstituted by the addition of ACP to a mixture containing all the individual enzymes catalyzing reactions (3)–(9). Even though the plant synthetases are included in this group, their inclusion should be considered tenuous since it is mainly based on the isolation of an ACP component and its requirement for fatty acid synthesis [31]. As will be seen later, the isolation of an ACP component may not serve as a reliable indicator as to the structure of the synthetase.

The type I synthetases are multifunctional proteins containing the 4'-phosphopantetheine moiety and catalyzing reactions analogous to those depicted in Eqs. (3)–(9). This group of FAS complexes can be subdivided into two groups. Type IA has a molecular weight of approximately 500,000. The type IB group, on the other hand, has molecular weights in the range of 1.0–2.3×10^6 and appears to consist of aggregates of lower molecular weight subunits. It should be noted that the molecular weight of many of these

TABLE 1

Classification of Fatty Acid Synthetase

Type	Source	Molecular weight	Major product	References
IA: Multifunctional proteins	Mammalian liver (rat; dog; man)	4–5×10^5	Palmitate	6; 7; 8; 9
	Avian liver (pigeon; chicken)	4–5×10^5	Palmitate	6; 7, 10
	Mammary gland (rat; rabbit[a]; cow; guinea pig)	4–5×10^5	Palmitate	11; 12; 13, 14; 15
IB: Aggregated multifunctional proteins	Fungi (yeast; *Neurospora*; *Penicillium*)	2.3×10^6	Palmitoyl-CoA	16; 17; 18
	Higher bacteria (*Myco-bacterium*; *Corynebacterium*; *Streptomyces*)	$>1 \times 10^6$	Acyl-CoA,[b] long-chain fatty acids	5; 19; 20
	Algae (etiolated *Euglena*)	$\sim 10^6$	Palmitate	21
II: Unassociated proteins	Lower bacteria (*E. coli*; *Clostridium* sp.; *Bacillus*; *Pseudomonas*)		Palmitate, acyl-ACP	22–24; 25; 5; 22
	Plants and algae (*Chlamydomonas*; avocado; lettuce; photosynthesizing *Euglena*)		Acyl-ACP	26; 27; 28; 29, 30

[a] The molecular weight of the synthetase of rabbit mammary gland was reported to be 910,000 [12].

[b] The products of many of the bacterial, plant, and algal FAS systems have not been completely characterized.

complexes has been estimated by gel filtration, sucrose gradient centrifugation, and sedimentation velocity, and therefore the value of the molecular weight reported is uncertain. The product of type IA FAS is free palmitic acid, whereas type IB synthetases yield primarily acyl-CoA derivatives and may require FMN for their activities.

With one possible exception, the synthetases that have been obtained from animal tissues are of type IA. The synthetase of rabbit mammary gland has been reported to have a molecular weight of 910,000 [12], which is in contrast to the value of about 500,000 obtained for the enzyme isolated from the mammary gland of the cow [13,14], rat [32], and guinea pig [15]. It seems that a more detailed study of these enzymes is warranted before a conclusion can be reached regarding any differences in their size.

Reaction Mechanism

The overall reaction catalyzed by the type I FAS complexes is essentially the same as that shown in reaction (2). Although the intermediate acyl derivatives and the enzymes involved in the individual reactions were not isolated and identified, the similarity between the overall reaction catalyzed by these complexes and that of *E. coli* is obvious. The similarities between the reactions catalyzed by the type II FAS of *E. coli* and the type I FAS have been determined by the use of model substrates. These studies have included the isolation of acyl enzyme intermediates and the determination of some of the functional groups at the active sites [3,4].

The FAS complexes were reported to contain about 1 mole of 4'-phosphopantetheine per mole of enzyme (yeast FAS contains 3.5–6.0 moles per mole of enzyme). The 4'-phosphopantetheine is apparently bound to the protein via a phosphodiester linkage to serine as in *E. coli* ACP [33].

Studies have been made to determine the binding sites of the type I synthetases using the pigeon liver and yeast complexes. The acetyl and malonyl enzymes have been isolated by reacting the pigeon enzyme with acetyl- and malonyl-CoA. Joshi, Plate, and Wakil [34] found from extrapolating the acetyl-CoA or malonyl-CoA to infinite concentration that the enzyme binds 1.4 and 0.64 moles of acetate and malonate per mole of enzyme based on a molecular weight of 450,000. They found that the bound acetate and malonate would participate in fatty acid synthesis. All the [14C]acetate or [14C]malonate derived from the acyl enzyme appeared in fatty acids after incubating the labeled enzyme with the appropriate CoA ester and NADPH. In order to distinguish between thiol and nonthiol binding, performic acid oxidation was performed on the acyl enzyme. Thirty-three and 43% of the acetyl and malonyl groups, respectively, were stable to performic acid oxidation. They proposed that two of the acetyl and one of the malonyl moieties are bound to the enzyme by a thioester linkage. That there are thiol and nonthiol binding sites is supported by the finding that N-ethylmaleimide could be used to selectively block the binding to the thiol sites without affecting the binding to the sites stable to performic acid oxidation.

An attempt was made to determine the groups to which the acetyl and malonyl groups are bound. Peptic digestion and Dowex chromatography of the [^{14}C]acetyl and [^{14}C]malonyl enzymes yielded one major peak in each case containing over 90% of the radioactivity. It was found that 94 and 83% of the radioactivity of the acetyl and malonyl peptides, respectively, were unstable to performic acid oxidation. Since both fractions contained acetate or malonate and pantothenate it was suggested that the 4'-phosphopantetheine is covalently linked to both of these moieties. In addition, a peptide containing a small amount of the total bound acetate (less than 5%) was isolated and on further proteolysis yielded O-acetylserine. From these results it was proposed that acetyl and malonyl groups form O-acylserine linkages with the transacylases and a thioester linkage with the 4'-phosphopantetheine.

Porter's group [35] has reached similar conclusions regarding the nature of the binding sites that comprise the pigeon liver enzyme. They found that the enzyme binds 3 moles of acetate and 1 mole of malonate [36]. The [^{14}C]acetyl- or [^{14}C]malonyl-labeled enzyme was subjected to pepsin hydrolysis followed by Dowex chromatography. A fraction containing a small amount of the initial radioactivity was further fractionated into two radioactive components by high-voltage paper electrophoresis and analyzed for amino acids, acetate, malonate, β-alanine, and taurine content. Acetate or malonate, β-alanine, and taurine were found in equal amounts and, since cysteine was absent, it was concluded that both acetate and malonate are bound to the 4'-phosphopantetheine. Moreover, when the enzyme was incubated with acetyl- and malonyl-CoA and subjected to the same manipulations, a peptide fraction containing 12% of the radioactivity in acetoacetate was isolated from this fraction [37]. Although the nature of the binding of acetoacetate to the peptide was not reported, it was proposed that it was bound to the 4'-phosphopantetheine. The proposal that acetate and malonate are bound to the 4'-phosphopantetheine is further supported by the isolation of S-carboxymethylcysteamine from HCl hydrolyzate of the peptide after treating the acetyl and malonyl peptide with iodoacetamide and then hydroxylamine followed by ^{14}C-labeled iodoacetamide [35]. Another peptide containing acetate which was performic acid labile was isolated and found to contain cysteine as its only thiol-containing moiety. Malonate was not found to be associated with this peptide.

Peptides containing acetate and malonate which were not labile to performic acid oxidation were also isolated and found to contain serine but no β-alanine or taurine. However, the residues to which the acyl group was bound were not identified. From these findings, it was proposed that acetate and malonate and the condensation product of these molecules (acetoacetate) are bound to the 4′-phosphopantetheine and that the acetyl and malonyl groups form covalent linkages with serine.

Lynen and co-workers [38,39] have reached a similar conclusion regarding the binding of acetyl and malonyl groups to the yeast enzyme. They blocked the free SH groups with Ellman's reagent without affecting the transacylase activities and then labeled the protein with acetate and malonate. After proteolysis of the labeled enzyme, the amino acid sequence of the acetyl peptides was determined. The acetyl group was shown to be bound to the amino group of serine. Since alkaline hydrolysis of the [14C]acetyl peptides and the model compound pyrrolidonyl(O-[14C]acetyl)serylglycine in 8 M urea were the same, it was concluded that the N-acetylserine had resulted from an O-acetyl → N-acetyl shift at the serine residue. A [14C]malonyl peptide stable to performic acid oxidation was also obtained from [14C]-malonyl enzyme. A comparison of the partial sequence of the [14C]acetyl peptide and the amino acid composition of the [14C]malonyl peptides indicated that the malonyl and acetyl groups are bound to distinct sites.

A malonyl peptide was also isolated from [14C]malonyl enzyme and found to contain equal amounts of malonate, β-alanine, cysteamine, and organic phosphate, but no cysteine.

It was also possible to isolate a palmityl enzyme by reacting the enzyme with palmitoyl-CoA [40]. A peptide was isolated from this enzyme which contained equal molar quantities of pantothenic acid, palmitate, and β-alanine. A palmitoyl peptide stable to performic acid oxidation was also isolated. Although this peptide had an amino acid composition similar to that of the malonyl peptide stable to performic acid oxidation, it is not possible to decide whether the two sites are the same.

It should be pointed out that the assignment of the binding of the acyl groups to certain amino acid residues is tenuous. As Lynen observed, there is the possibility that the acyl group may undergo a transacylation reaction during the isolation of the peptides. This possibility makes the proposal regarding the binding sites even more uncertain when the acyl peptide isolated represents only a small fraction of the labeled peptides.

On the basis of these observations, a mechanism for fatty acid synthesis (Scheme 1) by multienzyme complexes has been proposed. The acetyl transacylase transfers the acetyl group from CoA to the active seryl residue of the enzyme and then to the thiol group of the 4′-phosphopantetheine (ACP). The acetyl group is transferred to the SH group of the active cysteine residue of the condensing enzyme. The SH group of ACP is then charged with the malonyl group via the malonyl transacylase. Condensation of the acetyl and malonyl groups follows, yielding CO_2 and acetoacetyl-ACP. The latter is then

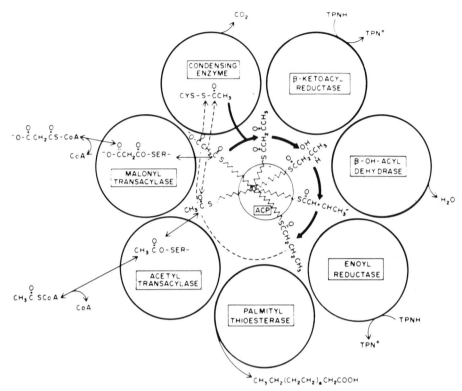

Scheme 1

reduced by NADPH, dehydrated, and reduced again to form butyryl-ACP. The butyryl group is transferred to the SH group of the cysteine residue, releasing the SH group of the pantetheine and making it available for the acceptance of another malonyl group via the malonyl transacylase. Again, condensation occurs and β-ketohexanoyl-ACP is formed. The latter acyl group is reduced, dehydrated, and reduced again to yield hexanoyl-ACP, which undergoes chain elongation by the same sequence of reactions to ultimately yield palmitoyl-ACP. In the case of yeast FAS, the palmitoyl group is transferred to CoA via palmitoyl transacylase or, as in the case of animal FAS, is hydrolyzed by the palmitoyl thioesterase to free palmitic acid. Factors determining chain elongation and chain termination are not understood. The condensing enzyme that leads to chain elongation and the thioesterase or palmitoyl transferase that terminates it may be important components in terminating fatty acids at the C_{16} and C_{18} chain length [41,42].

The primary features of this scheme are the juxtaposition of the enzymes in such a way as to carry out the catalyses sequentially and the involvement of a single 4'-phosphopantetheine moiety (ACP) and a thiol group of a cysteine residue, presumably that of the condensing enzyme.

STRUCTURE AND FUNCTIONAL ORGANIZATION OF TYPE I FAS COMPLEXES

Reversible Dissociation into Subunits

The animal synthetases readily dissociate into subunits of equal or nearly equal size. The factors affecting the dissociation of the animal synthetases are summarized in Table 2. Low ionic strength, low temperature (0°–5°C), and alkaline pH (> 8) favor the subunit form to a variable degree depending on the source of the individual synthetases [6,7,11,43–45].

The effect of NADPH on the dissociation of the complex has been investigated [46]. Apparently NADPH decreases the rate of dissociation of the pigeon liver enzyme since the enzyme at 3 mg/ml in the presence of 100 μM NADPH, pH 8.3, at 25°C exhibited no sign of dissociation after 5 hr, while the amount of complex had decreased 50% in the absence of NADPH. The enzyme was found to be less stable in the presence of either acetyl- or malonyl-CoA but the effect on the dissociation of the enzyme is unknown.

The subunits have a molecular weight that is approximately one-half the value of the complex [7,43–45], suggesting that a dimer–monomer equilibrium exists in animal FAS. However, it has not been reported that the equilibrium between the complex and its subunits is consistent with an $A_2 \rightleftharpoons 2A$ type of system [47]. Moreover, it is not possible to assess the relative effect of the factors causing dissociation of the complex since the dissociation constants have not been reported.

TABLE 2

Conditions Affecting Dimer–Monomer Transition of Animal Fatty Acid Synthetase

Dimer ($\sim$ 480,000 daltons, active) $\rightleftharpoons$ Monomer (240,000 daltons, inactive)

High ionic strength	Low ionic strength
pH 6.5–7.5	pH 8.3–9.5
Room temperature, 25°C	Low temperature, 0°–5°C
High enzyme concentration	Low enzyme concentration
NADPH	—

The dissociation studies of the fatty acid synthetase complexes indicate that the monomer is inactive and that the loss of activity precedes the dissociation of the enzyme [48]. Therefore, it cannot be concluded that dissociation per se results in loss of enzyme activity. For example, after the pigeon enzyme was incubated in 500 mM Tris–glycine (pH 8.35), approximately 85% of the enzyme was in the complex form while the activity had decreased to 10% of original [44].

The reassociation and the total restoration of enzyme activity of the complex may be affected by increasing the salt concentration, raising the temperature from 0° to 25°C, or decreasing the pH to near neutrality [48]. It appears that inactive complex formation may precede restoration of enzyme activity. When reactivation of the enzyme activity was performed at 6°C in 0.2 M KCl, 0.74 mM Tris, 35 mM glycine, 1 mM EDTA, 10 mM DTT (dithiothreitol) at pH 8.3, the rate constant of reactivation was first order and did not vary more than 2-fold over the protein concentration range 0.025–2 mg/ml [48].

Studies of the yeast synthetase [49] indicate that it is an aggregate of 500,000 molecular weight subunits which in turn consist of 250,000 molecular weight subunits. Thus, when the enzyme was subjected to freezing and thawing in the presence of 1 M NaCl or LiCl, sucrose density gradient centrifugation showed two peaks corresponding to an apparent molecular weight of approximately 500,000 and 250,000. These values were estimated using marker proteins, and more reliable methods are required to establish the size of these subunits. Further, it was found that by lowering the salt concentration by dilution both the high molecular weight aggregate and activity were restored.

Structural Properties and Enzymatic Activities of the Subunits

Even though it has been proposed that the subunits are not identical, until recently there has been little evidence to support this contention. Yun and Hsu [45] have reported that the chicken liver enzyme dissociates into two nonidentical subunits in Tris–glycine buffer based on the finding of two protein bands after Tris–glycine gel electrophoresis. Attempts to separate the subunits by chromatography have been unsuccessful. The bovine mammary fatty acid synthetase was reported to dissociate into nonidentical subunits. No evidence was presented to support this claim [43]. In contrast to the reported separation of the subunits of the chicken enzyme, Porter's group [50] has found that the pigeon liver enzyme dissociates into nonidentical subunits that are not separated by Tris–glycine gel electrophoresis. However, it was possible to separate these subunits using a column consisting of a pantetheine derivative of Sepharose. The enzyme preparation was first stored at $-20°C$ for 6 weeks and then subjected to chromatography. At $0°C$, pH 8.4, 45% of the protein was eluted from the column (fraction 1). Raising the temperature to $25°C$ and pH to 10 resulted in the elution of a second fraction consisting of 12% of the protein. Improved yields of fraction 1 (51%) and fraction 2 (33%) were obtained if the enzyme was incubated with acetyl-CoA before chromatography was performed. The relative mobility of the protein in the two fractions was the same on disc gel electrophoresis performed at pH 7. The physicochemical evidence that the two fractions consist of different proteins is based on the finding that fraction 1 contains 8 SH groups per mole while fraction 2 contains 60, as determined by Ellman titration. The native enzyme is reported to contain 67 SH groups per mole. Fraction 1 contained the 4'-phosphopantetheine moiety, the ketoacyl reductase [Eq. (6)], and the enoyl reductase [Eq. (8)], while fraction 2 contained the acetyl and malonyl transacylase activities [Eqs. (3) and (4)]. Both fractions contained the dehydratase [Eq. (7)] and palmitoyl thioesterase [Eq. (9)] activities. Fractions 1 and 2 were designated subunits I and II, respectively. The specific activities of the partial activities were twice that found for the dissociated complex, except in the case of the two activities common to both subunits, the dehydratase and thioesterase, which had the same specific activity found in the complex. This result supports the claim that the two subunits of nearly equal size have been separated. Recombination of the subunits in equimolar amounts resulted in a 64% restoration of the original fatty acid synthetase activity.

The recent finding that pigeon fatty acid synthetase preparations contain a complex that does not contain the 4'-phosphopantetheine (apoenzyme) [51] may obscure the interpretation of the results regarding the separation of the two fractions. Thus, the possibility exists that the isolated fractions 1 and 2 consist of the subunits of holoenzyme and apoenzyme. However, the unequal distribution of SH groups and partial enzyme activities found in the two fractions appear to negate this possibility. Even so, the interpretation of these studies would be more straightforward if the apoenzyme were removed from the mixture before chromatography is performed.

The dissociation of the pigeon fatty acid synthetase into subunits has little effect on specific activity and K_m of the individual activities except for the condensing enzyme [50,52], which exhibits no activity after the dissociation of the complex is complete. The condensation activity was measured by determining the incorporation of ^{14}C into malonyl-CoA (reverse reaction) in an assay containing $NaH^{14}CO_3$, malonyl-CoA, caproyl-CoA, and CoA. The conclusion that the condensation activity is lost on dissociation of the complex would be more convincing if the rate of condensation of acetyl- and malonyl-CoA were measured. The loss of the condensing enzyme activity is supported by the finding that the dissociated enzyme when incubated with

[^{14}C]acetyl-CoA and treated with pepsin did not yield a labeled peptide, which is proposed to contain the condensing enzyme binding site.

The bovine mammary gland enzyme was reported to have the enoyl reductase activity associated only with the complex [43]. This proposition is questionable since the time course of the loss of synthetase and enoyl reductase activities is not the same. After the enzyme was incubated for 29 hr in Tris–glycine buffer (pH 8.45) analytical ultracentrifuge studies indicated that the dissociation of the complex was virtually complete and only 7% of the synthetase activity remained. In contrast, the enoyl reductase activity was 4-fold higher. Perhaps the V_{max}, K_m, or both are partially altered on dissociation of the enzyme, but there is no evidence that the reductase activity is absent in the subunit.

The chicken liver enzyme has been reported to yield nonidentical subunits on preparative gel electrophoresis under dissociating conditions in Tris–glycine buffer [53]. Two fractions were obtained. Fraction 1 contained β-ketoacyl reductase, acetyl transacylase, and palmitoyl thioesterase activities, while fraction 2 was lacking only the β-ketoacyl reductase activity. The assay of the other partial activities was not reported. In the ultracentrifuge, fraction 2 was found to contain three different peaks with estimated sedimentation coefficient values of 9, 6.7, and < 3 S. On aging for 10 days, fraction 1 (9 S) yielded protein(s) that showed a broad peak, with a sedimentation coefficient of about 3 S. The authors interpreted these results in terms of dissociation of a complex into the component proteins, but these results can be equally explained in terms of proteolysis of the complex (see below). It is clear that further studies of the chicken liver enzyme are required before the question of structure as well as identity or nonidentity of the subunits can be resolved.

It is clear from the above-mentioned studies that the subunits of the animal synthetase have all the individual activities shown in Eqs. (3)–(8) except the condensing activity [Eq. (5)]. Until recently, the working hypothesis has been that the animal and yeast synthetases are complexes of the individual proteins which are related to these enzyme activities and are represented by the circles in Scheme 1. Lynen [54,55] has proposed that the yeast complex consists of seven enzymes. This conclusion was based on the finding of equal amounts of seven different N-terminal amino acids and on the presence of at least six different proteins in starch gels of the urea-treated enzyme. Consistent with the multiprotein complex hypothesis was the isolation of a 16,000 molecular weight peptide containing 4′-phosphopantetheine from the Gdn(guanidine)·HCl-treated synthetase [56,57]. Similar results were reported by Porter and his group [58] for the pigeon liver enzyme. They proposed that the complex consists of eight proteins based on the finding of five different N-terminal amino acids and eight protein bands in phenol-acetic acid–urea gels. The multiprotein concept was further supported by the isolation of a protein near 10,000 molecular weight containing the 4′-phosphopantetheine moiety from pigeon [59], dog [8], and chicken liver [53] enzymes. In addition, it has been reported that the chicken liver enzyme separates into six protein components after preparative Tris–glycine gel electrophoresis and that a β-ketoacyl-reductase component has been isolated [53].

In contrast to this model of an *E. coli*-like system in which the individual enzymes and the acyl carrier protein are tightly held together by noncovalent interactions, it has been proposed by Schweizer *et al.* [60] that the yeast complex is an aggregate of two polypeptides of 179,000 and 185,000 molecular weight. It was reported that the synthetase was subject to proteolysis during the isolation of the enzyme. If phenylmethane sulfonyl fluoride (PMSF) was included in the isolation buffers, the proteolysis was reduced and the molecular weight of the peptides could be estimated by SDS gel electrophoresis. Contrary to the previous findings, Gdn·HCl did not release a small peptide containing the 4′-phosphopantetheine. Instead, SDS gel electrophoresis indicated that the prosthetic group is associated with the 185,000 molecular weight subunit [60].

Our studies [7] of the subunit size of the rat, chicken liver, and yeast synthetases in SDS and 6 *M* Gdn·HCl indicate that the complexes consist of two polypeptide chains of the same or nearly the same size. In contrast to Schweizer *et al.* [60], we were unable to resolve the yeast subunits into two bands on SDS gels when the enzyme was isolated in the presence of PMSF. If PMSF is omitted we have observed two or more bands on SDS gel electrophoresis [61]. These results raise the possibility that the two bands Schweizer's group observed resulted from limited proteolysis of the enzyme.

The chicken liver fatty acid synthetase preparations contained a protease, and proteolysis was enhanced in the presence of SDS. The protease could not be removed by various chromatographic procedures, which included gel filtration, ion-exchange chromatography, isoelectric focusing, and preparative gel electrophoresis [62]. The number of bands obtained on SDS gels increased and the size of the polypeptides decreased as the time of incubation and the protein concentration in SDS was increased (Fig. 1). The proposal that the fatty acid synthetase preparations contain a protease that is active in SDS solutions was confirmed by the degradation of bovine serum albumin (Fig. 1). The proteolysis could be diminished by incubating the synthetase in SDS in the presence of the alkylating agents tosyllysine chloromethyl ketone, iodoacetamide, and iodoacetate and apparently eliminated by subjecting the protein–SDS solution to heat in a boiling-water bath (Fig. 2). In this manner, it was possible to obtain a meaningful estimate of the number and size of the subunits that comprise the chicken enzyme. Interestingly, the rat

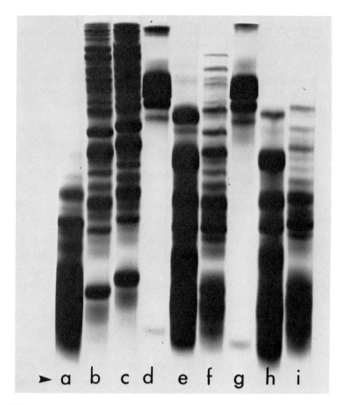

Fig. 1 Effect of protein concentration and time of incubation in SDS on SDS gel (10%) pattern of chicken FAS. Synthetase incubated at the indicated concentration for 24 hr (a–f) and 96 hr (g–i) in a solution containing 2.2% SDS, 1% HS–EtOH, and 0.1 M sodium phosphate, pH 7.0. Each gel was loaded with 50 μg of FAS or albumin or both. Preparations a–c contained FAS at 11.5, 2.3, and 1.1 mg/ml, respectively; (d) albumin (5.0 mg/ml), (e) FAS (5.8 mg/ml) and albumin (5.0 mg/ml); (f) FAS (5.8 mg/ml); (g) albumin (5.0 mg/ml); (h) albumin (5 mg/ml) and FAS (5.8 mg/ml); (i) FAS (5.8 mg/ml). The arrow indicates the approximate position of the dye front.

and yeast enzyme exhibited little or none of the time-dependent formation of the multiple lower molecular weight peptides when incubated in SDS at 10 mg/ml for 7 days. The molecular weights of the rat, chicken, and yeast synthetases were estimated by SDS gel electrophoresis to be 220,000, 220,000, and 200,000, respectively.

In addition, molecular weight determinations of the FAS by sedimentation equilibrium in Gdn·HCl were carried out. It was found that over 80, 90, and 90% of the protein in the cell consists of one molecular size with an estimated molecular weight of 240,000, 247,000, and 184,000 for the chicken, rat, and

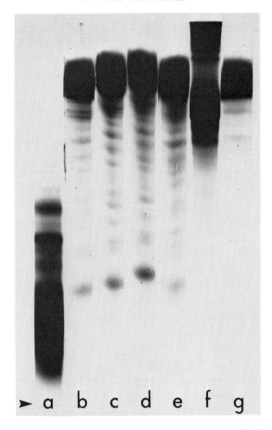

Fig. 2   The SDS gel pattern of chicken FAS after treatment with heat or alkylating agents. Synthetase (4.5 mg/ml) was heated or alkylated in the presence of SDS, and after 48 hr 50 μg was loaded on 5% gels (b–e and g) or 10% gels (a, f). The synthetase was treated as follows: (a) no treatment; (b) 10 mM iodoacetate; (c) 10 mM iodoacetamide; (d) 1.3 mM tosyllysyl-CH$_2$Cl; (e) heat; (f) 1.3 mM tosyllysyl-CH$_2$Cl plus albumin (5 mg/ml); (g) FAS dissociated in Tris–glycine buffer, pH 8.3, prior to treatment with heat. The arrow indicates the approximate position of dye front.

yeast enzymes, respectively. These values are in good agreement with those obtained from SDS gel electrophoresis.

These proteins apparently behave similarly to most other proteins in 6 M Gdn·HCl since their circular dichroism spectra show no absorbance over the 210–240 nm range [63]. These results indicate that the synthetases are a random coil in this solvent, and it would be expected that interactions between the subunits have been eliminated. Consequently, the molecular weight values obtained in 6 M Gdn·HCl correspond to the minimum size of the subunits that comprise the complex. In addition, the good agreement between the values of the molecular weights of the subunits determined in SDS and Gdn·HCl support the proposal that the 184,000–247,000 molecular weight peptides are not complexes of lower molecular weight subunits.

It follows from these findings and the measurements of the partial activities of the subunits that the two polypeptide chains contain multicatalytic sites and the 4'-phosphopantetheine. This conclusion is further supported by the finding that synthetase reacted with [^{14}C]acetyl-CoA or [^{14}C]malonyl-CoA and subjected to SDS gel electrophoresis showed that the radioactivity was associated with the 220,000 molecular weight subunits, suggesting that the acetyl and malonyl binding sites are associated with this protein [7]. In addition, chromatography of [^{14}C]pantothenate-labeled enzyme on a Sepharose 6B column in the presence of SDS, followed by SDS gel electrophoresis of the radioactive fractions, showed that the 4'-phosphopantetheine moiety was also associated only with the 220,000 molecular weight subunit. This observation suggests that the 4'-phosphopantetheine is an integral part of the 220,000 subunit. These findings explain why we have not been able to obtain ACP of low molecular weight from the FAS complex except after limited proteolysis [7,64]. It would appear, then, that the reported isolation of lower molecular weight proteins from FAS of pigeon, rat, dog, chicken, man, and yeast has resulted from proteolysis.

From these findings it is concluded that the ultimate subunit size of the animal complex is near 220,000–250,000 molecular weight and that fatty acid synthetase consists of two subunits. The yeast enzyme consists of subunits of a similar size (180,000–200,000 molecular weight), and physicochemically it differs from the animal complex in that the subunits form higher molecular weight aggregates. It will be of interest to learn if the large aggregates found in some bacterial systems (Table 1) are similar to the yeast enzyme. It will also be of interest to determine if the plant synthetases have been correctly assigned to the type II class. As can be seen from the results above, the isolation of an acyl carrier protein component from plants does not reliably distinguish between the type I and type II synthetases.

More recently, Alberts *et al.* [65] have shown that cultured Chang liver cells have FAS that gives a single band on SDS gels which corresponds to 240,000 molecular weight. A similar result was obtained with the rat liver synthetase. When the cells were cultured in the presence of [^{3}H]pantothenate and the labeled enzyme was subjected to SDS gel electrophoresis, 97% of the label was located in the 240,000 molecular weight subunit. Lower molecular weight peptides were observed if PMSF was not included in the SDS incubation mixture. These results support the studies of the chicken, rat, and yeast enzymes.

It is clear from these recent findings that reliable studies of the subunit composition and structure of fatty acid synthetase must take into consideration the possibility of proteolysis during the isolation and handling of the enzymes.

Genetic studies on the FAS from yeast (*Saccharomyces cerevisiae*) support the multifunctional protein model. Schweizer and his co-workers [66–69] have isolated an extensive series of palmitate-requiring mutants which map at two unlinked loci, *fas-1* and *fas-2*. These results have been independently confirmed by Henry and colleagues [70,71] and leave little doubt that there are but two structural genes for FAS in yeast. Mutants mapping at a third locus were also seen by both groups [67,70], but upon further study it was found that this gene is unrelated to FAS [72].

Detailed mapping of *fas-1* strongly suggests that this complex locus is a single cistron rather than a gene cluster or operon [69,73]. A large number of *fas-1* and *fas-2* mutants are pleiotropic and noncomplementing. In *fas-1*, at least, these map throughout the locus, while a high degree of clustering was observed for mutants deficient in single enzyme activities.

An examination of the activities missing in point mutants at both loci has permitted Schweizer *et al.* [73,74] to assign different enzymatic reactions to the putative products of these genes. The *fas-1* locus appears to code for the enoyl reductase, the dehydratase, and the malonyl and palmitoyl transferases. The *fas-2* locus, on the other hand, controls the condensation reaction, the β-ketoacyl reductase, and the region of the complex bearing the 4'-phosphopantetheine prosthetic group [60]. These results can be contrasted to those of Bratcher and Hsu [53] and Lornitzo *et al.* [50], who found that in the avian liver systems certain activities were common to two separate fractions.

Studies on the FAS isolated from wild-type and complementing mutants in general support the multifunctional protein model. Purified FAS has been resolved by SDS gel electrophoresis into two nonidentical subunits [75], one of which seems to be subject to proteolysis. This subunit carries the 4'-phosphopantetheine and in certain *fas-2* mutants either lacks the prosthetic group or is altered in its sensitivity to proteolytic degradation. Thus, it would appear that this component is the product of the *fas-2* gene.

So far at least, it has not been possible to isolate free subunits even from the pleiotropic, noncomplementing mutants and study the assembly of the complex *in vitro*. It has been suggested either that the unassociated subunits are rapidly degraded *in vivo* or that their synthesis is under reciprocal positive control by the *fas-1* and *fas-2* loci [75]. Once formed, the complex is reasonably stable; even temperature-sensitive alleles yield thermotolerant FAS *in vitro* [76]. It would be desirable to examine the enzymatic properties and thermostability of the FAS from intragenically complementing diploid strains for further confirmation of the multifunctional model. Nonetheless, the preponderance of the evidence suggests that yeast FAS consists of two nonidentical multifunctional proteins coded by two unlinked genes.

In contrast to the situation with yeast, comparatively little effort has been expended on the FAS of other organisms amenable to genetic analysis. In *Neurospora crassa* the nutritional requirements and lipid composition of a rather leaky saturated fatty acid mutant (*cel*) [77] have been described by Keith and others [78,79]. Recently, Elovson has purified FAS from both wild-type and this mutant [17]. His investigations indicate that the *Neurospora* FAS is similar to the yeast, although there remain difficulties with separating the putative subunits.

The Pantothenate Content of FAS Complex

As indicated earlier, the fatty acid synthetase of animal tissues was reported to contain 1 mole of 4'-phosphopantetheine per mole of enzyme [6,9,11,33,36,80]. The yeast synthetase is reported to contain 4.9 moles of 4'-phosphopantetheine per mole of the complex (molecular weight 2.3×10^6) [49,57], or approximately one prosthetic group per 500,000 molecular weight subunit. As a result of these findings, the proposed biosynthetic pathway and mechanism studies have been interpreted in terms of one 4'-phosphopantetheine (Scheme 1). The discovery [7] that the animal synthetases from chicken and rat liver and chicken oviduct consist of two polypeptide chains of the same or similar molecular weight led us to reconsider the problem regarding the prosthetic group content of the animal syntheses [81].

Table 3 lists the taurine and β-alanine content of five different chicken and Table 4 lists the taurine content of two different rat liver enzyme preparations. The β-alanine and cysteamine content was determined by isolating the partially purified 4'-phosphopantetheine after base treatment of the enzyme followed by gel filtration. Alternatively, the content of β-alanine and cysteamine associated with a tryptic peptide that was separated from the chicken synthetase by gel filtration and ion-exchange chromatography was determined. In addition, the taurine content of the chicken and rat preparations was estimated from total amino acid analysis of the protein (Table 4). As is apparent in Tables 3 and 4, both procedures give similar results.

These findings indicate that the chicken and rat enzymes contain two 4'-phosphopantetheine groups per mole of enzyme. The fact that different chicken preparations contain from 1.6 to 1.8 (but not 2) moles of the prosthetic group per mole of enzyme may indicate that these preparations contain enzyme that may have one or no prosthetic group. In this regard, Porter's group [51] has reported that the pigeon enzyme preparations consist of a mixture of synthetase with and without the prosthetic group (holo- and apoenzyme). Even though the relative amount of the apo- and holoenzymes is not reported, it can be estimated from their results that the preparation contains 10–20% apoenzyme.

TABLE 3

β-Alanine and Taurine Content of Chicken Liver Fatty Acid Synthetase[a]

Enzyme preparation[b]	Load synthetase (μmole)[c]	β-Alanine obtained (μmole)	Taurine obtained (μmole)	Yield radioactivity in β-alanine (%)[d]	Moles β-alanine or taurine per mole synthetase	HCl hydrolysis (hr)[e]
Obtained from alkali-treated synthetase						
I	0.0229	0.0383		104	1.7	24 + 50
I	0.00747		0.0122	—	1.6	24
II	0.0105		0.0173	—	1.6	72
III	0.0301	0.0513		101	1.7	48 + 48
III	0.00344	—	0.00582	—	1.7	24
IV	0.0273	0.0465		99	1.7	48 + 48
IV	0.00818		0.0150		1.8	48
IV	0.00850		0.0153		1.8	96
Obtained from trypsin-treated synthetase[f]						
V[g]	0.0199	0.0337		100	1.7	48
V[g]	0.0202	0.0334		100	1.7	48 + 48
V	0.00678		0.0109		1.6	48

[a] Sample subjected to performic acid oxidation unless stated otherwise.

[b] Specific activity (nanomoles NADPH oxidized per minute per mg protein) of the enzyme preparations determined at 23°C: I, 1200; II, 1200; III, 1100; IV, 1200; V, 1200.

[c] Calculated from initial radiospecific activity of the enzyme in disintegrations per minute per micromole obtained from the A_{280}-weight relationship and a molecular weight of 480,000 and the total disintegration per minute (or micromole of synthetase) in the sample analyzed for taurine or β-alanine.

[d] Corrected for losses due to decarboxylation by dividing the values obtained by 0.73 [81].

[e] In the case of retreatment with HCl, the sample was dried, then another aliquot of HCl was added, and the sample was heated for the time indicated. This procedure was found to be necessary in order to obtain the reported values for β-alanine.

[f] Tryptic peptide was not subjected to the alkali treatment.

[g] Protein was not subjected to performic acid oxidation.

TABLE 4

Molecular Weight of Chicken and Rat Fatty Acid Synthetases Based on 2 Moles of Taurine

Enzyme preparation	Molecular weight[a]		Moles taurine/mole synthetase[b]	
	Chicken	Rat	Chicken	Rat
I	547,000[c]	436,000[f]	1.8	2.3
II	512,000[d]	465,000[g]	1.9	2.1
III	540,000[e]		1.8	

[a] A summation of the total number of amino acid residues multiplied by their individual molecular weights corresponding to 2 moles of taurine.

[b] Calculated from molecular weights obtained from ultracentrifuge analysis (480,000 for the chicken and 494,000 for the rat) divided by values calculated in this table and multiplied by 2.

[c-g] The average of 5, 9, 7, 6, and 4 determinations, respectively. The maximum variation in these values was 10%.

Proposed Scheme for Fatty Acid Synthesis Catalyzed by Animal FAS

The finding of nearly two 4'-phosphopantetheine groups in the complex raises some relevant questions concerning the structure of the complex and the biosynthetic pathway for fatty acid synthesis. The FAS of chicken and rat livers is comprised of two polypeptide chains, each of which may contain one 4'-phosphopantetheine. Moreover, the finding that the two subunits are essentially the same size and contain blocked amino termini [7] suggests that the two polypeptide chains may be homologous. Clearly, further evidence is needed to ascertain the validity of these proposals.

In considering the pathway of fatty acid biosynthesis, a proposal was made to involve the two pantetheine groups (Scheme 1). The lack of synthetase activity of the monomer and the dependence of the condensing enzyme (β-ketoacyl synthetase) activity on the dimer are consistent with this hypothesis. Then the condensation reaction involves the participation of the acyl and malonyl thioesters of 4'-phosphopantetheine as substrates. This scheme is analogous to the condensation reaction of the *E. coli* system, in which both acyl and malonyl derivatives of the 4'-phosphopantetheine of ACP are substrates. The SH group of the pantetheine of one subunit (subunit 1) is acylated by the transfer of the acetyl group from acetyl-CoA via the *O*-serine residue of the acetyl transacylase (Scheme 2). As a result of this acylation, conformational changes may occur so that the second subunit (subunit 2) of the synthetase accepts the malonyl group more readily than before (i.e.,

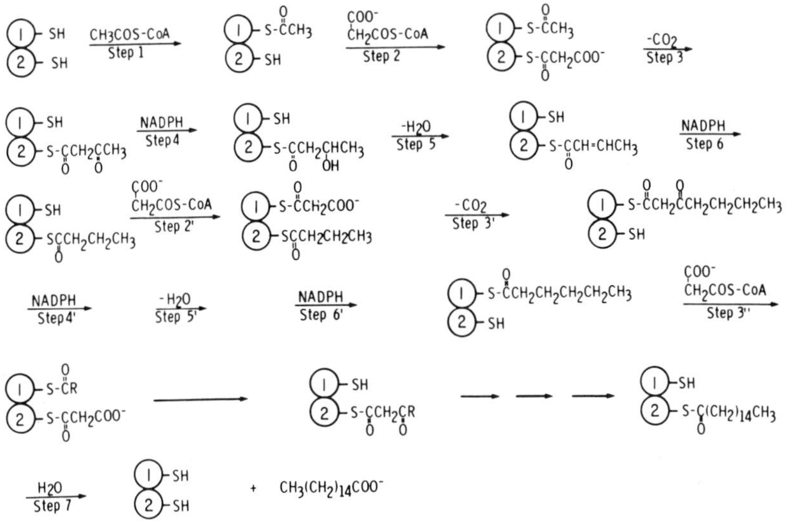

Step 1: (1)-SH, (2)-SH + CH3COS-CoA → (1)-S-$\overset{O}{\overset{||}{C}}$CH3, (2)-SH

Step 2: + $\overset{COO^-}{\underset{}{CH_2COS\text{-}CoA}}$ → (1)-S-$\overset{O}{\overset{||}{C}}$CH3, (2)-S-CCH2COO⁻

Step 3: $\xrightarrow{-CO_2}$

Step 4: (1)-SH, (2)-S-$\underset{O}{\overset{||}{C}}$CH2$\underset{O}{\overset{||}{C}}$CH3 $\xrightarrow{NADPH}$

Step 5: (1)-SH, (2)-S-CCH2CHCH3 (OH) $\xrightarrow{-H_2O}$

Step 6: (1)-SH, (2)-S-CCH·CHCH3 $\xrightarrow{NADPH}$

Step 2': $\xrightarrow{CH_2COS\text{-}CoA,\ COO^-}$ (1)-SH, (2)-SCCH2CH2CH3

Step 3': (1)-S-$\overset{O}{\overset{||}{C}}$CH2COO⁻, (2)-SCCH2CH2CH3 $\xrightarrow{-CO_2}$ (1)-S-CCH2CCH2CH2CH2CH3, (2)-SH

Step 4': $\xrightarrow{NADPH}$ Step 5': $\xrightarrow{-H_2O}$ Step 6': $\xrightarrow{NADPH}$ (1)-S-CCH2CH2CH2CH2CH3, (2)-SH

Step 3'': $\xrightarrow{CH_2COS\text{-}CoA,\ COO^-}$

(1)-S-$\overset{O}{\overset{||}{C}}$R, (2)-S-CCH2COO⁻ → (1)-SH, (2)-S-CCH2CR → — — — → (1)-SH, (2)-S-C(CH2)14CH3

Step 7: $\xrightarrow{H_2O}$ (1)-SH, (2)-SH + CH3(CH2)14COO⁻

Scheme 2

lower K_m), yielding the malonyl derivative of the pantetheine group of subunit 2. Condensation of the acetyl and malonyl groups then takes place through the involvement of the HS group of the cysteine residue of the condensation site. The acetoacetyl derivative of the pantetheine of subunit 2 is then reduced, dehydrated, and reduced again to form the butyryl derivative. Once butyrate is formed, a conformation change in subunit 1 of the synthetase takes place which is similar to the change that occurred when the acetyl group was bound to the enzyme, causing it to more readily accept a malonyl group and yielding malonyl thioester of the pantetheine moiety of the subunit. Condensation of the butyryl and malonyl groups then occurs and the ketohexanoyl derivative is formed. The cycle is then repeated until the palmitoyl derivative is formed, which is then hydrolyzed to free palmitate by the thioesterase site. One of the significant features of this scheme is that the growing fatty chain acid is transferred from one pantetheinyl group to the other in each round of elongation.

REGULATION OF FATTY ACID SYNTHETASE IN ANIMALS

The major homeostatic function of lipogenesis is to store the chemical energy of carbohydrates as triglycerides. The lipogenic process is precisely regulated in animals in response to their ever-changing energy needs. For instance, fatty acid synthesis is lowered in fasted or alloxan diabetic animals and in animals fed a high-fat diet. On the other hand, when fasted animals are refed a fat-free, high-carbohydrate diet, fatty acid synthesis is restored. Initially, the carboxylase was assigned a primary role in the control of fatty acid synthesis on the basis of its specific activity in the liver extracts obtained from animals in various nutritional states [82]. However, recent studies [83–85] suggest that, under optimum assay conditions, the carboxylase and synthetase activities are not significantly different, and therefore either one or both of the enzymes could be involved in the regulation of fatty acid synthesis. Confirmation of the latter postulate has been based on the adaptive changes in the amount of FAS and acetyl-CoA carboxylase observed under different nutritional conditions. The changes in the activities of the enzymes involved in fatty acid synthesis can be attributed to two major factors: (a) allosteric or metabolic control of the enzymes of fatty acid synthesis by metabolites derived from glucose and fatty acids and (b) control of the amount of the enzymes of fatty acid synthesis by modifying their rates of synthesis and degradation. Several reviews deal with the overall regulation of lipogenesis [86–88].

Metabolite Control

Studies in this laboratory have demonstrated that pigeon liver FAS is stimulated by phosphorylated sugars, especially fructose 1,6-diphosphate [89,90]. The rate is increased 4-fold by lowering the K_m for NADPH and is dependent on malonyl-CoA concentration. Other phosphorylated sugars (glucose 1-phosphate, glucose 6-phosphate) or phosphate can replace fructose 1,6-diphosphate, but at relatively higher concentration. In addition, Roncari [91] has reported that human liver FAS activity is stimulated 2- to 5-fold by fructose 1,6-diphosphate. Porter and co-workers [92], however, were unable to demonstrate any effect of fructose 1,6-diphosphate on the pigeon liver FAS. The reason for the discrepancy between these results remains unknown. The physiological role of phosphorylated sugars in the regulation of FAS activity appears doubtful, since high concentrations (10–30 mM) of these sugars are required to affect FAS activity.

It has been proposed that liver [91,93] and yeast [94] FAS activity is regulated by palmitoyl-CoA by feedback inhibition. The significance of these studies is questionable since palmitoyl-CoA irreversibly inactivates FAS, presumably through its detergent properties [95]. Moreover, this inactivation of FAS is in contrast to the reversible inhibition of acetyl-CoA carboxylase by palmitoyl-CoA [96,97]. In summary, there is no convincing evidence that metabolites play a significant role in the regulation of FAS activity.

Covalent Modification

The isolation of holo-ACP synthetase and ACP hydrolase from *E. coli* [98,99] and the participation of these enzymes in the rapid turnover of the 4'-phosphopantetheine prosthetic group of ACP in *E. coli* [100] raise the possibility that such turnover of the prosthetic group could provide a method of regulating FAS activity in animals as well. Larrabee and co-workers [101] studied the prosthetic group turnover in rat liver by pulse labeling with [³H]pantothenate and determining the specific radioactivity of the 4'-phosphopantetheine of the synthetase and CoA. The turnover rate of the prosthetic group of FAS was at least an order of magnitude faster than that of the whole FAS complex, which has a half-life ($t_{1/2}$, first order) of about 70 hr. Similar results have been obtained for the prosthetic group of synthetase from brain and adipose tissue [102]. The turnover of the prosthetic group of the liver enzyme decreased 4-fold as a result of starvation [102,103] and increased about 2-fold during fat-free feeding [102]. A 4'-phosphopantetheine hydrolase activity has been measured in crude extracts of livers of starved rats [91]. Parallel with the increase in FAS activity seen upon refeeding, Yu and Burton [104,105] have reported the presence of a 4'-phosphopantetheine transferase activity and a possible accumulation of apo-FAS.

More recently, Porter and co-workers [51] have separated apo- and holo-fatty acid synthetases by chromatography. The [¹⁴C]pantetheine-labeled FAS preparation was loaded on a column of Sepharose–ε-aminocaproylpante-theine. Aposynthetase, lacking the [¹⁴C]pantetheinyl moiety and containing all the partial activities except the β-ketoacyl synthetase, was eluted first, while the radioactive holosynthetase containing all of the partial activities and the synthetase activity eluted last. It is of interest that this order of elution is in contrast to that found for the elution of the dissociated synthetase, in which the pantetheine-containing protein was eluted first [50]. Fatty acid synthetase activity was generated when the aposynthetase was incubated with CoA, ATP, Mg^{2+}, and a high-speed supernatant preparation from livers of starved–refed rats. The latter supernatant was made free of synthetase by chromatography and served as a source of 4'-phosphopantetheine trans-ferase. Although the supernatant containing 4'-phosphopantetheine trans-ferase activity was free of synthetase activity, it was not demonstrated that such preparations lacked inactive synthetase. Furthermore, a relationship between enzyme activity and the incorporation of [¹⁴C]pantetheine from CoA into aposynthetase was not reported, nor was the proposed transferase isolated. Thus, it is clear from the preceding discussion that regulation of FAS by prosthetic group turnover remains to be elucidated.

Recently, attempts have been made to demonstrate phosphorylated and nonphosphorylated forms of enzymes involved in fatty acid biosynthesis.

Porter and co-workers have reported the separation of phosphorylated (holo-b) and dephosphorylated (holo-a) forms of avian fatty acid synthetase by chromatography on Sepharose–ε-aminocaproylpantetheine [106]. The conversion of dephosphorylated enzyme to the phosphorylated form was followed by the incorporation of ^{32}P from [γ-^{32}P]ATP. The two forms were separated by chromatography, and it was concluded that the nonphosphorylated and phosphorylated forms of FAS were active and inactive, respectively. However, this conclusion appears to be premature since the "dephosphorylated" FAS fraction exhibited 40% of the specific radioactivity of the "phosphorylated" FAS fraction. Furthermore, Rous [107] could not demonstrate incorporation of injected [^{32}P]orthophosphate, in vivo, into purified FAS from rat liver. Further studies are required to establish a role for phosphorylation in the regulation of FAS.

Adaptive Changes of Enzyme Content Involved in Nutritional and Hormonal Regulation of FAS

Although evidence for a physiological role for FAS in short-term regulation of fatty acid synthesis is not convincing, considerable information indicates that the synthetase is a critical enzyme in long-term regulation. The importance of regulation of FAS in fatty acid synthesis during long-term feeding with a diet high in carbohydrate is suggested by simultaneous measurements of the rate of fatty acid synthesis and malonyl-CoA concentration, since the utilization of malonyl-CoA by the synthetase is limiting under these circumstances [108]. The FAS of animal liver is one of the several lipogenic enzymes whose activity is affected in a coordinate manner by the nutritional status of the animal [87]. The liver FAS activity is reduced as much as 20-fold on starvation and is elevated to above normal levels after refeeding a high-carbohydrate, low-fat diet [87]. The mechanisms underlying the changes in hepatic FAS activity in various nutritional states have been studied. Craig et al. [109], using immunochemical procedures, demonstrated that the content of FAS paralleled the reduction in FAS activity in livers of fasted rats. The relative rate of synthesis of FAS decreased to one-fifth the normal value in the fasted state and increased to 14 times the normal value in rats refed a fat-free diet. Measurement of the degradation of the synthetase after administration of [^{14}C]leucine yielded an identical half-life value of 69 hr for normal animals as well as for animals refed a fat-free diet. It should be noted that reutilization of the tracer in the above-mentioned experiment could result in an overestimation of the half-life of the synthetase. However, after administration of nonreutilizable [*guanido*-^{14}C]arginine and following decay of label in purified synthetase, Tweto et al. [101] demonstrated that the hepatic enzyme in fed rats was degraded with a $t_{1/2}$ of 71 hr. Together these studies suggest that changes in the level of synthetase which are produced on re-feeding rats are primarily due to changes in the rate of synthesis rather than degradation of the enzyme. More recently, Volpe *et al.* [110], using immuno-chemical techniques, observed that the rates of synthesis and degradation of FAS in livers of starved rats were nearly 10 times lower and 4 times higher, respectively, than those in fed rats ($t_{1/2}$ in starved rats = 18 hr). Thus, an increase in the rate of degradation and a decrease in the rate of synthesis of FAS play a major role in determining the lower hepatic synthetase activity in starved rats.

The factors responsible for the regulation of FAS synthesis have been investigated using polysomal translation in a heterologous cell-free system [65,111]. Rat liver polysomes that synthesize FAS were identified by the binding of ^{125}I-labeled anti-FAS specific antibody. The binding of ^{125}I-labeled antibody to polysomes was found to correlate with the FAS activity in various nutritional conditions. Furthermore, the relative amounts of FAS synthesized by polysomes from livers of normal, starved, and starved-refed rats correlated well with the synthetase activities in these livers, suggesting that the control of the adaptive synthesis of FAS is mediated through changes in the rate of translation.

The nature of the physiological stimuli required for the induction of FAS on refeeding starved animals is not clear. The rapid rise in hepatic FAS activity upon refeeding after a fast is paralleled by a similar increase in levels of plasma insulin [87]. Hepatic lipogenesis [112] and FAS activity [113] are reduced in experimental alloxan diabetes and are restored by injection of insulin. The decrease in synthetase activity in diabetic rats is caused by a diminution in synthesis of the enzyme, and administration of insulin to diabetic rats restores activity and synthesis of the enzyme [114,115]. Similar observations have been made for the synthetase of adipose tissue [115,116]. Although insulin administration is known to correct the diminution of synthesis of FAS in diabetic rat liver, it is not required for the regulation of the synthetase activity since, in the absence of insulin administration, fructose feeding to diabetic rats increases the rate of synthesis of the enzyme [115]. These studies suggest that fructose, which can enter the glycolytic pathway in the absence of insulin in livers of diabetic animals, may induce the synthetase through changes in the concentration of certain intermediates of glycolysis or beyond. Further studies by crossover point analysis of the intermediates of glycolysis [117] in livers of diabetic and fructose-fed diabetic rats should identify the presumptive intermediate.

Marked changes in lipogenesis and activities of enzymes involved in fatty acid synthesis occur during the neonatal period. In contrast to the active lipogenesis in mammalian fetal liver [118], *de novo* fatty acid synthesis in chick embryo liver is low but detectable [119,120]. Hepatic lipogenesis and

FAS activity are stimulated when newly hatched chicks are fed a high-carbohydrate diet [120,121]. The FAS activity increased 3-fold on hatching and
thereafter 35- to 50-fold over the basal embryonic activity [120]. The factors
underlying the changes in FAS activity of developing avian liver have been
determined by immunochemical techniques [122]. The increase in hepatic
FAS activity after hatching was shown to be a consequence of an increase in
the rate of FAS synthesis, resulting in a net increase in the amount of FAS
[122,123].

In contrast to liver FAS, significant synthetase activity is associated with
the early embryonic avian heart [120]. The enzyme activity decreases in heart
with age of the embryo and is absent in newly hatched chicks and older
chickens. The loss of enzyme activity with development is possibly related to
differentiation and maturation of cardiac myocytes.

The possible involvement of hormonal factors in the induction of FAS
activity in neonatal chick liver has been investigated [122]. The liver synthetase
activity is prematurely induced 24- to 44-fold in 20-day-old chick embryos
by the administration of insulin, glucagon, or dibutyryl cyclic AMP. There
was a maximal induction of FAS activity 6–12 hr after injection of inducer
followed by a decline in the enzyme activity. The induction of FAS activity
by suboptimal doses of glucagon is potentiated by the phosphodiesterase
inhibitor theophylline, suggesting that elevated intracellular levels of cyclic
AMP may be involved in the action of glucagon. Quantitative precipitin
reaction using antibodies against FAS gave the same equivalence point for
FAS from adult and induced embryonic chicken liver, suggesting that the
increase in synthetase activity following administration of various inducers
results from an increase in the content of the enzyme and not from its activation. Immunochemical analysis of liver extracts after pulse labeling with
[³H]leucine showed that the increased level of synthetase in embryos treated
with insulin, glucagon, and dibutyryl cyclic AMP is due to a rapid increase
in the rate of synthesis of the enzyme as early as 2–4 hr after administration
of inducers. Sodium dodecyl sulfate gel electrophoresis of radioactive immuno-
precipitates obtained after addition of antibodies to [³H]leucine-labeled liver
extracts of induced embryos gave a single radioactive band corresponding to
the 240,000 molecular weight FAS subunit [123], indicating that all of the
[³H]leucine in the immunoprecipitate is associated with FAS. These observations established that the increased synthetase activity subsequent to the
administration of inducers results from an increase in the rate of FAS
synthesis [122]. It is particularly noteworthy that in induced embryos the rate of
FAS synthesis is 4–7.6% of the total overall rate of soluble protein synthesis.
A common feature of this type of premature induction is the rapid decay of
activity after elevation to higher levels. The $t_{1/2}$ of embryonic liver FAS
activity, as calculated from the first-order decay after elevation to higher

levels, was found to be 4 hr in contrast to the $t_{1/2}$ of 70 hr in the adult chicken. The decay of FAS activity appears to represent degradation of the enzyme protein since there is a coordinate decay of the component activities of FAS [122]. This increased degradation of the synthetase in chick embryo liver is reminiscent of the accelerated degradation of the synthetase in starved rats [110].

The inductive effect of glucagon and dibutyryl cyclic AMP on hepatic synthetase *in ovo* is contrary to their inhibitory effect on lipogenesis. In order to verify the physiological significance of the induction of FAS activity by glucagon and dibutyryl cyclic AMP, we have tested these compounds in cultured liver explants [123]. Addition of insulin (0.5 μg/ml) to embryo liver explants cultured in serum-free medium resulted in a 15-fold increase in FAS activity. The induction of FAS was inhibited by cycloheximide, suggesting that new protein synthesis is required for induction of synthetase. Furthermore, since immunochemical analysis gave similar equivalence points for FAS from control and insulin-treated liver explants, the increase in enzyme activity in insulin-treated explants resulted from an increase in enzyme content. Measurement of the rate of incorporation of [³H]leucine into FAS established that the increased enzyme content of insulin-treated explants resulted from at least a 5-fold increase in the relative rate of FAS synthesis. Glucagon (10 μg/ml) or dibutyryl cyclic AMP (1 mM) did not induce FAS activity in liver explants, nor did glucagon affect insulin induction of FAS activity. These results suggest that the administration of glucagon and dibutyryl cyclic AMP to chick embryos induces FAS activity by indirect means, possibly by eliciting endogenous insulin production. To our knowledge, the liver explant system described here is the only available primary culture system from liver that responds to physiological levels of insulin (10^{-8} M) by increasing FAS content. Thyroxine or triiodothyronine is required for the induction of lipogenesis and FAS activity in cultured avian hepatocytes [124]. Inability of insulin alone to increase synthetase activity in cultured avian hepatocytes [124] may be related to the presence of serum in the culture medium. Recently, Alberts *et al.* [125] have demonstrated a 2-fold increase in FAS activity in the established Chang liver cell line by the addition of insulin.

REGULATION OF FATTY ACID SYNTHETASE IN
Mycobacterium smegmatis

The FAS activity of *M. smegmatis* is stimulated by polysaccharides of two types: MMP (3-*O*-methylmannose-containing polysaccharide) and MGLP (lipopolysaccharide containing 6-*O*-methylglucose and glucose) [126,127].

The polysaccharides lower the K_m values for acetyl-CoA and malonyl-CoA 9-fold and 4-fold, respectively. However, they do not alter the partial enzyme activities (transacylases and β-ketoacyl synthetase) of FAS. In the presence of polysaccharides, FAS exhibits both an increase in total fatty acid synthesis and an increase in the relative proportion of palmitate. This change of the bimodal distribution of fatty acids suggested that the polysaccharide decreased end-product inhibition of the synthetase by palmitoyl-CoA and, by binding palmitoyl-CoA, made this compound less available for further elongation [128]. Since the intracellular concentration of palmitoyl-CoA is sufficiently high (200–300 μM) to cause substantial synthetase inhibition, the reversal of palmitoyl-CoA inhibition of FAS activity by mycobacterial polysaccharides may be of physiological significance [129].

SUMMARY

The fatty acid synthetases from eukaryotic cells have molecular weights 400,000–500,000 (avian and mammalian enzymes) or about 2,000,000 (yeast and other fungal enzymes). The synthetases are complexes which may be readily dissociated under nondenaturing conditions into subunits with 200,000–250,000 MW. In the presence of sodium dodecyl sulfate or guanidine hydrochloride, the polypeptide chains have 200,000–250,000 MW, indicating that the fatty acid synthetases from animal tissues consist of 2 polypeptides while those from fungi consists of aggregates of about 12 polypeptides. Since the subunits have many of the component activities involved in fatty acid synthesis, the polypeptides are multifunctional enzymes which contain several active sites on single polypeptide chains. This structural view of the fatty acid synthetase is in variance with the previous model of an *E. coli*-like system in which the individual enzymes are thought to be tightly held together by noncovalent interactions to form a multienzyme complex. The evidence for the multienzyme hypothesis was apparently derived from studies of synthetases which had undergone limited proteolysis during the various manipulations of the enzyme. We have been able to reduce or eliminate proteolysis and thereby determine this unique multifunctional enzyme structure of the fatty acid synthetase. These findings suggest that other multienzyme systems may also be multifunctional enzymes.

Recent studies indicate that animal synthetases contain two prosthetic groups (4'-phosphopantetheine) per mole of enzyme instead of the one prosthetic group previously reported. A scheme consistent with these structural findings is proposed for the fatty acid synthetases in which both the acyl and malonyl groups are bound to the prosthetic groups prior to condensation to form the β-ketoacyl derivatives. In this scheme, the growing fatty acid chain is transferred from one prosthetic group to the other in each round of elongation.

Fatty acid synthetases are tightly regulated. The enzyme is induced by insulin or by a carbohydrate diet after starvation. Even though the induction involves an increase in the rate of synthesis of fatty acid synthetase, the mechanism of regulation is not known.

ACKNOWLEDGMENTS

The investigations carried out in the authors' laboratory were supported in part by Grants GM-19091, HD-07516, and HL-17269 from the National Institute of Health, and Grant Q-587 from the Robert A. Welch Foundation.

REFERENCES

1. S. J. Wakil, *J. Am. Chem. Soc.* **80**, 6465 (1958).
2. R. Bressler and S. J. Wakil, *J. Biol. Chem.* **236**, 1643 (1961).
3. S. J. Wakil, *in* "Lipid Metabolism" (S. J. Wakil, ed.), p. 1. Academic Press, New York, 1970.
4. D. J. Prescott and P. R. Vagelos, *Adv. Enzymol.* **36**, 269 (1972).
5. D. N. Brindley, S. Matsumura, and K. Bloch, *Nature (London)* **224**, 666 (1969).
6. D. W. Burton, A. G. Haavik, and J. W. Porter, *Arch. Biochem. Biophys.* **126**, 141 (1968).
7. J. K. Stoops, M. J. Arslanian, Y. H. Oh, K. C. Aune, T. C. Vanaman, and S. J. Wakil, *Proc. Natl. Acad. Sci. U.S.A* **72**, 1940 (1975).
8. D. A. K. Roncari, *J. Biol. Chem.* **249**, 7035 (1974).
9. D. A. K. Roncari, *Can. J. Biochem.* **52**, 221 (1976).
10. R. Y. Hsu and S. Yun, *Biochemistry* **9**, 239 (1970).
11. S. Smith and S. Abraham, *J. Biol. Chem.* **245**, 3209 (1970).
12. E. M. Carey and R. Dils, *Biochim. Biophys. Acta* **210**, 371 (1970).
13. J. Knudsen, *Biochim. Biophys. Acta* **280**, 408 (1972).
14. S. K. Maitra and S. Kumar, *J. Biol. Chem.* **249**, 118 (1974).
15. C. R. Strong and R. Dils, *Int. J. Biochem.* **3**, 369 (1972).
16. F. Lynen, *Fed. Proc., Fed. Am. Soc. Exp. Biol.* **20**, 941 (1961).
17. J. Elovson, *J. Bacteriol.* **124**, 524 (1975).
18. K.-H. Holtermüller, E. Ringelmann, and F. Lynen, *Hoppe-Seyler's Z. Physiol. Chem.* **351** 1411 (1970).
19. H. W. Knoche and K. E. Koths, *J. Biol. Chem.* **248**, 3517 (1973).
20. A. Rossi and J. W. Corcoran, *Biochem. Biophys. Res. Commun.* **50**, 597 (1973).
21. J. Delo, M. L. Ernst-Fonberg, and K. Bloch, *Arch. Biochem. Biophys.* **143**, 384 (1971).
22. W. J. Lennarz, R. J. Light, and K. Bloch, *Proc. Natl. Acad. Sci. U.S.A.* **48**, 840 (1962).
23. P. W. Goldman, A. W. Alberts, and P. R. Vagelos, *J. Biol. Chem.* **238**, 1255 (1963).

24. E. L. Pugh, F. Sauer, M. Waite, R. E. Toomey, and S. J. Wakil, *J. Biol. Chem.* **241**, 2635 (1966).
25. P. H. W. Butterworth and K. Bloch, *Eur. J. Biochem.* **12**, 496 (1970).
26. R. Sirevåg and R. P. Levine, *J. Biol. Chem.* **247**, 2586 (1972).
27. P. Overath and P. K. Stumpf, *J. Biol. Chem.* **239**, 4103 (1964).
28. J. L. Brooks and P. K. Stumpf, *Arch. Biochem. Biophys.* **116**, 108 (1966).
29. M. L. Ernst-Fonberg and K. Bloch, *Arch. Biochem. Biophys.* **143**, 392 (1971).
30. I. Goldberg and K. Bloch, *J. Biol. Chem.* **247**, 7349 (1972).
31. R. D. Simoni, R. S. Criddle, and P. K. Stumpf, *J. Biol. Chem.* **242**, 573 (1967).
32. S. Smith and S. Abraham, *J. Biol. Chem.* **246**, 6428 (1971).
33. D. A. K. Roncari, R. A. Bradshaw, and R. P. Vagelos, *J. Biol. Chem.* **247**, 6234 (1972).
34. V. C. Joshi, C. A. Plate, and S. J. Wakil, *J. Biol. Chem.* **245**, 2857 (1970).
35. G. T. Phillips, J. E. Nixon, A. S. Abramovitz, and J. W. Porter, *Arch. Biochem. Biophys.* **138**, 357 (1970).
36. E. J. Jacob, P. H. W. Butterworth, and J. W. Porter, *Arch. Biochem. Biophys.* **124**, 392 (1968).
37. C. J. Chesterton, P. H. W. Butterworth, and J. W. Porter, *Arch. Biochem. Biophys.* **126**, 864 (1968).
38. E. Schweizer, F. Piccinini, C. Duba, S. Günther, E. Ritter, and F. Lynen, *Eur. J. Biochem.* **15**, 483 (1970).
39. J. Ziegenhorn, R. Neidermeier, C. Nüssler, and F. Lynen, *Eur. J. Biochem.* **30**, 285 (1972).
40. J. Ayling, R. Pirson, and F. Lynen, *Biochemistry* **11**, 526 (1972).
41. M. D. Greenspan, C. H. Birge, G. Powell, W. S. Hancock, and P. R. Vagelos, *Science* **170**, 1203 (1970).
42. F. Lynen, *Biochem. J.* **102**, 381 (1967).
43. S. K. Maitra and S. Kumar, *J. Biol. Chem.* **249**, 118 (1974).
44. S. Kumar, R. A. Meusing, and J. W. Porter, *J. Biol. Chem.* **247**, 4749 (1972).
45. S. L. Yun and R. Y. Hsu, *J. Biol. Chem.* **247**, 2689 (1972).
46. S. Kumar and J. W. Porter, *J. Biol. Chem.* **246**, 7780 (1971).
47. D. E. Roark and D. A. Yphantis, *Ann. N.Y. Acad. Sci.* **164**, 245 (1968).
48. R. A. Meusing, F. A. Lornitzo, S. Kumar, and J. W. Porter, *J. Biol. Chem.* **250**, 1814 (1975).
49. M. Sumper and F. Lynen, *Colloq. Ges. Biol. Chem.* **23**, 365 (1972).
50. F. A. Lornitzo, A. A. Qureshi, and J. W. Porter, *J. Biol. Chem.* **250**, 4520 (1975).
51. A. A. Quershi, M. Kim, F. A. Lornitzo, R. A. Jenik, and J. W. Porter, *Biochem. Biophys. Res. Commun.* **64**, 836 (1975).
52. S. Kumar, J. A. Dorsey, R. A. Meusing, and J. W. Porter, *J. Biol. Chem.* **245**, 4732 (1970).
53. S. C. Bratcher and R. Y. Hsu, *Biochim. Biophys. Acta* **410**, 229 (1975).
54. F. Lynen, *Prog. Biochem. Pharmacol.* **3**, 1 (1967).
55. F. Lynen, "New Perspectives in Biology" (M. Sela, ed.), Vol. 4, p. 132. Elsevier, Amsterdam, 1964.
56. K. Willecke, E. Ritter, and F. Lynen, *Eur. J. Biochem.* **8**, 503 (1969).
57. E. Schweizer, K. Willecke, W. Winnewisser, and F. Lynen, *Vitam. Horm. (N.Y.)* **28**, 329 (1970).
58. C. P. Yang, P. H. W. Butterworth, R. M. Bock, and J. W. Porter, *J. Biol. Chem.* **242**, 3501 (1967).
59. F. A. Lornitzo, A. A. Qureshi, and J. W. Porter, *J. Biol. Chem.* **249**, 1654 (1974).

60. E. Schweizer, B. Kniep, H. Castorph, and U. Holzner, *Eur. J. Biochem.* **39**, 353 (1973).
61. M. J. Arslanian and J. K. Stoops, unpublished results.
62. M. J. Arslanian, Y. H. Oh, and J. K. Stoops, unpublished results.
63. J. K. Stoops, unpublished results.
64. M. Arslanian, J. Stoops, J. McClure, H. Barakat, and Y. Oh, *Fed. Proc., Fed. Am. Soc. Exp. Biol.* **32**, 633 (1973).
65. A. W. Alberts, A. W. Strauss, S. Hennessy, and P. R. Vagelos, *Proc. Natl. Acad. Sci. U.S.A.* **72**, 3956 (1975).
66. E. Schweizer and H. Bolling, *Proc. Natl. Acad. Sci. U.S.A.* **67**, 660 (1970).
67. E. Schweizer, L. Kühn, and H. Castorph, *Hoppe-Seyler's Z. Physiol. Chem.* **352**, 377 (1971).
68. L. Kühn, H. Castorph, and E. Schweizer, *Eur. J. Biochem.* **24**, 492 (1972).
69. G. Burkl, H. Castorph, and E. Schweizer, *Mol. & Gen. Genet.* **119**, 315 (1972).
70. S. A. Henry and S. Fogel, *Mol. & Gen. Genet.* **113**, 1 (1971).
71. M. R. Culbertson and S. A. Henry, *Genetics* **75**, 441 (1973).
72. K. H. Meyer and E. Schweizer, *J. Bacteriol.* **117**, 345 (1974).
73. P. Tauro, U. Holzner, H. Castorph, F. Hill, and E. Schweizer, *Mol. & Gen. Genet.* **129**, 131 (1974).
74. A. Knobling, D. Schiffmann, H. S. Sickinger, and E. Schweizer, *Eur. J. Biochem.* **56**, 359 (1975).
75. G. Dietlein and E. Schweizer, *Eur. J. Biochem.* **58**, 177 (1975).
76. A. Knobling and E. Schweizer, *Eur. J. Biochem.* **59**, 415 (1975).
77. D. D. Perkins, M. Glassey, and B. A. Bloom, *Can. J. Genet. Cytol.* **4**, 187 (1962).
78. S. A. Henry and A. D. Keith, *J. Bacteriol.* **106**, 174 (1971).
79. A. D. Keith, B. J. Wisnieski, S. A. Henry, and J. C. Williams, *in* "Lipids and Biomembranes of Eukaryotic Microorganisms" (J. A. Erwin, Ed.), p. 259. Academic Press, New York, 1973.
80. I. P. Williamson, J. K. Goldman, and S. J. Wakil, *Fed. Proc., Fed. Am. Soc. Exp. Biol.* **25**, 340 (1966).
81. M. J. Arslanian, J. K. Stoops, Y. H. Oh, and S. J. Wakil, *J. Biol. Chem.* **251**, 3194 (1976).
82. J. Ganguly, *Biochim. Biophys. Acta* **40**, 110 (1969).
83. H. C. Chang, I. Seidman, G. Teebor, and M. D. Lane, *Biochem. Biophys. Res. Commun.* **28**, 682 (1967).
84. S. Smith and S. Abraham, *Arch. Biochem. Biophys.* **136**, 112 (1970).
85. J. J. Volpe and Y. K. Kishimoto, *J. Neurochem.* **19**, 737 (1972).
86. P. R. Vagelos, *Curr. Top. Cell. Regul.* **4**, 119 (1971).
87. D. M. Gibson, R. T. Lyons, D. F. Scott, and Y. Muto, *Adv. Enzyme Regul.* **10**, 187–204 (1972).
88. S. Numa and S. Yamashita, *Curr. Top. Cell. Regul.* **8**, 197 (1974).
89. S. J. Wakil, J. K. Goldman, I. P. Williamson, and R. E. Toomey, *Proc. Natl. Acad. Sci. U.S.A.* **55**, 880 (1966).
90. C. A. Plate, V. C. Joshi, B. Sedgewick, and S. J. Wakil. *J. Biol. Chem.* **243**, 5439 (1968).
91. D. A. K. Roncari, *Can. J. Biochem.* **53**, 135 (1975).
92. J. W. Porter, S. Kumar, and R. E. Dugan, *Prog. Biochem. Pharmacol.* **6**, 1 (1971).
93. P. K. Tubbs and P. B. Garland, *Biochem. J.* **89**, 258 (1963).
94. F. Lynen, I. Hopper-Kessel, and H. Eggerer, *Biochem. Z.* **340**, 95 (1964).
95. J. A. Dorsey and J. W. Porter, *J. Biol. Chem.* **243**, 3512 (1968).

96. S. Numa. E. Ringelmann, and F. Lynen, *Biochem. Z.* **343**, 243 (1965).
97. A. G. Goodridge, *J. Biol. Chem.* **247**, 6946 (1972).
98. J. Elovson and P. R. Vagelos, *J. Biol. Chem.* **243**, 3603 (1968).
99. P. R. Vagelos and A. R. Larrabee, *J. Biol. Chem.* **242**, 1776 (1967).
100. G. L. Powell, J. Elovson, and P. R. Vagelos, *J. Biol. Chem.* **244**, 5616 (1969).
101. J. Tweto, M. Liberti, and A. R. Larrabee, *J. Biol. Chem.* **246**, 2468 (1971).
102. J. J. Volpe and P. R. Vagelos, *Biochim. Biophys. Acta* **326**, 293 (1973).
103. J. Tweto and A. R. Larrabee, *J. Biol. Chem.* **247**, 4900 (1972).
104. H. L. Yu and D. N. Burton, *Arch. Biochem. Biophys.* **161**, 297 (1974).
105. H. L. Yu and D. N. Burton, *Biochem. Biophys. Res. Commun.* **61**, 483 (1974).
106. A. A. Qureshi, R. A. Jenik, M. Kim, F. A. Lornitzo, and J. W. Porter, *Biochem. Biophys. Res. Commun.* **66**, 344 (1975).
107. S. Rous, *FEBS Lett*. **44**, 55 (1974).
108. R. W. Guynn, D. Veloso, and R. L. Veech, *J. Biol. Chem.* **247**, 7325 (1972).
109. M. C. Craig, C. M. Nepokroeff, M. R. Lakshmanan, and J. W. Porter, *Arch. Biochem. Biophys.* **152**, 619 (1972).
110. J. J. Volpe, T. O. Lyles, D. A. K. Roncari, and P. R. Vagelos, *J. Biol. Chem.* **248**, 2502 (1973).
111. A. W. Strauss, A. W. Alberts, S. Hennessy, and P. R. Vagelos, *Proc. Natl. Acad. Sci. U.S.A.* **72**, 4366 (1975).
112. K. J. Mathes, S. Abraham, and I. L. Chaikoff, *J. Biol. Chem.* **235**, 2560 (1960).
113. A. Gellhorn and W. Benjamin, *Science* **146**, 1166 (1964).
114. M. R. Lakshmanan, C. M. Nepokroeff, and J. W. Porter, *Proc. Natl. Acad. Sci. U.S.A.* **69**, 3516 (1972).
115. J. J. Volpe and P. R. Vagelos, *Proc. Natl. Acad. Sci. U.S.A.* **71**, 889 (1974).
116. E. D. Saggerson and A. L. Greenbaum, *Biochem. J.* **119**, 221 (1970).
117. J. H. Exton, *in* "Methods in Enzymology" (B. W. O'Malley and J. G. Hardman, eds.), Vol. 37, p. 277. Academic Press, New York, 1975.
118. F. J. Ballard and R. W. Hanson, *Biochem. J.* **102**, 952 (1967).
119. W. E. Donaldson and N. S. Mueller, *Can. J. Biochem.* **49**, 563 (1971).
120. V. C. Joshi and J. B. Sidbury, Jr., *Dev. Biol.* **42**, 282 (1975).
121. A. G. Goodridge, *Biochem. J.* **108**, 655 (1968).
122. V. C. Joshi and J. B. Sidbury, Jr., *Arch. Biochem. Biophys.* **173**, 403 (1976).
123. V. C. Joshi, unpublished results.
124. A. G. Goodridge, A. Garay, and P. Silpanata, *J. Biol. Chem.* **249**, 1469 (1974).
125. A. W. Alberts, K. Ferguson, S. Hennessy, and P. R. Vagelos, *J. Biol. Chem.* **249**, 5241 (1974).
126. M. Ilton, A. W. Jevans, E. D. McCarthy, D. Vance, H. B. White, and K. Bloch, *Proc. Natl. Acad. Sci. U.S.A.* **68**, 87 (1971).
127. D. E. Vance, O. Mitsuhashi, and K. Bloch, *J. Biol. Chem.* **248**, 2303 (1973).
128. P. K. Flick and K. Bloch, *J. Biol. Chem.* **249**, 1031 (1974).
129. P. K. Flick and K. Bloch, *J. Biol. Chem.* **250**, 3348 (1975).

CHAPTER

15

Sulfane-Transfer Catalysis by Enzymes

John Westley

INTRODUCTION

The sulfanes comprise a class of inorganic sulfur compounds made up of linear chains of sulfur atoms all at the sulfenyl (—S—) level of oxidation and bonded only to each other, except for the chain-terminal atoms, which bear single hydrogens. By extension, the term "sulfane" is also used to designate sulfenyl sulfur atoms bonded only to sulfur in other compounds: the outer sulfur atom of inorganic thiosulfate and unsubstituted organic thiosulfonates, the internal chain atoms of polythionates and of organic polysulfides such as thiocystine (cystine trisulfide), and the terminal sulfur atom of organic hydrodisulfides (persulfides) [1,2].

Four enzymes widely distributed in nature have sulfane compounds as substrates or products. Two of these enzymes, rhodanese (thiosulfate sulfurtransferase, EC 2.8.1.1) and 3-mercaptopyruvate sulfurtransferase (EC 2.8.1.2), apparently can transfer sulfane sulfur directly by mechanisms involving sulfur-substituted enzymes. A substantial amount of kinetic, chemical, and other work relating to rhodanese action in particular has been carried out, permitting a number of inferences about the mechanism, some of them probably correct. Of the two enzymes which make sulfane sulfur by cleaving carbon-sulfur bonds, one, cystathionase (cystathionine γ-lyase, EC 4.4.1.1), functions by a typical pyridoxal phosphate-dependent mechanism to produce a persulfide as a product of disulfide cleavage but evidently does not contain a sulfane atom bound directly to the enzyme at any stage of catalysis [3]; the other (3-mercaptopyruvate sulfurtransferase) requires no comparable prosthetic group and probably functions by a mechanism similar to that of rhodanese. The remaining sulfane-related enzyme, the glutathione-dependent thiosulfate reductase [4–6] (EC unassigned) has not yet been investigated extensively as to mechanism. Brief reviews of the literature pertaining to cystathionase and thiosulfate reductase (probably better called sulfane reductase) are included below.

The two sulfurtransferases have attracted special interest for two principal reasons: (a) the likelihood that their chemical mechanisms might be relatively simple, facilitating analysis in familiar terms as well as permitting direct comparisons with uncomplicated models, and (b) the fact that the physiological functions of the sulfane sulfur which these enzymes can provide appear to be of great potential importance, although they are yet to be elucidated fully. In the reactions catalyzed by both of these enzymes a sulfur-donor substrate is cleaved, apparently with transfer of a single sulfur atom to the enzyme, which can then transfer it on to an acceptor. Since there is in neither case strict specificity for the sulfur acceptor, the physiological roles of the sulfurtransferases have been difficult to establish. However, recent developments in several fields of biochemistry suggest some essentially structural roles for this sulfur. These functions are discussed following the review of sulfurtransferase mechanisms.

RHODANESE (THIOSULFATE SULFURTRANSFERASE)

The sulfane transferase that has been called rhodanese through most of its history is a mitochondrial enzyme of mammalian liver and kidney tissue. The activity has also been detected in many other mammalian organs and in a broad variety of nonmammalian organisms over the entire phylogenetic range, including plants and prokaryotes. The history, biological distribution, and biochemistry of this enzyme through 1973 have been covered in recent reviews [1,2,7]. However, there has been a considerable amount of more recent work, which is discussed here along with enough of the background material to render the present status of the enzymatic mechanism intelligible.

Rhodanese has been prepared in apparently homogeneous condition from bovine liver [8–11], bovine kidney [12,13], culture filtrates of the fungus *Trametes sanguinea* [14], human liver [15], and rat liver [16]. It has also been purified to various degrees from several bacterial species. The enzyme is assayed most frequently by Sörbo's original procedure [17] based on colorimetric determination of thiocyanate produced in the catalyzed reaction between cyanide and thiosulfate. Methods based on determination of the other product, sulfite [18–20], or of the cyanide [21] or thiosulfate [9] substrates have also been used. Polarographic methods can be used to follow all of these species [22] as well as other substrates (and their products) that can be utilized in rhodanese-catalyzed reactions [23]. When dihydrolipoate is used as the sulfur-acceptor substrate, the reaction can be followed spectrophotometrically at the wavelength of the absorption maximum of oxidized lipoate, which is formed by the rapid, spontaneous decomposition of the immediate product lipoate persulfide. This method has been used extensively in kinetic studies of the enzyme [24,25]. When aromatic thiosulfonates are used as sulfur donors, the disappearance of the ultraviolet absorption maximum of these compounds can also be used to follow the reaction [26].

Studies Relating to Formal Mechanism

The double-displacement formal mechanism of rhodanese action was first inferred from the results of polarographic studies in which stoichiometric amounts of the crystalline enzyme were mixed with cyanide or sulfite [22]. These ions had been known previously to be acceptor substrates. In each case the result was the disappearance of substrate, with the production of an equivalent amount of thiocyanate or thiosulfate as product. The implication that the isolated enzyme contained bound, reactive substrate sulfur was clear. This conclusion was confirmed and extended by studies with radioactive sulfur [27]. Rhodanese became labeled when treated with $^{35}SSO_3{}^{2-}$ but not when treated with $S^{35}SO_3{}^{2-}$. The radioactive sulfur could be removed from the enzyme, apparently instantaneously, by treatment at 0°C with cyanide to form $^{35}SCN^-$ or with sulfite to form $^{35}SSO_3{}^{2-}$.

With thiosulfate and cyanide as substrates, therefore, the most probable general mechanism appeared to consist of reactions (1) and (2).

$$SSO_3{}^{2-} + \text{enzyme} \rightleftharpoons SO_3{}^{2-} + \text{enzyme—S} \tag{1}$$

$$\text{Enzyme—S} + CN^- \longrightarrow SCN^- + \text{enzyme} \tag{2}$$

In all such work, however, there is the chance that the substituted enzyme found is a side issue, not in the pathway of the main flux to products. As kinetic methods were developed strongly during the 1960's, it became clear that these provided a means for testing the relevance of such isolated intermediates to the actual catalysis. Application of steady-state kinetic methods to the rhodanese-catalyzed sulfur transfer between thiosulfate and dihydrolipoate showed unequivocally that the mechanism involved a substituted enzyme as intermediate [24]. Initial velocity patterns were all of the expected parallel variety, modified only by the double-competitive substrate inhibition that is characteristic of double-displacement enzymes. Later extension of studies to the thiosulfate–cyanide and thiosulfonate–cyanide sulfur-transfer reactions yielded essentially similar results [28]. Similar results have also been reported for the rat liver rhodanese [16]. Kinetic results confirming the same basic formal mechanism have also been obtained for the human liver rhodanese, although in this case an isomerization of the substituted enzyme also appears to be involved in determining the velocity [15].

Since a mechanism involving a sulfur-substituted enzyme intermediate has been established as the basic form of the rhodanese catalytic cycle, the next question logically concerns the number and positions of kinetically significant intermediates in the cycle. This question appears to have been settled by kinetic studies employing comparisons of alternate substrates. Since rhodanese-catalyzed sulfur transfer from thiosulfate to cyanide at substrate saturation was two orders of magnitude slower than similar transfer from organic thiosulfonates (RSO_2S^-), it was concluded that the enzyme must form a kinetically significant complex with thiosulfate [29]. In addition, with thiosulfate as the donor, it was found that dihydrolipoate gave a lower maximal velocity at pH 8.6 than did cyanide, yielding the conclusion that a kinetically significant complex is formed with dihydrolipoate [24]. Thus, with thiosulfate and lipoate as substrates, the formal mechanism can be represented by reactions (3)–(7), the first four of which involve the enzyme. With organic

$$SSO_3^{2-} + \text{enzyme} \rightleftharpoons (\text{enzyme} \cdot SSO_3^{2-}) \qquad (3)$$

$$(\text{enzyme} \cdot SSO_3^{2-}) \rightleftharpoons SO_3^{2-} + \text{enzyme—S} \qquad (4)$$

$$\text{enzyme—S} + RCH(SH)CH_2CH_2SH \rightleftharpoons (\text{enzyme—S} \cdot RCH(SH)CH_2CH_2SH) \quad (5)$$

$$(\text{enzyme—S} \cdot RCH(SH)CH_2CH_2SH) \rightleftharpoons \text{enzyme} + RCH(SSH)CH_2CH_2SH \quad (6)$$

$$RCH(SSH)CH_2CH_2SH \longrightarrow H_2S + RCH(\overline{S)CH_2CH_2S} \qquad (7)$$

thiosulfonates as sulfur donors no kinetically significant substrate complexes with the enzyme have been established [28]. Similarly, no cyanide substrate complexes significant for initial velocity have been established definitely, although product inhibition studies have shown the formation of a complex of free enzyme with thiocyanate, which may indicate the occurrence of a significant intermediate in the reaction of cyanide with the sulfur-substituted enzyme [30].

A recent study of the pH dependencies of the steady-state kinetic parameters in the rhodanese-catalyzed thiosulfate–cyanide reaction has resulted in an expansion of the formal mechanism along a pH dimension [31]. These results show which protonation states of the free enzyme, enzyme–thiosulfate complex, and sulfur-substituted enzyme are catalytically competent species and lead to inferences concerning the chemical mechanism of action (see below).

Enzyme Conformational Changes in the Catalytic Cycle

The occurrence of significant changes of conformation in rhodanese during the course of its catalysis was first suggested by the results of studies on the apparent thermodynamic and activation parameters of substrate thiosulfate binding and scission [10]. The subsequent studies of Wang and Volini established that the true thermodynamic and activation parameters based on bisubstrate kinetic analysis, with extrapolation of all inhibiting anion concentrations to zero, support the suggestion of major conformational change in this reaction [30]. Further, these results indicate that successive changes to more constrained conformational states occur as substrate thiosulfate is bound to the enzyme and the first product, sulfite, is released from the sulfur-substituted enzyme. Relaxation back to the unconstrained state occurs only with release of the last product, thiocyanate.

Importantly, these findings were followed up and confirmed by direct spectroscopic observations of the enzymatic forms in the catalytic cycle [32] as well as of some inert conformers [33]. The magnitude of the changes seen by observation of the absorption band of the peptide chromophore with optical rotatory and absorption techniques corresponds to what would be seen on converting some 20 amino acid residues all the way from random coil to β structure as thiosulfate is bound and sulfite is discharged. It is striking that the association of sulfate with the enzyme, a binding competitive with that of substrate thiosulfate, occurs unaccompanied by conformational changes.

The foregoing meticulously documented conformational mobility of rhodanese in response to the presence of substrates has also led to further developments. When human liver rhodanese was isolated and subjected to steady-state kinetic examination, it was found to yield double-reciprocal plots for cyanide as varied substrate that were concave from below when and only when thiosulfate was the donor substrate [15]. One formal mechanism consistent with this behavior would involve the occurrence of a sulfur-substituted rhodanese that undergoes a conformational transition at a rate slow enough to be significant in the steady-state velocity. Application of this model to the kinetic data obtained yielded a good fit [15], but other explanations for the nonlinearity are, of course, still possible.

A second consequence of the demonstration of substantial conformational mobility in rhodanese in response to substrates has been a kinetic study of a different sort. The question examined concerned the possibility that the enzyme might "remember" the structure of the substrate from which it received the transferable sulfur atom for a significant period after the first product, which contains that structure, has been discharged from the enzyme. The data established that the second-order rate constant for the cyanide

reaction of sulfur–enzyme obtained in the presence of ethanethiosulfonate
was two and one-half times that for the same reaction of the sulfur–enzyme
obtained in the presence of thiosulfate [34]. The simplest interpretation of these
results appears to be in terms of the occurrence of a conformation-based
"enzymic memory." It is not known how general such a phenomenon may
be among double-displacement enzymes.

Chemical Mechanism of Bond Cleavage

The evident simplicity of the inorganic reaction between thiosulfate and
cyanide which rhodanese catalyzes has invited efforts to elucidate the chemical mechanism of the enzyme-catalyzed reaction. The thiosulfate anion is a
simple tetrahedral structure with all bond lengths and angles known [35].
The structure is exceedingly like that of sulfate, with the minor distortion
that the S—S bond at 2.00 Å is almost 40% longer than the S—O bonds. The
outer sulfur atom is in the sulfenyl (-2) oxidation state, while the central
sulfonyl sulfur atom has an oxidation number of $+6$ [36]. Both nucleophilic
attack on such sulfenyl atoms and electrophilic catalysis of S—S bond
scissions are general phenomena in sulfur chemistry that are rather well
understood at present [37].

The essential chemical functions of rhodanese are (a) cleavage of a S—S
bond between a sulfane sulfur atom and a second sulfur atom which may be
at any oxidation level, in a donor substrate, (b) stabilization of the sulfane
atom after discharge of the product containing the second sulfur atom, and
(c) acceptance of a thiophilic acceptor substrate which then reacts to remove
the sulfane atom.

Preliminary to cleavage of the S—S bond of thiosulfate, rhodanese forms
a kinetically significant complex with the sulfane donor, as mentioned in the
foregoing section. Studies employing variation of ionic strength and dielectric
constant yielded results suggesting that the enzyme contains a cationic
binding site for this substrate [29]. Later studies indicated that the thiosulfate binding site has a positive charge of 2 [10,30]. At present, general
agreement among workers in this field concerning the chemical identity of the
cationic site has not been established (see the section on structure below).

The cationic binding site appears also to serve as an electrophilic catalyst
assisting nucleophilic cleavage of the S—S bond of the donor substrate.
Replacement of one of the negatively charged oxygen atoms of thiosulfate
ion with an alkyl or aryl group greatly facilitates nucleophilic attack on the
sulfenyl sulfur. Such replacements also result in rhodanese donor substrates
(RSO_2S^-) for which the rate constants for S—S bond cleavage are much
greater than is that for thiosulfate [38]. The inference was made that scission
of the S—S bond of thiosulfate by the enzyme involves electrophilic catalysis

of a nucleophilic attack. Juxtaposition of an electrophilic group would achieve the same result as the organic replacement cited above: conversion of the sulfonyl fragment into a better leaving group [37].

A mechanism of this kind requires a nucleophile. Data relating to the activation parameters for the bond cleavage step in the enzyme-catalyzed reaction led to the inference that, despite the electrophilic catalysis, a strong sulfur nucleophile, one not markedly weaker than CN^-, was probably required to achieve the rate increase seen [10]. Sörbo had earlier proposed the occurrence of a reactive sulfhydryl group in rhodanese to fulfill the role of a nucleophile in the mechanism [38]. Wang and Volini subsequently reported a variety of further evidence establishing the existence of an essential sulfhydryl group in rhodanese [39], and there appears to be general acceptance of this sulfhydryl group as the nucleophile active in the catalytic mechanism. Further, in the nucleophilic cleavage of thiosulfate as well as of the thiosulfonates, the nucleophilic attack must almost certainly be on the sulfenyl rather than the sulfonyl sulfur [37]. This formulation thus satisfies the requirement for discharge of the sulfite or sulfinate as first product, with formation of a sulfur-substituted enzyme.

One recent study has been interpreted in a way that is substantially at odds with the above-mentioned kind of mechanism. From an investigation of the sulfhydryl groups of rhodanese, Man and Bryant [40] proposed a mechanism in which two substrate thiosulfate ions reacted with a disulfide bond in the enzyme, producing two enzymatic sulfhydryl groups and a tetrathionate ion. These products were presumed to be present as a complex stabilized by unspecified forces. The complex reacted with two cyanide anions to regenerate the disulfide bond and produce two ions each of thiocyanate and sulfite.

This hypothesis has several weaknesses. The mechanism does not contain a stable substituted enzyme form capable of reacting with cyanide to produce only thiocyanate or with sulfite to produce thiosulfate, both reactions that are known to occur [22,27]. Further, Sörbo previously demonstrated convincingly that the freshly prepared enzyme contains no disulfide bond [38], a result repeatedly confirmed in other laboratories [39,41,42]. The proposed mechanism is also clearly inconsistent with the steady-state kinetic data that established a double-displacement formal mechanism for the rhodanese-catalyzed thiosulfate–cyanide reaction [15,16,28,31].

A principal basis for the Man and Bryant mechanism was the observation that rhodanese is inactivated by treatment with dithiothreitol [2,40], the presumption being that reduction of a disulfide bond was the mode of inactivation. However, Kim and Horowitz have recently proposed a quite different mechanism for this inactivation [43]. The experimental studies of these investigators indicate that the dithiothreitol inactivation of rhodanese requires the participation of oxygen and involves the formation of a mixed disulfide between the reagent and an enzymatic sulfhydryl group. Failure of the disulfide to cleave in the manner considered normal for mixed disulfides of dithiothreitol is dependent on maintenance of the native conformation of the enzyme. The inference is drawn that the similarity of dithiothreitol to the rhodanese acceptor substrate dihydrolipoate is involved in the constraints to cleavage of the mixed disulfide. Recently, this interpretation has been confirmed and somewhat extended [43a].

The result of all these considerations is that there is little experimental evidence available to support the Man and Bryant hypothesis, while there is much that refutes it and favors the nucleophilic mechanism. It thus appears that nucleophilic cleavage assisted by electrophilic catalysis is the most probable fundamental mechanism by which rhodanese attacks the S—S bond of thiosulfate and thiosulfonates.

A completely explicit chemical mechanism of this kind has been proposed in an attempt to correlate all of the available information [31]. It was established kinetically that the binding of substrate thiosulfate to free rhodanese requires that an enzymatic group with a pK' of 9.9 be protonated. The mechanism relating to this electrophilic group is proposed in terms of the ionization of a water ligand of a metal ion at the active site, but the data would also be consistent with direct protonation of the nitrogen of basic residue side chains. The reaction that results in S—S bond cleavage in the donor substrate requires that an enzymatic group with a pK' of 6.5 be deprotonated. This group is proposed to be the essential sulfhydryl group, the pK' lowered by interaction with the divalent cationic site. The proposed mechanism constitutes a detailed qualitative summary, a working hypothesis consistent with a considerable body of evidence. It must be remembered, however, that any attempt to take quantitative account of all the factors in this reaction will have to deal with the large enzyme conformational changes that must enter into the overall energetics [30,32].

The Sulfur-Substituted Enzyme

There has been much discussion of the nature of the sulfur-substituted intermediate form of rhodanese. The first suggestion of such an intermediate invoked a trisulfide structure [44]. When Sörbo showed that the free enzyme contains no disulfide bond, however, he suggested formulating the intermediate as a persulfide [38]. Villarejo and Westley were unable to find the expected persulfide absorption band in the ultraviolet spectrum of the sulfur–rhodanese [23], and Davidson and Westley proposed the possibility of a binding by charge-transfer interaction with a tryptophyl side chain [45]. The basis for this suggestion was activity loss correlated with tryptophyl loss on treatment with N-bromosuccinimide, inhibition by pyridinium compounds, and fluorescence quenching of the sulfur–enzyme relative to the free enzyme. Nevertheless, the mechanism in which a sulfhydryl nucleophile cleaves the sulfur-donor substrate does clearly require generation of a covalent persulfide intermediate at some stage of the reaction.

Consideration of the stability of the sulfur–enzyme appears to contain the essential clue to this puzzle. The sulfur–rhodanese is enormously more stable than small molecular persulfides, which are not stable enough to be isolated, and the factors affecting its stability must also affect its spectral properties. Probably the persulfide sulfur occurs in the sulfur–enzyme as a ligand of the electrophilic site [31] as well as in hydrophobic association with apolar side chain groups [41]. Finazzi Agrò et al. have demonstrated a small absorption increment in the sulfur–rhodanese spectrum at wavelengths where persulfides absorb [46], and this probably represents the diminished absorption of the stabilized persulfide. Further, Guido and Horowitz have recently reported an investigation of the quenching of rhodanese fluorescence by bound substrate sulfur [47]. The results of this study, which employed cesium ion-resolved emission spectra to distinguish solvent accessibility of tryptophyl residues, indicate that the quenching is probably not caused by direct contact of the sulfur with a tryptophyl residue but by nonradiative energy transfer. The same workers have also reported a reinvestigation of the N-bromosuccini-mide inactivation of rhodanese, which demonstrated sulfhydryl loss rather than tryptophyl loss as a result of treatment with this reagent [48]. Another recent study by the same research group indicates that the binding of pyridin-ium inhibitors for rhodanese [49] is stabilized by ionic interaction rather than charge transfer. By eliminating the charge-transfer binding hypothesis, these conclusions also favor a persulfide structure for the sulfur–enzyme.

The reaction of the sulfur-substituted rhodanese with many good sulfur nucleophiles is rather slow. That is, not all thiophiles can serve well as acceptor substrates, and there is no general production of elemental sulfur or sulfide as side product. With dihydrolipoate as acceptor substrate, however, the re-action proceeds smoothly to sulfide by way of a kinetically significant inter-mediate enzyme–substrate complex and the immediate product lipoate persulfide [23,24]. In contrast, the reaction with cyanide anion as acceptor substrate to produce thiocyanate appears to involve no intermediate longer lived than a transition-state complex. At pH values below a pK' near 6, the second-order rate constant for the cyanolysis approaches values expected for diffusion-controlled reactions and the enthalpy of activation is appropriately very small [31].

Structure

Bovine liver rhodanese is a small protein containing no residues other than amino acids [50,51]. It is available in substantial quantities as purified to crystallinity from bovine liver by current methods [11]. The apparently homogeneous protein has been studied by a great variety of procedures. Nevertheless, two aspects of rhodanese structure continue to be topics of controversy in the current literature. These are the question of the molecular weight of the smallest functional form occurring naturally and the question of the identity of the electrophilic site required by the mechanism.

Sörbo, who was the first to purify rhodanese to homogeneity, also determined its molecular weight by sedimentation velocity–diffusion measurements [8], obtaining a value of 37,500 daltons. The sedimentation values were confirmed repeatedly in other laboratories [12,39]. However, it was found in subsequent studies utilizing column chromatography [52] and fluorescence polarization techniques [41] as well as in sedimentation studies of the enzyme in solutions containing dodecyl sulfate [39] that, if care was taken to prevent oxidation, the enzyme could be dissociated into two polypeptide chains of equal size, about 19,000 daltons. Peptide mapping analysis indicated that the two chains were probably identical [52], a result subsequently confirmed in another laboratory [51].

These conclusions agreed reasonably well with some analytical data for the number of substrate sulfur atoms that could be bound in the sulfane-substituted enzyme. In one laboratory, polarographic determinations had yielded a value of about 1.9 sulfane atoms per 37,000 daltons [22]; ^{35}S tracer techniques gave a value of 1.5 atoms [27]. In another laboratory, a colorimetric method gave 1.1–1.3 atoms [38,50]. The colorimetric procedure subsequently gave 1.35 atoms in a third laboratory [51]. An unspecified method has recently given an average value of 1.5 atoms per 37,000 daltons for the bovine kidney enzyme in yet another laboratory [53]. A value of 1.4 for the ratio of bound sulfur to rhodanese as determined by a polarographic method has also been reported [43a]. Since all of the values are between 1 and 2, it was suggested that there are two active sites per dimeric molecule, the bound sulfur being slowly dissociable on repeated washing and recrystallization [2]. This notion was also in accord with the particularly clear demonstration in rhodanese of two fully equivalent essential sulfhydryl groups per 37,000 daltons [39] and the demonstration of two equivalent binding sites for divalent cations per 37,000 daltons [52,54,55].

Recently, the dimeric structure of the rhodanese of 37,000 daltons has been called into question. Several research groups have reported being unable to demonstrate the occurrence of an 18,000-dalton species of bovine liver rhodanese under denaturing conditions [56–60]. These reports give the molecular weight of the enzyme variously at values ranging from 32,000 to 35,000 daltons.

In the meantime, work on the primary structure of bovine liver rhodanese has proceeded, but progress has been slow and beset with low peptide yields and other unusual difficulties. There has continued to be controversy regarding even the amino acid composition [61]. However, the current version of the total covalent structure [62] adds up to 32,828 daltons (293 residues).

Work on the three-dimensional structure as inferred from X-ray crystallographic data has also been carried out. The most recent published interpretation (3 Å resolution) consists of 266 residues distributed into two domains of equal size and exceedingly similar conformation despite having dissimilar primary structures [58]. The crystallographers assume that their two-domain, single polypeptide structure is a product of gene duplication followed by extensive separate evolution of the two domains. However, this interpretation appears to be inconsistent with Oi's recent finding that native rat liver rhodanese has a molecular weight of 17,500 daltons, emerging from calibrated Sephadex G-200 columns at virtually the same elution volume as sperm whale myoglobin [16].

It is conceivable that many of these observations could be explained by the occurrence of cleavage of a double-domain structure in some preparations, the cleaved enzyme remaining fully active. It has been shown that rhodanese partially degraded by trypsin retains activity (actually has increased activity per weight of protein), but these preparations were reported to contain no active species smaller than about 29,000 daltons [56]. On the other hand, the X-ray crystallographic work evidently has been done on a preparation containing an active protease that cleaved the rhodanese to 11,000-dalton pieces upon denaturation in 2% sodium dodecyl sulfate at pH 8.0 [58].

A logical alternative explanation would be the artifactual linkage of the active rhodanese polypeptide to another polypeptide in some preparations. It has been shown, for example, that denaturation of rhodanese with dodecyl sulfate or urea results in a sulfhydryl oxidation that is dependent on the detailed kinetic course of the denaturation [61a].

The fact is that neither of these possible general explanations appears to be capable of explaining all of the published data. Final evaluation of the identity of the smallest functional form of rhodanese occurring naturally will have to await the full publication of sequence and X-ray crystallographic evidence and quite likely further experimental studies.

The second unresolved structure-related question has to do with the obligatory occurrence of a metal ion as a part of the fully functional active site of rhodanese. Unaware of Sörbo's earlier report that rhodanese contains no heavy-metal ions [63], Volini et al. reported finding two zinc ions per 37,000 daltons in a series of rhodanese preparations [52]. Although the zinc could be removed without loss of activity [52], a result also confirmed in two other laboratories [64,65], the occurrence of the metal binding site appears to be well established [52,64,66]. Further, since solutions of the common substrates for rhodanese all contain measurable concentrations of divalent cations, it has remained unclear whether the enzyme is metal ion activated. Volini's recent preliminary reports of experiments relating to this question [55,67] indicate that binding of a cyanide complex of cupric ions results in an active but altered rhodanese; the complex with nickel is totally inhibitory; and those of alkaline-earth ions, which associate very strongly with the enzyme, yield the activity parameters regarded as normal. It is interesting that in all of this work there are two fully equivalent cation binding sites per 37,000 daltons which affect the activity. Publication of the detailed evidence relating to the metal ion involvement is awaited with a good deal of interest.

Model Catalytic Systems

Some work has been done with nonenzymatic catalysts for the rhodanese reaction. The chemical features of bovine rhodanese that have reasonably established importance for catalysis are (a) an electrophilic cationic site, (b) a sulfhydryl nucleophile, and (c) an apolar site capable of stabilizing the sulfur-substituted catalyst with respect to simple dissociation to yield elemental sulfur and with respect to reaction with solvent or buffer-derived nucleophiles other than acceptor substrates.

Although simple, small molecular thiols are poor catalysts for the thiosulfate–cyanide reaction, the present mechanism for rhodanese action suggests that substrates having electron distributions more like the inferred enzyme–thiosulfate complex (e.g., thiosulfonates) might be readily cleaved by simple sulfhydryl compounds, forming persulfides (alkyl disulfanes). It has in fact been shown that cysteine, mercaptoethylamine, glutathione, and N-acetylcysteine are all catalysts for the reaction between alkyl thiosulfonates and cyanide [28,68,69]. The reaction mechanism involves a persulfide type of substituted catalyst that is demonstrable both spectrophotometrically and kinetically [28]. As might be expected, these catalysts are far less efficient than the enzyme, particularly in their failure to protect the substituted catalyst intermediates from side reactions.

Some evidence possibly relating to analogs for the known apolarity of the active-site microenvironment of rhodanese has been sought in recent studies of micellar catalysis of the thiosulfonate–cyanide reaction. Micelles of cetyltrimethylammonium bromide were found to provide substantial catalysis for this reaction only when the donor substrates were aromatic thiosulfonates [70]. The mechanism of this catalysis is probably very similar to that described by Bunton for cationic micellar catalysis of other reactions of aromatic anions [71].

3-MERCAPTOPYRUVATE SULFURTRANSFERASE

Like rhodanese, 3-mercaptopyruvate sulfurtransferase forms a sulfur-substituted enzyme that can transfer its sulfane sulfur to a variety of strong thiophiles, including cyanide, sulfite, sulfinates, and thiols.* Unlike rhodanese, however, this enzyme cleaves a C—S bond rather than an S—S bond in the donor substrate. 3-Mercaptopyruvate sulfurtransferase also occurs in the cytosol rather than the mitochondria of liver and kidney tissue and is present in practically all tissues, including erythrocytes, as well as in some bacteria.

The literature relating to 3-mercaptopyruvate sulfurtransferase is much less voluminous than that relating to rhodanese. It was very ably reviewed by Sörbo [7] in 1975, and there is no large volume of new material since that time. Therefore, the review that follows is selective and quite brief.

Studies Relating to Mechanism

Highly purified 3-mercaptopyruvate sulfurtransferase from either mammalian [72] or bacterial [73] sources is not a markedly stable enzyme. Probably for this reason and because of complexities inherent in assay procedures for this activity, no bisubstrate kinetic studies of this enzyme have been reported. Nevertheless, there is reason to suppose that the mechanism is basically a double displacement.

Among the earliest observations having to do with this activity was the finding that a progressive turbidity, which consisted of a mixed precipitate of protein and sulfur, appeared in direct proportion to the amount of pyruvate formed from mercaptopyruvate on incubation with liver preparations or *Escherichia coli* extracts at neutral pH [74]. Sörbo subsequently proposed the occurrence of a persulfide enzyme intermediate [75], and Hylin and Wood demonstrated the accumulation of protein-bound polysulfide in the absence of sulfur-acceptor substrates [76]. Kun and Fanshier, who had earlier formulated the intermediate as an enzymatic trisulfide structure, presuming the occurrence of a disulfide bond at the active site [77], showed that the enzyme highly purified from rat liver acetone powder contained only a single half-cystine residue and so modified the proposed intermediate to a persulfide [72]. There are two half-cystine residues in the *E. coli* enzyme but one of these is a readily accessible nonessential cysteinyl residue [73], again eliminating the trisulfide structure for the sulfur-substituted intermediate from consideration.

All of the foregoing suggestions presume a double-displacement formal

---

* As a note of caution, it must be pointed out that the participation of the 3-mercaptopyruvate sulfur–enzyme in the main flux of the catalytic cycle has not been established definitely, as it has been for the sulfur–rhodanese. This point will not be settled fully until the necessary kinetic studies are reported.

mechanism for mercaptopyruvate with a sulfur-substituted enzyme intermediate [Eqs. (8) and (9)]. With cyanide as acceptor substrate, then, this

$$\text{HSCH}_2\text{C(O)COO}^- + \text{enzyme} \rightleftharpoons \text{CH}_3\text{C(O)COO}^- + \text{enzyme—S} \qquad (8)$$

$$\text{Enzyme—S} + \text{acceptor} \rightleftharpoons \text{S—acceptor} + \text{enzyme} \qquad (9)$$

mechanism appears to have much in common with the reaction mechanism for rhodanese catalysis summarized in reactions (1) and (2). No detail at the level of reactions (3)–(7) is available at present for the 3-mercaptopyruvate sulfurtransferase reaction.

Despite the limitations on data available, however, the strong analogy to the more easily studied rhodanese reaction invites some comparisons that may serve to illuminate the mechanism. One point of comparison has to do with the stability of the sulfur–enzyme intermediates. The sulfur-substituted rhodanese, as noted above, is a stable species that is poorly reactive with even such powerful thiophiles as mercaptoethanol. In contrast, 3-mercaptopyruvate sulfurtransferase transfers sulfur to mercaptoethanol or to protein sulfhydryl groups generally to form persulfides. Depending on the pH and availability of additional sulfhydryl groups, these persulfides may give rise to polysulfide accumulation or the discharge of either elemental sulfur or HS^-. Even in the presence of CN^- as acceptor substrate, thiocyanate formation is by no means stoichiometric with pyruvate formation, and the cleavage of mercaptopyruvate is stimulated by inclusion of mercaptopyruvate as a more facile acceptor [73]. It will be of great interest to compare the rhodanese and 3-mercaptopyruvate sulfurtransferase active-site microenvironments when this information becomes available.

The second area in which a comparison of the two sulfurtransferases appears to be useful has to do with the fundamental mechanism of bond cleavage. By analogy to the rhodanese mechanism, as well as because of the occurrence of a persulfide intermediate, it might be anticipated that 3-mercaptopyruvate sulfurtransferase would cleave the C—S bond of the donor substrate by an electrophile-assisted nucleophilic attack. Data establishing the electrophilic catalysis directly have not been obtained, partly because mercaptopyruvate is the only known donor substrate [78]. However, various bound metal ions have been reported to occur in highly purified preparations of the enzyme. Kun and Fanshier found cupric ion in the rat liver enzyme at a ratio of 1 copper per 40,000 daltons [77] or 1 copper per 10,000 daltons [72]. On the other hand, Van Den Hamer et al. were able to remove virtually all of the copper from their preparations of the rat liver and erythrocyte 3-mercaptopyruvate sulfurtransferases without inactivating the enzymes [79]. Vachek and Wood found 1.1 zinc and 0.5 copper ions per 24,000 daltons in the E. coli enzyme [73].

Perhaps the most convincing argument for an electrophilic participation in the bond cleavage proceeds from the finding of Vachek and Wood that Mn^{2+} catalyzes the decomposition of mercaptopyruvate to yield persulfide sulfur [73]. This highly suggestive model catalysis invites mechanistic speculation based on an electrophilic mechanism and the evidence for the occurrence of mercaptopyruvate primarily in the enol form in aqueous solutions [80].

Structure

The 3-mercaptopyruvate sulfurtransferase purified extensively from rat liver acetone powders was reported to have a molecular weight of 35,000–40,000 daltons, calculated from sedimentation velocity and diffusion data [77]. Subsequent studies in which the enzyme was separated by chromatography on DEAE-cellulose into unstable active subunits yielded 10,000 daltons as the molecular weight of the subunit. With the possible exception of a metal ion, the enzyme appears to contain no cofactor or prosthetic group. There is one cysteinyl residue per subunit [72].

The enzyme purified to homogeneity from *E. coli* has a molecular weight of about 24,000 daltons, as determined by sedimentation equilibrium measurements [73]. Chromatography on DEAE-cellulose also dissociates this enzyme into active fragments, apparently nonidentical but of the same size, around 12,000 daltons. There are two nonequivalent sulfhydryl groups per 24,000 daltons, accounting for all the half-cystine residues in the molecule. Like the rat liver enzyme, the *E. coli* 3-mercaptopyruvate sulfurtransferase appears to contain no prosthetic group. Unlike the results reported for the rat liver enzyme, however, chelators fail to inhibit the *E. coli* enzyme [73].

THIOSULFATE REDUCTASE

There seem to be at least two potential mechanisms for the biological reduction of the planetary sulfur atom of thiosulfate and probably of sulfane sulfur generally. One of these involves the reaction of sulfur-substituted rhodanese with dithiol acceptor substrates such as dihydrolipoate to produce unstable persulfide products that undergo spontaneous decompositions to inorganic sulfide and the intramolecular disulfide of the substrate, as shown in reactions (3)–(7). The other mechanism, which involves a separate enzyme, utilizes glutathione as an electron donor for the reduction and appears to be present in many bacteria and fungi as well as in animal tissues. Both Sörbo [18] and Koj [5,19] have provided evidence indicating that the glutathione-dependent enzyme is responsible for most of the thiosulfate reduction in liver. On the other hand, Leinweber and Monty have provided evidence indicating that the glutathione-dependent enzyme, although present, is *not* responsible for most of the thiosulfate reduction in the bacterium *Salmonella tryphimurium* [81]. The substantial literature on the assimilatory and dissimilatory thiosulfate reduction processes of microorganisms has recently been reviewed by Siegel [82] (see also the recent paper by Chambers and Trudinger [83]).

A glutathione-dependent thiosulfate reductase was purified from yeast by Kaji and McElroy, who proposed a mechanism of action involving a sulfenyl thiosulfate intermediate formed by attack of thiosulfate ion on a presumed disulfide bond in the active site of the enzyme [4]. Available chemical evidence on the reactions of sulfenyl thiosulfates suggests, however, that such a mechanism is probably untenable [44]. Instead, quite possibly the S—S bond of thiosulfate is cleaved by an electrophilic–nucleophilic mechanism similar to that of rhodanese, followed by transfer to the glutathione sulfhydryl group to form a persulfide that discharges sulfide on reaction with a second glutathione sulfhydryl group. The extraordinary instability and elusiveness of this enzyme in animal tissues have thus far prevented systematic studies of the mechanism.

CYSTATHIONASE (CYSTATHIONINE γ-LYASE)

The enzyme responsible for cleavage of a C—S bond in the thioether cystathionine has also been found to be capable of producing sulfide from cysteine under some conditions [84–88]. Cystathionase is of relevance here because it appears that the actual substrate in this reaction is the disulfide cystine and the immediate product is the disulfane cysteine persulfide. This reaction, then, might be no less a source of sulfane sulfur than is the 3-mercaptopyruvate sulftransferase reaction. Szczepkowski and Wood have demonstrated that a coupled system consisting of cystathionase and rhodanese can degrade cystine with production of thiosulfate when sulfite is available [89].

The two enzymes that have in common the ability to cleave C—S bonds with the production of sulfane sulfur, 3-mercaptopyruvate sulfurtransferase and cystathionase, might be expected to have much in common mechanistically. It is well established, however, that the detailed chemical mechanisms for these two enzyme-catalyzed reactions must differ. Cystathionase is a pyridoxal phosphate enzyme [90,91]. As with all pyridoxal phosphate-dependent enzymatic reactions, the mechanism involves formation of a Schiff base between the substrate and the pyridoxal phosphate. This Schiff base involving the amino group of substrate cystine is understood to assist in the formation of an enzyme-stabilized carbanion intermediate which could in general cleave in any of several ways. The cystathionase apoenzyme, however, directs the occurrence of the cleavage to the C—S bond, resulting in the elimination of cysteine persulfide, followed by hydrolysis of the amino-acrylate Schiff base to yield pyruvate and ammonia [87,91]. Information on the specific properties of the different cystathionases is available in the recent reviews of Greenberg [3] and Flavin [92].

PHYSIOLOGICAL FUNCTIONS OF SULFANE SULFUR

There is a considerable variety of sulfane-containing compounds in biological materials. Injection of $^{35}SSO_3^{2-}$ into rats gives rise very rapidly to radioactive polythionates and radioactive elemental sulfur associated with serum proteins [93]. Organic and inorganic per- and polysulfides are known products of enzyme-catalyzed reactions, as noted in foregoing sections of this review. Thiosulfate is also well known as a biological product. Neither the quantities nor the functions of this sulfane sulfur seem to have been much investigated, but it has been shown that radioactive sulfur equilibrating among these forms *in vivo* is metabolically active and can be incorporated into the cysteine synthesized by rats [93]. This mixed pool of sulfane sulfur is also the source of the sulfur that can achieve detoxication of cyanide and sulfide where this occurs to a significant extent [2,7,89,94]. Some discussion and evaluation of these several roles appear in recent reviews [2,7,95].

The recent research literature suggests at least three additional possible roles for sulfur contributing to the sulfane pool. (a) The sulfur of mercapto-pyruvate has been found to be a superior precursor for the sulfur of the thionucleotides of tRNA [96]. These thionucleotides, which receive their sulfur after formation of the RNA, appear to have an essential structural role in the apparatus of protein synthesis, either at the level of the detailed secondary structure of the tRNA or in the recognition site itself. (b) The regular occurrence of sulfane sulfur proposed to be present as a persulfide has been reported in both of the iron–sulfur flavoprotein hydroxylases, xanthine oxidase and aldehyde oxidase [97]. The report that cystine tri-sulfide is an activator of the photosynthetic bacterial enzyme 5-amino-levulinate synthetase [98] may be a further indication of a structural role for sulfane sulfur in protein chemistry, as may the finding that the reconstitutive capacity of succinate dehydrogenase can be restored by treatment with rhodanese [99]. (c) The sulfane pool is very likely the source of the inorganic sulfur used in synthesis of the iron–sulfur proteins. This may occur directly, as in the reconstitution of ferredoxins by treatment with rhodanese and thiosulfate [100] or 3-mercaptopyruvate and its sulfurtransferase [101]. Alternatively, it may be that sulfane sulfur formed and transferred in the liver by some of the mechanisms discussed in this review is exported to peripheral tissues essentially as elemental sulfur in combination with plasma proteins. Reduction to the sulfide level of oxidation at which it occurs primarily in the iron–sulfur proteins could then be closely coordinated with synthesis of the iron–sulfur centers, a clear necessity in view of the extraordinary toxicity of inorganic sulfide.

ACKNOWLEDGMENTS

Work in this laboratory on topics related to this review is supported by National Science Foundation research grant BMS74-18144 and United States Public Health Service research grant GM-18939.

REFERENCES

1. A. B. Roy and P. A. Trudinger, "The Biochemistry of Inorganic Compounds of Sulphur," pp. x–xiii. Cambridge Univ. Press, London and New York, 1970.
2. J. Westley, *Adv. Enzymol.* **39**, 327 (1973).
3. D. M. Greenberg, *Metab. Pathways, 3rd Ed.* **7**, 505–528 (1975).
4. A. Kaji and W. D. McElroy, *J. Bacteriol.* **77**, 630 (1959).
5. B. Sido and A. Koj, *Acta Biol. Cracov., Ser. Zool.* **15**, 97 (1972).
6. C. P. Barton and J. M. Akagi, *J. Bacteriol.* **107**, 375 (1971).
7. B. Sörbo, *Metab. Pathways, 3rd Ed.* **7**, 433–456 (1975).
8. B. H. Sörbo, *Acta Chem. Scand.* **7**, 1129 and 1137 (1953).
9. B. Davidson and J. Westley, *J. Biol. Chem.* **240**, 4463 (1965).
10. K. R. Leininger and J. Westley, *J. Biol. Chem.* **243**, 1392 (1968).
11. P. Horowitz and F. DeToma, *J. Biol. Chem.* **245**, 984 (1970).
12. J. Westley and J. R. Green, *J. Biol. Chem.* **234**, 2325 (1959).
13. C. Cannella, L. Pecci, and G. Federici, *Ital. J. Biochem.* **21**, 1 (1972).
14. S. Oi, *Agric. Biol. Chem.* **37**, 629 (1973).
15. R. Jarabak and J. Westley, *Biochemistry* **13**, 3233 (1974).
16. S. Oi, *J. Biochem. (Tokyo)* **78**, 825 (1975).
17. B. H. Sörbo, *in* "Methods in Enzymology" (S. P. Colowick and N. O. Kaplan, eds.), Vol. 2, p. 334. Academic Press, New York, 1955.
18. B. H. Sörbo, *Acta Chem. Scand.* **18**, 821 (1964).
19. A. Koj, *Acta Biochim. Pol.* **15**, 161 (1968).
20. A. J. Smith and J. Lascelles, *J. Gen. Microbiol.* **42**, 357 (1966).
21. R. A. Llenado and G. A. Rechnitz, *Anal. Chem.* **45**, 826 (1973).
22. J. R. Green and J. Westley, *J. Biol. Chem.* **236**, 3047 (1961).
23. M. Villarejo and J. Westley, *J. Biol. Chem.* **238**, 4016 (1963).
24. M. Volini and J. Westley, *J. Biol. Chem.* **241**, 5168 (1966).
25. S. Oi, *J. Biochem. (Tokyo)* **76**, 455 (1974).

26. B. Sörbo, *Acta Chem. Scand.* **16**, 243 (1962).
27. J. Westley and T. Nakamoto, *J. Biol. Chem.* **237**, 547 (1962).
28. J. Westley and D. Heyse, *J. Biol. Chem.* **246**, 1468 (1971).
29. R. Mintel and J. Westley, *J. Biol. Chem.* **241**, 3381 and 3386 (1966).
30. S.-F. Wang and M. Volini, *J. Biol. Chem.* **248**, 7376 (1973).
31. P. Schlesinger and J. Westley, *J. Biol. Chem.* **249**, 780 (1974).
32. M. Volini and S.-F. Wang, *J. Biol. Chem.* **248**, 7386 (1973).
33. M. Volini and S.-F. Wang, *J. Biol. Chem.* **248**, 7392 (1973).
34. R. Jarabak and J. Westley, *Biochemistry* **13**, 3237 and 3240 (1974).
35. P. H. Laur, *Sulfur Org. Inorg. Chem.* **3**, 224–225 (1972).
36. M. Schmidt, *Sulfur Org. Inorg. Chem.* **2**, 92–94 (1972).
37. J. L. Kice, *Sulfur Org. Inorg. Chem.* **1**, 154–207 (1971).
38. B. Sörbo, *Acta Chem. Scand.* **16**, 2455 (1962).
39. S.-F. Wang and M. Volini, *J. Biol. Chem.* **243**, 5465 (1968).
40. M. Man and R. G. Bryant, *J. Biol. Chem.* **249**, 1109 (1974).
41. P. Horowitz and J. Westley, *J. Biol. Chem.* **245**, 986 (1970).
42. K. M. Blumenthal and R. L. Heinrikson, *J. Biol. Chem.* **246**, 2430 (1971).
43. S. K. Kim and P. M. Horowitz, *Biochem. Biophys. Res. Commun.* **67**, 433 (1975).
43a. L. Pecci, B. Pensa, M. Costa, P. L. Signini, and C. Cannella, *Biochim. Biophys. Acta* **445**, 104 (1976).
44. T. W. Szczepkowski, *Acta Biochim. Pol.* **8**, 251 (1961).
45. B. Davidson and J. Westley, *J. Biol. Chem.* **240**, 4463 (1965).
46. A. Finazzi Agrò, G. Federici, C. Giovagnoli, C. Cannella, and D. Cavallini, *Eur. J. Biochem.* **28**, 89 (1972).
47. K. Guido and P. M. Horowitz, *Biochem. Biophys. Res. Commun.* **67**, 670 (1975).
48. K. Guido and P. M. Horowitz, *Biophys. J.* **16**, 203a (1976).
49. R. D. Baillie and P. M. Horowitz, *Biochim. Biophys. Acta* **429**, 402 (1976).
50. B. Sörbo, *Acta Chem. Scand.* **17**, 2205 (1963).
51. K. M. Blumenthal and R. L. Heinrikson, *J. Biol. Chem.* **246**, 2430 (1971).
52. M. Volini, F. De Toma, and J. Westley, *J. Biol. Chem.* **242**, 5220 (1967).
53. S. Pagani, C. Cannella, P. Cerletti, and L. Pecci, *FEBS Lett.* **51**, 112 (1975).
54. R. L. Bryant and S. Rajender, *Biochem. Biophys. Res. Commun.* **45**, 532 (1971).
55. B. Van Sweringen and M. Volini, *Fed. Proc., Fed. Am. Soc. Exp. Biol.* **33**, 1379 (1974).
56. B. L. Trumpower, A. Katki, and P. Horowitz, *Biochem. Biophys. Res. Commun.* **57**, 532 (1974).
57. L. M. Ellis and C. K. Woodward, *Biochim. Biophys. Acta* **379**, 385 (1975).
58. J. Bergsma, W. G. J. Hol, J. N. Jansonius, K. H. Kalk, J. H. Ploegman, and J. D. G. Smit, *J. Mol. Biol.* **98**, 637 (1975).
59. J. Russell, L. Weng, P. S. Keim, and R. L. Heinrikson, *Biochem. Biophys. Res. Commun.* **64**, 1090 (1975).
60. J. M. Crawford and P. M. Horowitz, *Biochim. Biophys. Acta* **429**, 173 (1976).
61. R. D. Baillie and P. M. Horowitz, *Biochim. Biophys. Acta* **427**, 594 (1976).
61a. R. D. Baillie and P. M. Horowitz, *Biochim. Biophys. Acta* **429**, 353 (1976).
62. R. L. Heinrikson, J. Russell, L. Weng, and P. S. Keim, *Fed. Proc., Fed. Am. Soc. Exp. Biol.* **35**, 1622 (1976).
63. B. Sörbo, *Proc. Int. Pharmacol. Meet., 1st, 1961* Vol. 6, pp. 121–128 (1962).
64. R. C. Bryant and S. Rajender, *Biochem. Biophys. Res. Commun.* **45**, 532 (1971).
65. C. Cannella, L. Pecci, and G. Federici, *Ital. J. Biochem.* **21**, 1 (1972).
66. D. W. Bolen and S. Rajender, *Arch. Biochem. Biophys.* **161**, 435 (1974).

67. M. Volini, B. Van Sweringen, S.-F. Wang, and J. Leung, *Fed. Proc., Fed. Am. Soc. Exp. Biol.* **34**, 495 (1975).
68. J. Westley, *Bioinorg. Chem.* **1**, 245 (1972).
69. J. Westley and H. Taylor, *Adv. Enzyme Regul.* **13**, 339 (1975).
70. J. Westley, unpublished observations.
71. C. A. Bunton, *in* "Reaction Kinetics in Micelles" (E. H. Cordes, ed.), pp. 73–97. Plenum, New York 1973.
72. D. W. Fanshier and E. Kun, *Biochim. Biophys. Acta* **58**, 266 (1962).
73. H. Vachek and J. L. Wood, *Biochim. Biophys. Acta* **258**, 133 (1972).
74. A. Meister, P. E. Fraser, and S. V. Tice, *J. Biol. Chem.* **206**, 561 (1954).
75. B. Sörbo, *Biochim. Biophys. Acta* **24**, 324 (1957).
76. J. W. Hylin and J. L. Wood, *J. Biol. Chem.* **234**, 2141 (1959).
77. E. Kun and D. W. Fanshier, *Biochim. Biophys. Acta* **32**, 338 (1959).
78. H. Fiedler and J. L. Wood, *J. Biol. Chem.* **222**, 387 (1956).
79. C. J. A. Van Den Hamer, A. G. Morell, and I. H. Scheinberg, *J. Biol. Chem.* **242**, 2514 (1967).
80. W. D. Kumler and E. Kun, *Biochim. Biophys. Acta* **27**, 464 (1958).
81. F.-J. Leinweber and K. J. Monty, *J. Biol. Chem.* **238**, 3775 (1963).
82. L. M. Siegel, *Metab. Pathways, 3rd Ed.* **7**, 217–286 (1975).
83. L. H. Chambers and P. A. Trudinger, *J. Bacteriol.* **123**, 36 (1975).
84. F. Binkley and D. Okeson, *J. Biol. Chem.* **182**, 273 (1950).
85. D. Cavallini, B. Mondovi, C. DeMarco, and A. Scioscia-Santoro, *Arch. Biochem. Biophys.* **96**, 456 (1962).
86. D. Cavallini, B. Mondovi, C. DeMarco, and A. Scioscia-Santoro, *Enzymologia* **24**, 253 (1962).
87. M. Flavin, *J. Biol. Chem.* **237**, 786 (1962).
88. M. P. Roisin and F. Chatagner, *Bull. Soc. Chim. Biol.* **51**, 481 (1969).
89. T. W. Szczepkowski and J. L. Wood, *Biochim. Biophys. Acta* **139**, 469 (1967).
90. Y. Matsuo and D. M. Greenberg, *J. Biol. Chem.* **230**, 545 and 561 (1958).
91. H. C. Dunathan, *Adv. Enzymol.* **35**, 79 (1971).
92. M. Flavin, *Metab. Pathways, 3rd Ed.* **7**, 457–503 (1975).
93. J. F. Schneider and J. Westley, *J. Biol. Chem.* **244**, 5735 (1969).
94. M. Auriga and A. Koj, *Bull. Acad. Pol. Sci., Ser. Biol.* **23**, 305 (1975).
95. B. Sörbo, *Sulfur Org. Inorg. Chem.* **2**, 143–169 (1972).
96. T.-W. Wong, M. A. Harris, and C. A. Jankowicz, *Biochemistry* **13**, 2805 (1974).
97. V. Massey, *in* "Iron-Sulfur Proteins" (W. Lovenberg, ed.), Vol. 1, p. 301. Academic Press, New York, 1973.
98. E. A. Wider de Xifra, J. D. Sandy, R. C. Davies, and A. Neuberger, *Philos. Trans. R. Soc. London, Ser. B* **273**, 79 (1976).
99. S. Pagani, C. Cannella, P. Cerletti, and L. Pecci, *FEBS Lett.* **51**, 112 (1975).
100. A. Finazzi Agrò, C. Cannella, M. T. Graziani, and D. Cavallini, *FEBS Lett.* **16**, 172 (1971).
101. T. Taniguchi and T. Kimura, *Biochim. Biophys. Acta* **364**, 284 (1974).

Index

A 7
B 8
C 9
D 0
E 1
F 2
G 3
H 4
I 5
J 6